3/06

AGENTS OF BIOTERRORISM

■ ■ ■ ■ ■ ■ ■ ■ ■ ■ ■ ■ ■ ■ ■ ■ ■

AGENTS OF BIOTERRORISM

■ ■ ■ ■ ■ ■ ■ ■ ■ ■ ■ ■ ■

PATHOGENS AND THEIR WEAPONIZATION

Geoffrey Zubay et al.

Columbia University Press
New York

COLUMBIA UNIVERSITY PRESS

Publishers Since 1893

New York   Chichester, West Sussex

Copyright © 2005 Columbia University Press

Library of Congress Cataloging-in-Publication Data

Agents of bioterrorism: pathogens and their weaponization / Geoffrey Zubay et al.

p.   cm.

Includes bibliographical references and index.

ISBN 0–231–13346–4 (alk. paper)

1. Pathogenic microorganisms. 2. Biological weapons. I. Zubay, Geoffrey.

QR175.M55   2005

579.165—dc22   2005045565

∞

Columbia University Press books are printed on permanent
and durable acid-free paper.

Printed in the United States of America

c 10 9 8 7 6 5 4 3 2 1

CONTENTS

PREFACE

Geoffrey Zubay

Concerns about bioterrorism led me to organize a college seminar course on microbiology and bioterrorism. This book, written mainly by students who participated in the seminar, follows the general structure of the course. Pathogens considered to be major biothreats are the central topic. They are examined from the standpoint of their biology and the steps that might be taken to defend against them. The information complements what is usually presented in a microbiology text, and there is little overlap.

The first chapter deals with general aspects of bioterrorism. Each of the 12 chapters that follow deals with a specific pathogen. In selecting the pathogens to be covered, we were guided by the Centers for Disease Control and Prevention (CDC), which has classified most pathogens into Categories A, B, or C, depending on their prevalence, ease of access and use, and potentially lethal effects if used in a bioterrorist attack. Six pathogens in Category A, three in Category B, and one in Category C are considered. In addition, two pathogens are discussed that have not been ranked by the CDC: the influenza virus and the virus that causes severe acute respiratory syndrome (SARS). Each pathogen is considered from the standpoint of its history, molecular biology, pathology, clinical presentation, diagnosis, weaponization characteristics, and specific defenses.

The four appendixes deal with rapid drug discovery, strategies for developing and manufacturing vaccines, protective measures for individuals, and sources of information on bioterrorism. The book concludes with a glossary.

Bioterrorism is a provocative subject that disturbs many people to the extent that they would rather not discuss it. Consider the fear generated by an incident involving anthrax that occurred shortly after September 11, 2001. A few letters that had been posted from a mailbox in Trenton, New Jersey, were found to contain powdered anthrax, and a few individuals were exposed to lethal doses. The investigation that followed resulted in the closing of the Senate office building for several months and considerable anxiety among postal workers, who wanted the entire postal service to be shut down. I do not mean to trivialize this event or to say that a devastating bioterrorist event involving anthrax could not occur. However, we should keep a sense of proportion. Nature is still far ahead of the bioterrorists. Our planet is currently populated by about 6 billion humans. Roughly 120 million die each year from various causes, and about 10 million of these deaths are from infectious diseases. We have yet to see a bioterrorist event result in the loss of more than several thousand lives, and perhaps we never will. The more we know about the subject, the more likely we are to avoid disaster.

One good thing that has come out of all this concern is an awakening to the realization that research on infectious diseases has fallen behind the needs of a burgeoning population and a world in which globalization has eliminated boundaries that once appeared to limit the spread of pestilence.

I acknowledge the generous assistance of Anthony Gomez throughout the preparation of this text.

This book is an outgrowth of a course in bioterrorism that was given three times after 9/11. The students gave seminars that were revised and put into book form. We consider this course to be very authoritative mainly because Kathleen Kehoe, the author of appendix 4, helped us to obtain the latest information from the vast literature on this subject. She was a major source of information to all of the participating students and to me.

LIST OF CONTRIBUTORS

Rian Balfour, Columbia University, New York, New York, U.S.A.

Barbara Chubak, Johns Hopkins University School of Medicine, Baltimore, Maryland, U.S.A.

William Edstrom, University of Southern California, Los Angeles, California, U.S.A.

Maria E. Garrido, Columbia University, New York, New York, U.S.A.

James Hudspeth, Washington University School of Medicine, St. Louis, Missouri, U.S.A.

Kathleen Kehoe, Biological Sciences Library and Physics/Astronomy Library, Columbia University, New York, New York, U.S.A.

Anuj Mehta, Weill Medical College of Cornell University, New York, New York, U.S.A.

Kira Morser, Kings College, London, United Kingdom

Rohit Venkat Puskoor, Columbia University, New York, New York, U.S.A.

Payal Shah, Columbia University, New York, New York, U.S.A.

Salwa Touma, Columbia University, New York, New York, U.S.A.

Joseph Patrick Ward, Columbia University, New York, New York, U.S.A.

Geoffrey Zubay, Department of Biological Sciences, Columbia University, New York, New York, U.S.A.; visiting scholar, Scripps Institute of Oceanography, La Jolla, CA.

AGENTS OF BIOTERRORISM

CHAPTER 1

■ ■

TERRORISM AND FEAR

HOW TO COPE

Geoffrey Zubay

Fear is the currency of terrorism. It comes in a variety of forms: fear of dying, fear of being crippled, fear of the unknown. To reassure Americans in a day of confusion after the attack on Pearl Harbor, President Roosevelt said they had nothing to fear but fear itself. Because ignorance and insecurity are prime ingredients of fear, the more we know about a particular terrorist event the lower its fear potential is likely to be and the sooner we are likely to find a constructive solution.

This book is dedicated to presenting a comprehensive picture of what bioterrorism is all about and how we can defend ourselves against it. Our belief is that knowing what we are confronting will help to minimize fear and encourage productive behavior.

An effective terrorist knows how to plan his acts of terror so that they have maximum impact. He must be highly motivated and willing to make personal sacrifices. It is hard to think of a more effective act of terror than was committed at the World Trade Center in New York City on September 11, 2001, when a handful of terrorists brought down the symbols of American economic power and killed over 3000 people of different nationalities. More were killed by this small group than in the Pearl Harbor attack conducted by a naval battle group. Furthermore, the September 11 event produced enormous fear in the aftermath as we came to realize how vulnerable our society is to attacks of this nature, and we have since lived in fear of another attack. The World Trade Center terrorists were very effective; they had a workable plan and they were willing to die for it.

AMERICA AS A TARGET FOR TERRORIST ACTIVITIES

There are numerous reasons why the United States has become a terrorist target. Some are obvious, others more subtle. Among the factors are our great wealth, which breeds resentment leading to hatred; our use of military power; our support for totalitarian governments; our support for democracy; our liberal culture; and our strategic dominance after the end of the Cold War. The United States is thus a likely target of terrorists who may disapprove of U.S. policy or culture or who are directly opposed to purported Western ideals. It should be noted that not all terrorists are foreigners—the "Unabomber" and Timothy McVeigh's 1995 bombing of the Federal Building in Oklahoma City are examples of home-grown terrorists. Nor may they have any intelligible reason for terrorist acts: the Aum Shinrikyo cult's nerve-gas attack in the Tokyo subway reflected a hopelessly muddled apocalyptic world view.

BIOLOGICAL AGENTS IN TERRORISM

A terrorist's goals may be met with various agents. Explosives are the most common type of weapon used by terrorists. Terrorist activities that employ biological agents have been largely ignored, probably because the use of explosives has been so successful. Another factor is that the handling of biological agents requires new kinds of skills. Whatever the explanation in the past, it seems highly likely that the use of biological agents will increase in the near future because they are easier to hide, cheaper, and far more devastating. More attention must be given to the likelihood of increased use of biological agents by terrorist groups.

We might ask whether bioterrorism is something new or something recently conceived. The answer is that it is as old as recorded history, as seen in subsequent chapters. What distinguishes contemporary bioterrorism are the scientific advances that have been made in our understanding of how pathogens produce their toxins and how we can protect ourselves against them. On the one hand, this has also resulted in the discovery of a wide range of new bioweapons. On the other hand, this has resulted in the discovery of procedures for protecting against serious pathogens.

NATURE: AN EXQUISITE BIOTERRORIST

Nothing engenders more fear than the thought that when you go outside you may encounter a deadly pathogen that will result in an agonizing illness followed by death. One should keep a sense of proportion. Thus far, at least, the agents of bioterrorism look rather puny in their impact. In terms of kill power, only about 10% of all deaths result from infectious disease, either directly or indirectly. That means that, in a world of 6 billion people with an average life expectancy of 60 years, about 10 million people will die every year from an infectious disease. Currently the three major killers are the human immunodeficiency virus (HIV), tuberculosis, and malaria, which account for, respectively, 3 million, 2 million, and 1 million deaths per year. It should be noted that none of these is considered an agent of bioterrorism and that bioterrorist events have never resulted in more than a few thousand deaths in a single episode.

A major reason that HIV, tuberculosis, and malaria are not considered likely agents for bioterrorism is that their action takes many months, or even years, to be felt. Most agents used by bioterrorists are effective in a matter of hours or days. Thus, time to impact appears to be an important component of bioterrorism.

THE RANKING OF BIOWEAPONS

The Centers for Disease Control and Prevention (CDC) is the agency within the U.S. Department of Health and Human Services whose job it is to protect American citizens from infectious diseases. They have ranked biological agents based on the dangers they pose and the likelihood that they will be used as agents of bioterrorism. Category A agents are easily disseminated and/ or highly infectious and are characterized by high mortality rates. Because of their speed of action, they are likely to strain the public health infrastructure if released. In addition, many are untreatable or difficult to treat and could cause public panic. Category B agents are considered lower in priority because of their lower mortality rates and the ease with which they can be treated. Category C agents are emerging pathogens considered to have a potential for weaponization due to their potentially high morbidity and mortality rates and their availability.

Classification schemes may change as we learn how to cope with existing pathogens or as terrorists learn how to make available pathogens more deadly. For example, *Vibrio cholerae*, which is now classified as a Category B agent, would once have been considered a Category A agent until it was learned that simple hydration therapy lowers its lethality from 50% to 1% On the other hand, anthrax might have been classified as a Category B agent until procedures for producing powdered spores gave it greater stability and potency.

Subsequent chapters deal with six pathogens classified as Category A threats: anthrax, smallpox, Ebola virus, *Francisella tularensis, Yersinia pestis*, and *Clostridium botulinum*. In addition, three Category B agents—*V. cholerae,* salmonella, and viral encephalitis—and one Category C threat—hantavirus—are discussed. Finally, two agents that have not yet been classified—the viruses that cause influenza and the severe acute respiratory syndrome (SARS)—are examined. The latter seems to have come out of the rural area of western China.

THE LITERATURE ON BIOTERRORISM
SHOULD NOT BE CENSORED

There is a movement by scientific journals to censor submissions dealing with bioterrorism-related research. The idea that such articles may pose more risk than benefit to our security is one that must be considered seriously as it could have a significant impact on the kind of research that is conducted. It is my view that such a censorship program would do more harm than good. It seems unlikely that it would inhibit the clandestine efforts of terrorists to weaponize and distribute deadly pathogens. But it would surely bring an abrupt halt to most of the efforts of well-intentioned researchers to find ways of dealing with such pathogens. The unrestricted flow of scientific ideas and information is critical to the advancement of science. Therefore, if science in the area of bioterrorism is to advance, scientists involved in this research must be free to conduct and publish their results.

The vast majority of these scientists do not conduct research relating to bioterrorism in order to make weapons but rather to devise ways to defend against them. An official policy of suppressing publication of this work would only put a damper on potentially life-saving research. If we are not encouraged to study these organisms, we are unlikely to learn how to counter them. I am very optimistic that, given the opportunity, we will find ways to counter the deleterious effect of weapons that might be used by bioterrorists. Perhaps this is because I believe that good scientists far outnumber scientists with evil intentions, and that by and large good scientists are smarter.

CHAPTER 2

■ ■

VIRAL ENCEPHALITIS (FLAVIVIRUSES)

Salwa Touma

The viral genus *Flavivirus* is a very serious public health threat. Twenty-two of the 34 mosquito-borne flaviviruses cause human disease.[2] In some strains, the flavivirus enters the brain's blood vessels and nerves and causes brain inflammation, which is known as encephalitis.[3] In the most severe cases, this inflammation may cause debilitating irreversible nerve damage, brain tumors, and death.[4] The lethality of encephalitis can be as high as 37%, although it varies by strain.[3]

There are some prophylactic measures against viral encephalitis and one synthetically created acyclic nucleoside analog, Acyclovir,[1–3] and many companies are working to develop a vaccine. However, since such a vaccine does not currently exist for the most common strains of encephalitis in the United States, it is a major bioterrorist threat. Its danger is increased by the ease of transmission from mosquitoes (mainly the *culex* species) and ticks and by their many breeding grounds.[1,5] However, because of its moderate to low mortality rate, its moderate ease of dissemination, and the fact that it requires specific enhancements of the Centers for Disease Control and Prevention's (CDC) diagnostic capacity and enhanced disease surveillance, the CDC categorizes viral encephalitis as a Category B bioweapon threat and a Biosafety Level 3 pathogen.[3]

Encephalitis is derived from the Greek word *encephalo* ("brain") and the Greek word *itis*, a term used pathologically to indicate inflammation of an organ or an abnormal state or condition. Thus, taken together, *encephalitis* literally means "inflammation of the brain," which is precisely what advanced encephalitis is and what causes its potential lethality. However, unlike meningitis, which is limited to inflammation of the meninges of the brain, encephalitis involves inflammation of both the meninges and the parenchyma[1,7,12,18] (figure 2.1A and B).

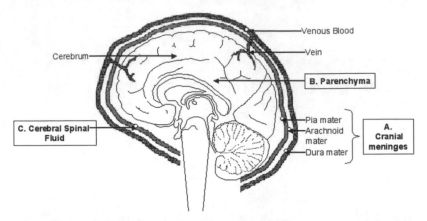

FIGURE 2.1. Encephalitis involves inflammation of the meninges (**A**) and the parenchyma (**B**). Neurons are surrounded by cerebral spinal fluid (CSF) (**C**). Encephalitis enters the CSF from the blood through the blood–brain barrier, which is normally closed to prevent the flow of materials from the blood into the brain. However, the barrier is opened by nasopharyngeal mucous membrane congestion when the body wages a general immune attack against encephalitis.

STRAINS OF ENCEPHALITIS AFFECTING THE UNITED STATES

Reported cases of different strains of viral encephalitis are genetically conserved and have been confirmed globally. The focus of this chapter is the viral genus *Flavivirus* within the viral family *Flaviridiae*. *Flavivirus* comprises two of the most widely occurring strains of viral encephalitis in the United States: St. Louis encephalitis (SLE) and West Nile encephalitis (WNE)[6,8,10,13,16] (figure 2.2).

Viral encephalitis in the United States can be divided into five strains: SLE, WNE, Eastern equine encephalitis (EEE), Western equine encephalitis (WEE), and La Crosse encephalitis (LCE). The names of the five strains are based on the location of the first identified outbreak and in some cases are based on the region of the resulting outbreaks. SLE, first seen in Paris, Illinois, has been reported throughout the United States. WNE, initially in the West Nile province of Uganda, first seen in New York City in 1999, has been reported throughout the United States. Cases of EEE, first seen in Massachusetts, have been reported elsewhere in the eastern United States. WEE, first identified in California, has been reported in the western and central United States. Finally, LCE, first seen in Wisconsin, has been reported in the eastern United States. Clearly, regardless

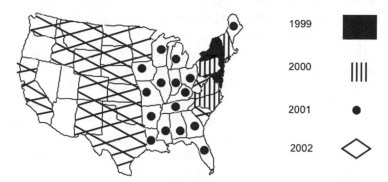

1999	■
2000	||||
2001	●
2002	◇

FIGURE 2.2. Flavivirus comprises the strains that cause West Nile virus (WNE) and St. Louis encephalitis (SLE), two of the most widely occurring types of viral encephalitis in the United States. The first reported human cases of WNE were in New York City. WNE quickly spread from east to west between 1999 and 2002.

of the location of the first reported case, each strain quickly spreads throughout a region. In the case of WNE, the strain rapidly moved from east to the west over the course of 4 years, from 1999 to 2002[6,8,10,13,16] (figure 2.2).

Flaviridiae comprises SLE and WNE. However, two other viral families contain strains of encephalitis: *Togaviridae* comprises EEE and WEE and *Bunyaviridae* comprises LCE.[6,8,10,13,16]

HISTORY

Meningoinflammatory diseases such as rabies and poliomyelitis were recognized centuries ago. Encephalitis was discovered much later, in the early twentieth century, and the flavivirus was not implicated as a cause until the 1950s.[6,8,10,13,16]

Rabies was known to be transmitted through bites, via the saliva of infected animals. Poliomyelitis was most commonly found in the contents and walls of human small intestines and was observed to be transmitted from person to person. The viral nature of these two diseases was discovered in the early 1900s.[6,8,10,13,16]

In the 1930s and the 1940s, many new meningoinflammatory diseases were appearing worldwide. However, unlike rabies and poliomyelitis, these diseases were most commonly found in and transmitted by mosquitoes and ticks. As a result, these new meningoinflammatory diseases were called arthropod-borne encephalitides; soon any virus carried by arthropods, even if it did not cause encephalitis, was called an arthropod-borne virus or an arbovirus.[6,7,8,10,13,16]

The first identifiable outbreaks of SLE, WNE, EEE, and WEE in the United States all occurred in the 1930s. However, SLE was the first flaviviral arthropod-borne encephalitis to be isolated between 1931 and 1937. In the 1930s and 1940s, the neutralization test, the complement fixation test, and the new hemagglutination test were used to specifically isolate and examine the relationships between SLE and WNE. The neutralization test inoculates an animal with a mixture of serum and virus to determine the antimicrobial activity of the serum. The complement fixation test is based on the binding of complement and the proteins of the complement system collectively by antigen–antibody complexes. The hemagglutination test combines an antibody with the virus to prevent red blood cells from agglutinating.[6,8]

Using these tests, researchers in the 1950s divided arboviruses into three distinct serological groups: A, B, and C. Casals and Brown are credited not only with discovering the flavivirus but also with distinguishing group B from the other groups. This differentiation was based on the study of the viral genes, structure, and replication cycle. This created a context and a framework within which the International Committee for the Taxonomy of Viruses could differentiate the other viral encephalitides.[6,8]

ENCEPHALITIS-CARRYING MOSQUITO BREEDING GROUNDS

Since mosquitoes are very active in warm weather, most cases of encephalitis occur between June and September. In warmer states, though, cases can occur throughout the year. The many mosquito breeding sites include lakes and ponds, plants, discarded tires, and trash. Since these environments are very difficult to control, encephalitis is further implicated as a biohazard.[3,9,10,13,16]

THE TRANSMISSION CYCLE

The transmission cycle of encephalitis begins with birds and other primarily infected animals being infected but unaffected by viral encephalitis (figure 2.3). Mosquitoes then feed on the infected birds and other animals. The virus infects the mosquitoes, replicates, and infects the mosquitoes' salivary glands. Like the birds and other primarily infected animals, the mosquitoes are infected with encephalitis but do not get sick.[3,9,16]

A mosquito then takes a blood meal from a human or horse. In the process, the virus in the mosquito's saliva is transmitted. The virus that is extravascular

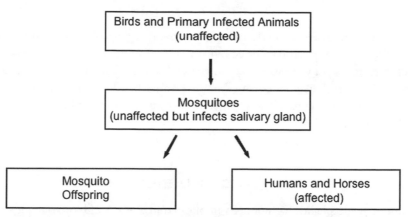

FIGURE 2.3. Direct pathway of encephalitis transmission from birds to mosquitoes to humans. 1) Birds and other primarily infected animals are infected but unaffected. 2) Mosquitoes then feed on the infected animals. 3) Encephalitis infects the mosquitoes, replicates, and infects the mosquitoes' salivary glands. Although the mosquitoes are infected, they do not get sick. 5) A mosquito takes a blood meal from a human or horse. 5) Encephalitis secreted in the mosquito's saliva is transmitted to the human or horse. 6) Extravascular encephalitis remains, but intravascular encephalitis is sucked up by the mosquito and can therefore be passed from the female mosquito to its offspring.

remains with the human or horse, but the virus that is intravascular is sucked back up by the mosquito with the blood and can therefore be passed from the female mosquito to its offspring.[3,9,16]

Interestingly, there is a salivary enzyme, apyrase, in the mosquito's saliva that inhibits ADP-dependent platelet aggregation and coagulation. This allows the virus to replicate at the site of infection and slows the initial spread of the virus throughout the body. Once the virus spreads, it can then potentially make the human or horse very sick with encephalitis in a process to be delineated below. Nonetheless, not all infected humans are symptomatic.[3,9,16]

OTHER CAUSES

Other causes of encephalitis include herpes simplex types 1 and 2 (HSV-1 and HSV-2), enteroviruses (primarily in the gastrointestinal tract), Coltivirus, measles, mumps, rabies, rubella, lymphatic choriomeningitis virus, cytomegalovirus, Epstein-Barr virus (EBV), human immunodeficiency virus (HIV), Varicella-zoster virus (VZV), and influenzas A and B.[3,9]

CURRENT STATUS

Most recently, SLE and WNE have affected many regions of the United States. In 2002, there were 4156 reported human cases of WNE—and 284 deaths—in 44 states. This is a vast increase compared with the 4478 reported human cases of SLE in the United States between 1964 and 1998. Because a vaccine does not exist for the strains of encephalitis most common in the United States, it is a bioterror threat.[6,8,16]

ENCEPHALITIS AND BIOWARFARE

Although encephalitis has not been implicated in any acts of biowarfare, it still poses a very dangerous bioterrorist threat. In fact, when WNE was first discovered in the Western Hemisphere at the Bronx Zoo, the first suspicion was that it was the work of terrorists. Although this was eventually disproven, it is not completely unfounded as a potential cause of an encephalitis outbreak. In fact, in the United States some arboviruses have already been developed for aerosolized release, enabling transmission without the use of contaminated mosquitoes or ticks.[6]

There are many features of encephalitis that make it an especially dangerous biological weapon, including, as noted earlier, the ease of transmission from mosquitoes and ticks and the ubiquity of their many breeding sites.[3,15]

MOLECULAR BIOLOGY

The flavivirus is a spherical, positive-sense-stranded, monopartite linear, RNA genomic retrovirus with a diameter of 40–50 nm and an electron-dense core. It is 6% RNA, 66% protein, 9% carbohydrate, and 17% lipid. It infects horses and humans. The viral species have been categorized into more than 80 genetically distinct groups. Some of these groups cause meningoencephalitis, which is the most significant and lethal form of viral encephalitis.[1,7,12,18]

PHYSICAL CHARACTERISTICS

Viral encephalitis is sensitive to acidic environments since it has an optimal pH range between 8.4 and 8.8. It is also unable to withstand heat above 40°C. The lipid envelope protein causes the virus to be destroyed by organic solvents, detergents, and formaldehyde. Encephalitis is inactivated by ultraviolet light, gamma radiation, and disinfectants.[1,7,12,18]

GENOMIC CHARACTERISTICS

The flavivirus is approximately 11 kb long. The genomic RNA for translation is in a single open reading frame. The genome contains a 5′ cap ($m^7G5′ppp5′A$) but does not have a polyadenylate tail. The 5′ and the 3′ ends correspond to a 100-nucleotide and a 400–700-nucleotide noncoding region, respectively. The genome contains ten distinct proteins: three structural and seven nonstructural. The proteins are present in the following order from the 5′ end to the 3′ end: capsid protein (C), membrane protein (prM or, more specifically, pr + M), envelope protein (E), nonstructural protein 1 (NS1), nonstructural protein 2a (ns2a), nonstructural protein 2b (ns2b), nonstructural protein 3 (NS3), nonstructural protein 4a (ns4a), nonstructural protein 4b (ns4b), and nonstructural protein 5 (NS5) (figure 2.4). The hydrophobicity of the viral protein suggests that the endoplasmic reticulum membrane may appear as shown in figure 2.5.[1,8,12,18]

THE THREE STRUCTURAL PROTEINS

The first structural protein is the nucleocapsid core protein (C protein) (11 kd). The two ends of the C protein are especially basic and are thought to form a

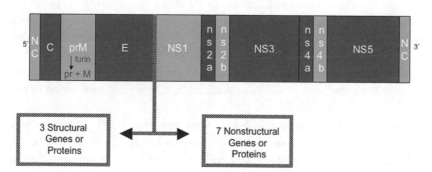

FIGURE 2.4. The genome contains 10 distinct proteins: three structural and seven nonstructural. From the 5′ end to the 3′ end are: a 100-nucleotide noncoding region with a cap (m7G5′ppp5′A) lacking a polyadenylate tail, capsid protein (C), membrane protein (prM or, more specifically, pr + M), envelope protein (E), nonstructural protein 1 (NS1), nonstructural protein 2a (ns2a), nonstructural protein 2b (ns2b), nonstructural protein 3 (NS3), nonstructural protein 4a (ns4a), nonstructural protein 4b (ns4b), nonstructural protein 5 (NS5), and a 400–700-nucleotide noncoding region. The flavivirus is a spherical, positive-sense-stranded RNA genomic retrovirus approximately 11 kb long. The genomic RNA for translation is in a single open reading frame.

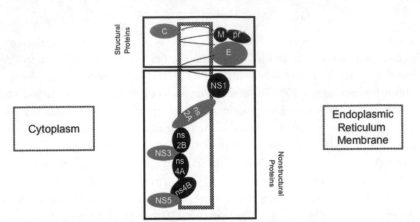

FIGURE 2.5. Flaviviral hydrophobicity suggests the location of the three structural and seven nonstructural proteins relative to the endoplasmic reticulum (ER) membrane: in the regions of the cytoplasm (CY), the transmembrane (TR), and the ER membrane itself. Capsid protein (C), CY; membrane protein (prM or, more specifically, pr + M), ER; envelope protein (E), ER; nonstructural protein 1 (NS1 at membrane), ER; nonstructural protein 2a (ns2a), TR; nonstructural protein 2b (ns2b), CY on membrane; nonstructural protein 3 (NS3), CY; nonstructural protein 4a (ns4a), CY on membrane; nonstructural protein 4b (ns4b), CY on membrane; and nonstructural protein 5 (NS5), CY at membrane.

ribonucleoprotein complex with packaged RNA; the basic regions conceivably allow the C region to bind to the genomic RNA. The center of the C protein is hydrophobic and has been suggested to assist in the assembly of the virion since the hydrophobic region can interact with the cell membrane. Supportive of this hypothesis is a protein called anchC protein, which has a basic region and is a signal sequence for the translocation of prM into the endoplasmic reticulum lumen. This anchC protein is then processed by cytoplasmic viral serine protease.[1,7,12,18]

The second structural protein is the membrane protein (M protein) (8 kd), which causes the maturation of the virus and its lethal differentiation. In the endoplasmic reticulum, host signal peptidases stimulate the N terminus. In the trans-Golgi, the enzyme furin cleaves prM into pr and M. Therefore, the M protein is only present in mature virions and contains three domains. The pr protein has N-linked glycosylation sites and cysteine residues, both of which are involved in intramolecular disulfide bridges.[1,7,12,18]

The envelope protein (E protein) (53 kd) is the third structural protein. It serves as the primary antigen when encephalitis infects the body by 1) interacting with viral receptors and 2) stimulating virus–cell fusion. It has transmembrane domains that allow for NS1 translocation. Like the pr protein, it has cysteine residues which are also involved in intramolecular disulfide bridges. The E gene has appreciable antigenic variability and is one of the most variable genes in the flavivirus. Given the function and variability of the E gene, if the E protein is mutated, the lethality of the virus can be modified.[1,7,12,18]

The Structure of the Envelope Protein The envelope protein is divided into three structurally distinct domains: domain III (or B), domain I (or C), and domain II (or A). Domain III contains 100 amino acids that form a β–barrel composed of seven antiparallel β-strands. This entire domain independently folds and has a single disulfide bond on which the antibodies depend. There is a flexible region that attaches domain III to domain I, which contains 50 amino acids that form a β-barrel composed of eight up-and-down strands. Domain I is attached to domain II, which is composed of two long loops involved in dimer contact.[1,7,12,18]

Association of the Three Structural Proteins On the surface of the spherical virion are hundreds of trimers including E1 and E2 heterodimers that associate with the central C monomer. The interactions of the three major proteins in encephalitis—C, prM (pr + M), and E—are interesting. Specifically, the prM and E proteins (figure 2.6) associate differently with the C protein in

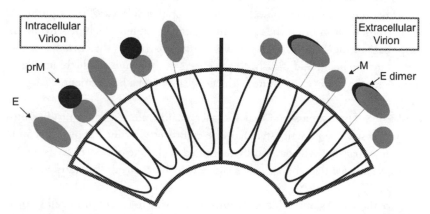

FIGURE 2.6. Two of the three flaviviral structural proteins in extracellular and intracellular virions: envelope protein (E) and membane protein (prM or, more specifically, pr + M). The prM causes the maturation of the virus and its lethal differentiation, and E protein serves as the primary antigen, interacting with viral receptors and stimulating virus–cell fusion when encephalitis infects the body. The prM and E proteins associate differently with the capsid protein (C) in extracellular and intracellular virions. In extracellular virions, E dimerizes or forms heterodimers whereas M does not. In intracellular virions, however, M dimerizes with pr whereas E is not dimerized.

extracellular and intracellular virions. In extracellular virions, E dimerizes or forms heterodimers while M does not. In intracellular virions, however, E is not dimerized. Instead, M dimerizes with pr, which has N-linked glycosylation sites and cysteine residues, both of which are involved in intramolecular disulfide bridges.[1,7,12,18]

THE SEVEN NONSTRUCTURAL PROTEINS

The seven nonstructural proteins are NS1, ns2a, ns2b, NS3, ns4a, ns4b, and NS5. Nearly all are involved in RNA replication as described below. However, some of these proteins have other very specific functions: most notably, NS1 is secreted only by mammal cells, not by mosquito cells. Thus, NS1 induces a humoral immune response in mammals. ns2b is a cofactor for NS3 function; NS3 has enzyme activities that make it very important in polyprotein processing. NS5 is the most conserved flaviviral protein that has RdRP sequence homology.[1,7,12,18]

THE EIGHT STEPS OF SCHEMATIC VIRAL EFFECT

Eight major steps are necessary for the schematic entrance of encephalitis from an insect bite to the effect of the virus on the brain: 1) the bite, 2) the local replication of the virus on the skin, 3) transient viremia, or the transient entrance of the virus from the skin into the blood through the epithelial layer, 4) the replication of the virion in the host, 5) the re-entrance of the virion into the bloodstream, 6) the passage of the virus through the blood–brain barrier, 7) the entrance of the virion into the brain, and, finally, 8) the effects of the virion on the brain. Each of these steps requires a mosquito with encephalitis and a host patient.[11]

Viral Entry First, we assume that the encephalitis-infected mosquito has bitten the prospective patient (step 1 of the eight steps outlined above). Second, we assume that the virus has locally replicated at the site of skin entry (step 2). Third, we assume that the virus has entered the bloodstream by transient viremia (step 3).

Once the virus is in the blood, it must find a way to enter a host cell so that it can replicate itself (figure 2.7). In order to do this, the virus accumulates in pits near the surface of the host cell. Then the virus attaches to hypothesized cell-surface receptors that are yet to be determined. It then enters the host cell by receptor-mediated endocytosis[1,7,12,18] (figure 2.7A).

Viral Replication Inside the host cell, the virion must replicate (step 4). As previously noted, the viral envelope has an optimal pH range of 8.4–8.8 while the

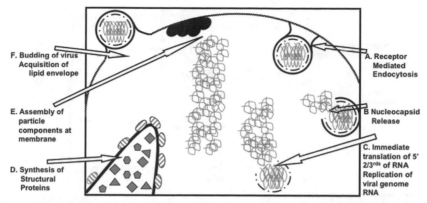

FIGURE 2.7. (A) The virus enters the host cell by receptor-mediated endocytosis. (B) Inside the host cell, the difference in pH between the virion and the cytoplasm causes the virion to undergo a conformational change, attach to the cell membrane of the host cell, and release its nucleocapsid into the cytoplasm of the host cell. (C) The structural protein's RNA in the first two-thirds of the 5′ positive-strand RNA is replicated first. (D) The nonstructural proteins are synthesized because they are required for the translation and the processing of all the virion's proteins. (E) Particle components are assembled, including the E protein and the other structural proteins, C and M. (F) The virus then buds from the host cell and, while exiting, acquires a lipid envelope.

host cell cytosol has a more acidic pH. As a result, when inside the host cell, the virion undergoes a conformational change and in doing so attaches to the cell membrane of the host cell. The cholesterol and sphingolipid in the host-cell membrane cause the fusogenic domain in the virion to be exposed. The nucleocapsid is then released into the cytoplasm of the host cell (figure 2.7B). It remains unclear whether the nucleocapsid releases its RNA first, followed by the nucleocapsid binding to the ribosomes, or whether the nucleocapsid binds to the ribosomes first, followed by the ribosome's release of RNA. Whatever the sequence of events, the RNA is released and the virion must make more of its RNA.[1,7,12,18]

In addition to the reasons previously outlined for the selection of the flavivirus as a likely agent of bioterrorism, it has been extensively studied. Moreover, it is highly conserved compared with its predecessors. When replicated, it has an open reading frame; because there are no terminal codons it can potentially translate as a polypeptide chain.[6,8,10,13]

The virus has positive-sense RNA (figure 2.8). This means that when the virus enters the host cell, it acts as an mRNA template and is read 5′ to 3′ (figure 2.8A). From the viral mRNA template strand, the nonstructural proteins

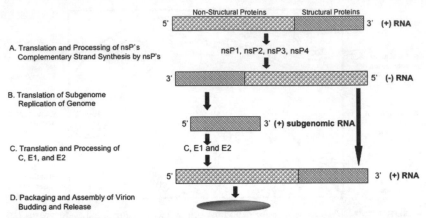

FIGURE 2.8. (A) The virus has positive-sense RNA; i.e., when the virus enters the host cell, it acts as an mRNA template and is read 5′ to 3′. (B) From the viral positive-sense mRNA template strand, the nonstructural proteins are translated. These are necessary for the synthesis of the complementary negative strand. The negative strand is the template for the genome-length final positive strand. The final positive 5′-3′ strand is translated in steps. (C) First, the three structural proteins are translated and processed. (D) Second, the virion's whole genome, including both structural and nonstructural proteins as a positive-sense RNA, is translated and processed.

are translated. These proteins are necessary for the translation of the complementary negative strand, the structural proteins, and the final positive strand as well as for the processing of the final virion. After the nonstructural proteins are synthesized, a complementary negative strand is synthesized (figure 2.8B). The resulting negative strand is the template for the structural proteins and for the genome-length final positive strand, which is 5′–3′ (figure 2.7C). The final positive 5′–3′ strand is translated in steps. First, the three structural proteins are translated and processed (figure 2.8C). Second, the virion's whole genome, including both structural and nonstructural proteins as a positive-sense RNA, is translated and processed[1,7,12,18] (figure 2.8D).

The nonstructural proteins help the virus enter the host cell by mediating receptor binding of the virus to the extracellular matrix of the host cell. Once inside the host cell, the nonstructural proteins assist in synthesizing the negative strand, creating the 5′ cap of the genome, and cleaving the polyprotein. If the nonstructural proteins are mutated, RNA synthesis is affected.[1,7,12,18]

Of the three structural proteins, E, M, and C (figure 2.7D), only the E protein bears the antigenic section of the virus and thus informs the host cell when a virus has invaded. Antibodies to the virus are the major component of immunity in natural infections. As indicated earlier, the structure of the E protein induces neutralizing antibodies at epitopes on all three structural envelope domains. This protects the body from future encephalitis infections.[1,7,12,18]

The E protein is necessary to the virion because it is essential for viral entry into the host cell; it allows viral receptor binding to the extracellular matrix of the host cell. Once inside the host, the virus must fuse with the host's cell membrane. The E protein mediates this pH-dependent fusion activity. While inside the host cell, the E protein protects the virion genome from cellular nucleases and protects the naked virion nucleocapsids from being degraded by ribonucleases.[1,7,12,18]

Viral Exit and Resulting Immune Response Once the virion RNA has been replicated and cleaved into the structural and nonstructural proteins, the new virus must leave the host cell with a new protective coating (step 5 of the eight steps). This coating is created from particle proteins at the host's cell inner plasma membrane. The virus buds from the host cell and while exiting acquires a lipid envelope (figure 2.7E and F). After more viruses reenter the bloodstream, the host cell dies by apoptosis.[4]

Once in the bloodstream, the virus induces a general immune response. The immune system responds not only to the presence of the virus in the blood but also to the host cells that were killed in the process of viral replication. The general immune response includes the responses of macrophages, natural killer T cells, T cells, CD4, and CD8 T lymphocytes, IgG, and immunoglobulin/virus complexes. While this general immune response is induced by the presence of the virus in the blood, ultimately the virus will enter the brain.[4]

Viral Entry to the Brain There are blood vessels that enter the brain to feed and nourish it. However, these vessels are not in contact with the neurons in the brain. Instead, the neurons are surrounded by cerebral spinal fluid (CSF) (figure 2.1C. The virus enters the CSF from the blood through the blood–brain barrier (step 6), a network of capillaries, arteries, and ganglia at the base of the brain. The barrier is normally closed to prevent the flow of substances into the brain.[1,4,7,12,17,18]

When the body wages the general immune attack described earlier in response to encephalitis, there is nasopharyngeal mucous membrane congestion, which causes the brain's end arteries and subcortical ganglia to dilate. The capillaries from the arteries and the ganglia are normally closed and do not allow viruses to enter the brain. However, when the ganglia dilate and the capillaries and the arteries are inflamed, they become more permeable and allow the viruses to enter the CSF[4,17] (step 7).

An interesting finding is that the virus usually lodges and has its effects on the left side of the brain. The explanation may be that the blood pressure on the left side of the brain is slightly higher than it is on the right side, because blood flows directly from the heart to the brain. However, the difference in blood pressure would be very small, as the carotid artery is the central transporter of blood from the heart to the brain and it splits off to the right and left sides almost at the very base of the brain. Since the virus has been found to lodge in the left side of the brain, many of the symptoms are observable preferentially on the right side of the body—the left side of the brain is known to control the right side of the body and vice versa.[17]

Brain Inflammation The effects of encephalitis on the right side of the body stem from the virus's two main targets in the brain: the blood vessels and nerves (step 8). Each target is the source of different side-effects, but both contribute to general brain inflammation.[4,14,17]

First, since the blood travels to the blood vessels in the brain, the virus attaches to the surfaces of those vessels. However, when the virus is attached to the brain's blood vessels, mononuclear cells such as plasma cells, lymphocytes, and macrophages aggregate around the virus particles and the underlying small veins of the brain. When aggregated, these cells form rings, or "cuffs," around the small veins. This causes ring bleeding—hemorrhaging into the cuff—which leads to aneurysms. Aneurysms are very dangerous in their own right. Ring bleeding and hemorrhaging can contribute to overall brain inflammation.[4,14,17]

Second, when the virus enters the CSF after it has crossed the blood–brain barrier, it comes into contact with nerves. Various immune cells invade the area near the nerves. Such immune cells as the polymorphocytes and lymphocytes, for example, become embedded in the cytoplasmic vacuoles of nerve cells and contribute to the apoptosis of nerves. When the nerves undergo apoptosis and immune cells invade the CSF, there is overall brain inflammation.[4,14,17]

Effects of Brain Inflammation There are several manifestations of inflammation of the brain. These include distention—expansion in volume and circumference—of blood vessels and regional congestion in the form of edematous, hemorrhaging neurophagocytosis as well as tissue disintegration and absorption. These manifestations of inflammation may lead to dangerous scarring in the brain. A scar, called a cyst, is harmful because it may grow and develop into a lethal brain tumor.[4,14,17]

CLINICAL DIAGNOSIS AND RESPONSE

In a clinical setting, the incubation period for encephalitis is 1–15 days. The severity of the illness increases with age, as is indicated by the 2% incidence of severe manifestations in young patients as opposed to the 22% incidence in those above 65 years old. Reasons for this trend include the higher incidence of hypersensitivity and arteriosclerosis disease as well as diabetes in elders. Additionally, immunocompromised individuals such as patients with HIV and cancer have been found to be more susceptible to encephalitis.[2,3]

SYMPTOMATIC CHARACTERIZATION

The encephalitis disease state is characterized by symptoms that may be divided into four categories according to the stage of the disease: initial, between 1 and 4 days, up to 3 years, and persisting. At first, all encephalitis patients experience malaise, fever, chills, headache, drowsiness, nausea, anorexia, and a sore throat and cough. Between 1 and 4 days, in 50% of cases, encephalitis patients experience acute or subacute neurological symptomatic signs. In 25% of cases, patients experience urinary-tract symptoms. For up to 3 years, in 30–50% of cases, patients experience convalescence, irritability, tremulousness, sleepiness, depression, memory loss, and headaches. Persisting symptoms include muscular stiffness, abnormal dragging of one foot when walking, speech disturbances, sensory-motor impairments, psychoneurotic complaints, and tremors. These persisting symptoms can be attributed to irreversible nerve damage.[2,3]

Nonclassic symptoms have been observed as well. In Brazil, jaundice has been reported to accompany encephalitis, which suggests that encephalitis may affect the liver in some fashion. In the tropics, hepatitis has been observed. A Dengue-like rash has been seen in encephalitis cases in different parts of the world. Thus, the spectrum of the effects of encephalitis has yet to be fully identified and understood.[2,3]

DIAGNOSIS

Proper diagnosis of encephalitis is critical because in the most severe cases it is characterized by debilitating and irreversible nerve damage, brain tumors, and death. Differential and elimination diagnosis are imperative. In differential diagnosis, the age of the patient, the season, the place of residence and patient exposure, and community occurrence of encephalitis should be taken into account. In elimination diagnosis, it is important to rule out treatable infections such as bacterial, mycobacterial, spirochetal, fungal, and herpes infections. It is also essential that a stroke not be misdiagnosed as encephalitis—a stroke can easily be confused with encephalitis since strokes occur mainly in the elderly and often, like many encephalitis infections, affect only one side of the body.[2,3,5]

DIAGNOSTIC EXAMINATION

In encephalitis, radionuclide brain scans, similar to computerized axial tomographic (CAT) scans, and computed tomography, similar to x-rays, appear to be normal. Effective tools for diagnosing encephalitis include serological and virus-detection assays. Serological assays (performed on samples of serum from the CSF) include IgM ELISA, IgG ELISA, PRNT, CF, HI, IFA, and dipsticks. Virus-detection assays (on virus isolated in cell culture from mosquito pools or tissues) include IFA, TaqMan reverse transcriptase–polymerase chain reaction (RT-PCR), Ag-captured ELISA, RT-PCR sequencing, dipsticks, and NASBA. Of all these diagnostic tools, the most popular serological CSF test is the IgM ELISA and the most popular viral-recovery test is the IFA.[3,5]

The serological CSF IgM ELISA test will reflect a heightened immune response in encephalitis, with an increase in IgM 3–5 days after onset, a peak at 7–14 days, and a drop by 60 days. The implications of this are twofold. First, if the test is not performed during the 7–14-day period that is normally the time of peak activity, the patient may be misdiagnosed. Second, in about 50% of encephalitis patients, there is a decrease in antibody titer for 1 year after the mosquito bite, which confuses the diagnosis.[3,5]

The viral-recovery IFA test entails virus recovery by intracerebral inoculation. In this procedure, suspensions of brain tissue, liver, spleen, lung, or kidney are isolated and immunofluorescence is performed. Electron microscopy indicates flavivirus-like particles. PCR to the blood, CSF, or tissues can also be performed in the future.[3,5]

TREATMENT

Although there are currently no known vaccines against WNE and SLE, some effective treatments are being implemented and other, potentially effective treatments are being developed. These include preventive and supportive care, an acyclic nucleoside analog, and an antiviral vaccination.[2,3,5,14]

PREVENTIVE AND SUPPORTIVE CARE

Until the cause of a case of meningoinflammation has been identified, antibiotics and a third-generation cephalosporin such as cefotaxime sodium (Claforan) and ceftriaxone sodium (Rocephin) should be administered as precautionary measures.[2,3,5,9,14]

If flaviviral encephalitis is implicated as the cause of infection, the three major effects—brain swelling, seizures, and breathing difficulty—should be treated. It is imperative that intracranial pressure (ICP), fever, and fluids be closely monitored to detect brain swelling, which is reduced through standard measures. Many (more than 50%) encephalitis patients experience seizures, which can be treated with standard anticonvulsants. Respirators and oxygen administration will help patients who experience breathing difficulty.[2,3,5,9,14]

Conditions associated with seriously ill patients include aspiration pneumonia, stasis ulcers, decubiti, contractures, deep-vein thrombosis, and infections at the sites of indwelling lines and catheters. Prophylactic measures should be used whenever possible to avoid these conditions, but if they develop, standard therapies should be used.[2,3,5,9,14]

POTENTIAL DEFENSES

ACYCLIC NUCLEOSIDE ANALOG

Acyclovir (amino-2[hydroxy-2-ethoxy]methyl-9-1 H-9 H purinone-6), also spelled aciclovir, is a synthetically created acyclic nucleoside analog (figure 2.9). Originally intended for herpes simplex, varicella, and herpes zoster viral infections, in 1982 it was approved for a wider range of viruses.[1,3,7,12,18]

Acyclovir can be administered either orally with a 200-mg capsule or as an intravenous preparation. The capsule is taken five times daily for 5 days; the intravenous preparation is administered 10 mg/kg every 8 hours for 10–14 days. The bioavailability ranges between 15% and 30%. If used within the first 4 days of contracting encephalitis, the likelihood of survival is 65–100%.[3]

The mechanism by which acyclovir works is based on the fact that flaviviruses, and therefore flavivirus-infected cells, encode the enzyme deoxypyrimidine

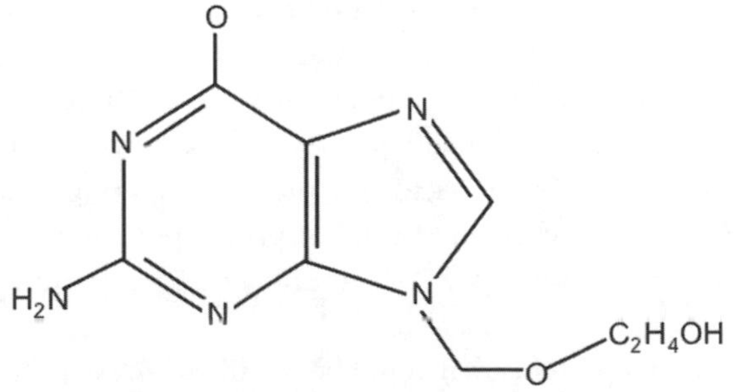

FIGURE 2.9. The chemical structure of acyclovir (amino-2[hydroxy-2-ethoxy]methyl-9–1 H-9 H purinone-6), a synthetically created acyclic nucleoside analog.

(thymidine kinase). Deoxypyrimidine phosphorylates acyclovir to acyclovir-5`-monophosphate, which is further phosphorylated to a triphosphate derivative. This triphosphate derivative causes premature termination of viral DNA chains and inhibits the synthesis of viral DNA by competing with a 2'-GTP as a substrate for viral DNA polymerase. The metabolized product of acyclovir is 9-carbomethoxymethylguanine.[3]

One of the advantages of acyclovir is its effectiveness in immunocompromised patients when 200–400-mg capsules are given four times daily. If administered within the first 4 days of the contraction of encephalitis, the likelihood of survival in immunocompromised patients, just as in immunocompetent patients, is increased to 65–100%. However, it is important that immunocompromised patients on acyclovir be closely monitored since acyclovir can delay or obscure the diagnosis of encephalitis or precipitate a toxic encephalopathy.[3]

ANTIVIRAL VACCINES

Although an effective vaccine for SLE and WNE does not exist, the development of one is very promising given that a highly (91%) effective inactive vaccine for Japanese encephalitis derived from an infected mouse has been available since 1954.[1,7,9,12,17]

To design an effective vaccine for SLE or WNE, it is important to have a full understanding of the molecular biology of the flavivirus genome and the epitopes involved in immunity. At this time, there is a complete genomic sequence for at least one member of each group of flaviviruses. In addition, cells that have RNA transcripts of viral cDNAs contain flaviviruses that could be isolated through reverse genetics. These flaviviruses could be altered to express foreign genes or to be used as live vaccines or in other ways. Ideally, noninfectious genetically engineered viruses should be used as vaccines. Effective vaccines can be created by modifying either structural or nonstructural proteins.[1,7,9,12,17]

The structural protein that can be modified is the prM/E protein. When encephalitis infects the body, the E protein serves as the primary antigen by interacting with viral receptors and stimulating virus–cell fusion. A new vaccine could be created using recombinant prM/E and baculovirus—a virus that is used as a eukaryotic expression vector for proteins requiring posttranslational modifications. However, when an SLE DNA prM/E protein vaccine was tested in mice, the survival rate was 60–75% and did not increase with a higher DNA dosage.[1,7,9,12,17]

The nonstructural proteins that could be modified are NS1, NS3, and NS5. NS1 is secreted only by mammalian cells and induces a humoral immune response in mammals. Therefore, if NS1 is altered, immunization with NS1 can elicit a humoral response from the body—if an individual is infected by encephalitis, the body will be protected. NS3 has enzyme activities that make it very important in polyprotein processing, while NS5 is the most conserved flavivirus protein that has RdRP sequence homology. If NS3 or NS5 is modified, immunization with NS3 and NS5 could affect viral replication and virulence of encephalitis in an infected individual.[1,7,9,12,17]

WEAPONIZATION

Three main strategies are involved in the weaponization of the WNE and SLE strains may also be extended to the other strains of encephalitis. These approaches involve 1) exploiting the ordinary mosquito viral transmission cycle, 2) increasing the virulence of the flavivirus, and 3) creating a new aerosolized viral transmission pathway.

THE ORDINARY VIRAL TRANSMISSION CYCLE

In the ordinary mosquito viral transmission cycle, as described previously, each infected mosquito has the ability to inject the flavivirus into several humans and/or horses. Also, every infected female mosquito may pass the virus to its offspring thus increasing the number of mosquitoes capable of spreading the disease. Because an artificial augmentation of the number of infected mosquitoes can exponentially increase the spread of the disease, it can serve as an effective form of weaponization.

One way to accomplish this would be to expose a flavivirus-infected animal or individual to mosquitoes. When the mosquitoes feed on the infected body, they will contract the virus. To collect the newly diseased mosquitoes, the temperature of the room should be lowered, which will induce a mosquito hibernation state, allowing the mosquitoes to be gathered in inconspicuous containers. The containers would then be transported to a target site and opened to release the mosquitoes. If this form of weaponization were carried out in warmer months, suspicion would be decreased; hospitals would be contaminated and the number of cases would be uncontrollable by the time terrorism could be implicated.[6,10,13,16]

THE LETHALITY OF THE VIRUS

Since the flavivirus is a Category B and Biosafety Level 3 pathogen, in weaponization both its lethality and the length of time that the virus stays alive may be exploited. In order to pursue this, a virulent strain must be isolated and then additional, previously sequenced virulent genes such as those from *Legionella pneumophila* can be cloned into the flavivirus genome. The resulting form is more lethal and can live longer than the original flavivirus.[3]

AEROSOLIZATION

Putting the newly formed flavivirus into a liquid medium makes aerosolization possible, maximizing its effectiveness as a weapon. The aerosolized virus can be disseminated by parachuting bomblets and by palm-sized magnet aerosolizers. Among the many advantages of aerosolization is the greater chance of precipitation of a virus from the air. Another is the higher likelihood that the pathogens will be inhaled and thus lodge more deeply in the lungs. Aerosolized liquid would be more effective than aerosolized powder because the virus is more likely to aggregate in liquid.[6,10,13,16]

In the United States, work has already been done on aerosolization of the standard flavivirus.

POTENTIAL DEFENSES

The potential defense against a potential outbreak of viral encephalitis may be divided into prevention, detection, and response. All defenses certainly involve a significant economic cost. Quick and effective communication between the general public; government at the local, state, and federal levels; and individuals involved in health care is very important.[3]

PREVENTION

Prevention includes measures such as recognizing, reporting, and properly addressing such mosquito breeding grounds as stagnant water, trash piles, and discarded tires as well as certain plants. It is important to use an insect repellent that contains a small amount of DEET, which has been shown to be very effective against mosquitoes (although DEET is known to be dangerous, in minimal concentrations it does not seem to have significant harmful effects).[3]

DETECTION

Detection involves reporting cases of dead birds, horses, and humans. Swift detection and open-air tests are imperative. Case reports play a major role in creating awareness of heightened danger, which can lead to increased preventive measures (e.g., vaccinations) and treatments in the area.[3]

RESPONSE

Immediately upon the detection of a case or outbreak of viral encephalitis—whether natural or terrorist-related—it is essential that the patients be treated and that the number of cases be contained. Patients should be treated with supportive care along with an effective drug such as acyclic nucleoside analog. In the effort to limit the number of cases, an emergency biohazard suit and mask are very important because they protect the skin in a mosquito attack and the mucosal linings in an aerosolized attack. Also, an effective vaccine, preferably one that does not contain a live pathogen, could be administered not only to residents in the immediate location but also to those in the surrounding area to contain the outbreak.[3]

REFERENCES

1. Abrahamson, I. 1935. *Lethargic Encephalitis.* New York: privately published.
2. Booss, J. and Esiri, M. M. 1986. *Viral Encephalitis: Pathology, Diagnosis, and Management.* Boston: Blackwell Scientific Publications.
3. Centers for Disease Control and Prevention. Online: http://www.cdc.gov.
4. Chambers, T. J. 1998. Flavivirus, infection and immunity. In *Encyclopedia of Immunology,* 2nd ed., pp. 926–932. London: Academic Press.
5. Chaudhuri A., Kennedy P. G. E. 2002. Diagnosis and treatment of viral encephalitis. *Postgrad Med J* 78: 575–583.
6. Day J. F. 2001. Predicting St. Louis encephalitis virus epidemics: Lessons from recent, and not so recent, outbreaks. *Annu Rev Entomol* 46: 111–138.
7. *Fields Virology,* 4th ed. 2001. D. M. Knipe and P. M. Howley, eds. Philadelphia: Lippincott Williams and Wilkins.
8. Fields, W. S. and R. J. Blattner. 1958. Viral Encephalitis: A Symposium. Fifth Annual Scientific Meeting of the Houston Neurological Society Texas Medical Center, Houston Texas. Bannerstone House, Springfield, Ill.
9. Gutierrez, K. M. and Prober, C. G. 1998. Encephalitis: Identifying the specific cause is key to effective management. *Postgrad Med* 103(3).
10. Hall, A. J. 1924. *Epidemic Encephalitis.* New York: William Wood.
11. *Harrison's Principles of Internal Medicine,* 14th ed. 1998. New York: McGraw-Hill.
12. Johnson, R. T. 1998. *Viral Infections of the Nervous System,* 2nd ed. New York: Lippincott-Raven.
13. Leslie T. and G. Fite. 1933. *A Virus Encountered in the Study of Material from Cases of Encephalitis in the St. Louis and Kansas City Epidemics of 1933. Science* (new series) 78(2029):463–465.
14. Matheson Commission. 1932. *Epidemic Encephalitis: Etiology, Epidemiology, Treatment Second and Third Reports.* New York: Columbia University Press, p. 2.
15. Mori, A., T. Tomita, O. Hidoh, Y. Kono, and D. W. Severson. 2001. Comparative linkage map development and identification of an autosomal locus for insensitive acetylcholinesterase-mediated insecticide resistance in *Culex tritaeniorhynchus. Insect Mol Biol* 10: 197–203.

16. Rayburn, C. R. 1932. *Epidemic Encephalitis*. Oklahoma: E. A. Johnson.
17. Strauss, I., T. K. Davis, A. M. Frantz. 1932. *Infections of the Central Nervous System: An Investigation of the Most Recent Advances*. Baltimore: Williams and Wilkins.
18. Wolstenholme, G. E. W and M. P. Cameron. 1961. *Virus Meningo-Encephalitis*. Boston: Little, Brown.

CHAPTER 3

■ ■

BOTULISM

(CLOSTRIDIUM BOTULINUM)

Rian Balfour

The bacteria *Clostridium botulinum* produces a toxin (botulinum neurotoxin) that causes the neuroparalytic disease known as botulism. Botulinum neurotoxin (BoNT)—the agent of botulinum—is the most poisonous substance known. It has been cited as 100,000 to 3 million times more potent than the nerve gas sarin, which is considered one of the most dangerous chemical weapons and was the predominant chemical weapon used in World War II. The Centers for Disease Control and Prevention (CDC) categorize BoNT as a Category A bioweapon threat because of its tremendous potency and lethality, relative ease of production and transport, and the need for prolonged intensive care among infected individuals.

HISTORY

DISCOVERY AND ISOLATION

Botulism was first discovered in 1793 by Justinius Kerner, a German physician. This case was of foodborne botulism, and Kerner cited a substance, endemic to spoiled sausage, that would later be discovered and termed *botulinum neurotoxin*. At that time, sausage was crudely produced and preserved by filling a pig's stomach with meat and blood, boiling it in water, and storing it at room temperature for days. These processes allowed spores of *C. botulinum* to survive and infect individuals with foodborne botulism. In 1897, Emile von Ermengem isolated *C. botulinum* from a foodborne outbreak in Ellezelles, Belgium. Von Ermengem is credited with identifying *C. botulinum* as the cause of botulism.[5]

Botulism got its name from those first discovered cases that implicated spoiled sausage as the causative agent of the disease. The term is derived from the Latin word *botulus*, which means "sausage." The derivation was particularly significant in Europe until the early 1900s, when the initial cases of botulism were linked to improperly cured and/or processed sausages. This connection is no longer significant—plant products, rather than animal products, are now the more common vehicles of *C. botulinum*.

USE IN BIOWARFARE

The use of BoNT as a bioweapon was first recorded in the 1930s during Japan's occupation of Manchuria. Manchurian prisoners were fed cultures of *C. botulinum* by the Japanese Biological Warfare group (Unit 731), and lethal effects were observed on the prisoners. The U.S. biological weapons program began producing BoNT during World War II. The production of BoNT during the war was so rampant in opponent countries, such as Germany, that more than 100 million toxoid vaccine doses were prepared for Allied troops involved in the D-Day invasion of Normandy.[1]

Although the executive orders of President Nixon ended the U.S. biological weapons programs in 1970 and the 1972 Biological and Toxin Weapons Convention prohibited bioweapon research and production, Iraq and the Soviet Union continued to produce BoNT as a potential bioweapon.

Following the Persian Gulf War in 1991, Iraq admitted to UN inspectors that it had produced 19,000 liters of concentrated BoNT, which is about three times the amount needed to kill the entire human population by inhalation. In 1990, Iraq filled missiles with a range of 13–600 km and bombs weighing 100–400 lb with BoNT and "prepared" them for immediate use. (Thankfully, these destructive weapons were never released on the world.)

Between 1990 and 1995, the Japanese cult Aum Shinrikyo obtained *C. botulinum* from soil collected in northern Japan and dispersed BoNT aerosols in Tokyo and against U.S. military installations in Japan. The group released BoNT on at least three occasions, but these attacks failed because of faulty microbiological techniques, poor aerosolization equipment, and possible internal sabotage.

The development of bioweapons, specifically BoNT, is still believed to be a problem today. Of the seven countries listed by the U.S. government as "state sponsors of terrorism," four—Iran, Iraq, North Korea, and Syria—have developed or are developing BoNT as a weapon. The threat of BoNT is real; "it is estimated that a point-source aerosol release of BoNT could incapacitate or kill 10% of persons within 0.5 km downwind" (William C. Patrick, unpublished data). Another likely terrorist use of BoNT is deliberate contamination of food, which could produce either a single, concentrated epidemic or episodic, widely separated outbreaks.

MOLECULAR BIOLOGY

DESCRIPTION

Clostridium botulinum is a rod-shaped, gram-positive, spore-forming, anerobic and motile bacterium that infects mammals and birds. The species is separated into four genetically diverse groups that share the ability to produce botulinum toxin, which is the most significant and lethal feature of *C. botulinum*. The toxin, also known as botox, is the most lethal and poisonous substance known.

The bacteria measure 0.5–2.0 µm in width and 1.6–22.0 µm in length. They grow naturally in soil from which they can be easily isolated, and they are present in the gastrointestinal (GI) tracts of both animals and humans. *C. botulinum* depends on water for its growth, and the bacteria are unable to grow in dehydrated foods or foods that contain high concentrations of sugars or salts.

The limiting factors for *C. botulinum* growth in foods are temperature, pH, water activity, redox potential, food preservatives, and competing microorganisms. *C. botulinum* strains, broadly grouped as *proteolytic* or *nonproteolytic*, possess varying growth conditions. (Proteolytic strains self-activate their neurotoxin by cleaving the inactive holotoxin, which forms the active dichain toxin, whereas nonproteolytic strains rely on the host proteases to active their toxin.) Both proteolytic and nonproteolytic strains grow optimally at 40°C, but nonproteolytic strains are able to thrive in harsher growth conditions. The minimum temperature for growth of proteolytic strains is 10°C, whereas that for nonproteolytic strains is 3.3°C.

Normally, the minimum pH is 4.6 for proteolytic strains and 5.0 for nonproteolytic strains. Some food proteins appear to have protective effects on the bacteria and allow them to grow at pH levels below 4.6. Furthermore, certain food preparations may contain low-acid "pockets," which also support the growth of *C. botulinum* and the production of BoNT.

Since *C. botulinum* bacteria are anaerobes, the optimal redox potential (E) for bacteria growth is −350 mV. This low $E_{-350 \text{ mV}}$ is consistent with a low O_2 environment, which the bacteria prefer. (As a consequence to *C. botulinum* normal growth conditions, vacuum-packing foods may promote their growth and toxin production.) However, BoNT production has been observed at $E_{+250 \text{ mV}}$. This finding explains why *C. botulinum* growth and BoNT production is possible in foods with high O_2 levels. Food preservatives such as nitrite, sorbic acid, parabeus, phenolic antioxidants, polyphosphates, and ascorbates inhibit *C. botulinum* growth and toxin production.

As is often the case, *C. botulinum* spores are capable of surviving much harsher conditions than the vegetative bacteria. Many *C. botulinum* spores are thermally resistant and less susceptible to higher pH levels and lower NaCl contents. Spores of some *C. botulinum* strains must be heated above the boiling point to ensure destruction.

THE FOUR GROUPS

As noted earlier, there are four genetically diverse groups of *C. botulinum* (table 3.1). The bacteria are considered the same species only because of their shared ability to produce neurotoxins—BoNTs—with similar pharmacological effects. Each strain of *C. botulinum*, belonging to a specific group, produces one of the seven forms of botulism, which differ in their severity on affected hosts. BoNTs are alphabetically classified as BoNT/A to BoNT/G according to their absence of cross-neutralization. For instance, anti-A antitoxin does not neutralize BoNTs/B–G, anti-B antitoxin does not neutralize BoNTs/A and C–G, etc. Furthermore, the serotypes also serve as epidemiological markers.

TABLE 3.1 CLASSIFICATION OF *C. BOTULINUM*

	I	II	III	IV
Toxin types	A, F, and proteolytic B	E, nonproteolytic B and F	C1, C2, and D	G
Disease in humans	+	+	−	?
Disease in mammals	Rare or ?	Rare or ?	+	−
Disease in birds	+	+	+	−
NaCl inhibition	10%	5%	3%	>3%
Minimum growth temperature	10–12°C	3.3°C	15°C	>3°C
	(50–54°F)	(38°F)	(59°F)	(>37°F)
Optimal growth temperature	37°C	30°C	37–40°C	25–45°C

Source: Adapted from E. R. First, Neuropharmacology Laboratory, Boston University School of Medicine (unpublished data).

Note: The bacteria can be grouped according to the BoNT toxin type they produce or their metabolic activity. Also shown are the toxin types and the animals they affect. It should be noted that both proteolytic and nonproteolytic strains produce botulism in humans and animals.

In addition to their serotypical classification, BoNTs are subdivided according to whether they are proteolytic or nonproteolytic. Recall that proteolytic strains activate their own BoNT internally, whereas the nonproteolytic strains require host enzymes. This is exemplified by a protease such as trypsin within a victim's stomach that could activate his own BoNT. The proteolytic types of BoNT include A, B, F, and G, and the nonproteolytic types include B, E, and F. BoNTs C_1, C_2, and D vary from being slightly proteolytic to nonproteloytic. (Note that both proteolytic and nonproteolytic forms of BoNTs B and F exist.) In addition, some BoNTs, such as C_1 and D, are bacteriophage-induced while others do not require a bacteriophage.

BoNT types A, B, E, and F have been linked to diseases in humans, whereas types C and D cause diseases in birds and nonhuman mammals. Type G has been isolated from Argentinean soil, but it has not yet been linked to any human or nonhuman cases of botulism. Between 1980 and 1996, 135 outbreaks of foodborne botulism were reported in the United States. Of those outbreaks, the sources were as follows: 54.1% type A, 14.8% type B, 26.7% type E, and 1.5% type F; in 3.0% the type of BoNT was unknown. Although types C and D have not yet been linked to human botulism cases, humans are believed to be susceptible to these toxin types because they cause foodborne botulism in primates. Furthermore, and perhaps most relevant to bioterrorism, aerosol studies have shown that primates are susceptible to BoNTs/C, D, and G when inhaled.[4]

MECHANISM OF ACTION

Botulinum neurotoxins are produced by *C. botulinum* as 150-kd single-polypeptide chains that consist of approximately 1290 amino acids (figure 3.1). The toxins are proteolytically activated by the cleavage of the holotoxin to form a heavy (H) chain (100 kd) and a light (L) chain (50 kd) linked by a single disulfide bond and noncovalent bonds.[6,9] Therefore, the active form of BoNT is a two-chain structure. Proteolytic *C. botulinum* strains release, on bacterial lysis, the active two-chain toxin, whereas nonproteolytic *C. botulinum* strains release the single-chain toxin, which is activated (cleaved into the H and L dichain structure) by host proteases, such as trypsin.[2] Although the amino acid sequences and the three-dimensional structures of the seven BoNT serotypes differ, the general protein domains as well as the toxins' actions are similar. Therefore, in the following sections, the actions of BoNTs are discussed based on their general domain properties. This explanation is followed by a discussion of the differences between BoNTs, which are primarily associated with specific active sites and substrates.[8,9]

In BoNT's mechanism of action, the H chain is believed to play an accessory role, whereas the L chain is believed to contain the toxic activity. The H

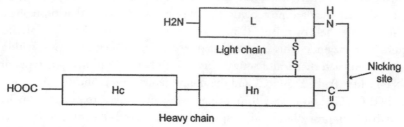

FIGURE 3.1. Schematic drawing of BoNT/B showing the organization of the toxin and modeling the functional domains of all BoNTs, which mediate cell intoxication by the toxin. The two chains and the three domains of the heavy (H) chain are shown. The light (L) chain functions as a zinc-dependent endopeptidase that cleaves the SNARE proteins. The carboxy-terminal domain (Hc) of the heavy chain binds to the surface of target nerve cells. The amino terminal (Hn) of the heavy chain facilitates the translocation of the L chain across the membrane.

Source: Adapted from Ref. 9.

chain is composed of two domains: a 50-kd amino terminal (H_n) and a 50-kd carboxy terminal (H_c). The H_c domain binds to gangliosides on the surface of target nerve cells (which specific one is unknown; this remains one of the least understood aspects of the mechanism of BoNT invasion), and the H_n is involved in translocating the L chain across the cell membrane. The L chain is a zinc-dependent endopeptidase, which confirms the toxicity of BoNTs, as this is the protein that cleaves the target substrate. The L chain cleaves the proteins known as SNARE proteins, which form the synaptic vesicle and fusion complex. Therefore, the toxic action of BoNT is dependent on three distinct protein domains and hence involves three steps: cell-surface binding, productive internalization, and intracellular expression of catalytic activity resulting in substrate cleavage.[9]

Vegetative cells produce the neurotoxin intracellularly, and the toxin is released during autolysis of the *C. botulinum* bacterium cell. (This action is sometimes mediated by a bacteriophage in the serotypes C_1 and D.) The neurotoxin causes the paralyzing effects of botulism by suppressing and blocking the presynaptic release of acetylcholine (ACh), the neurotransmitter responsible for initiating muscle contractions.

Under normal neurotransmitter release, the arrival of the nerve impulse at the presynaptic terminal causes Ca^{2+} ions to enter the terminal (figure 3.2). The influx of Ca^{2+} in turn releases vesicles containing neurotransmitters, which in this case contain ACh. At the same time, the SNARE complex of proteins—SNAP-

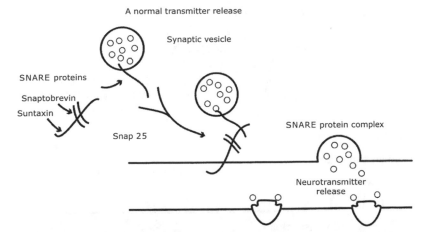

A normal transmitter release

Synaptic vesicle

SNARE proteins

Snaptobrevin

Suntaxin

Snap 25

SNARE protein complex

Neurotransmitter release

FIGURE 3.2. Normal neurotransmitter (NT) release. The arrival of the nerve impulse at the presynaptic terminal causes Ca^{2+} ions to enter the terminal. The influx of divalent cations releases vesicles containing Ach at the neuromuscular junction. The SNARE complex of proteins (synaptobrevin and syntaxin) forms; this complex enables synaptic vesicles containing Ach to dock with the plasma membrane. Once the vesicles have docked, membrane fusion occurs, and Ach is released into the synaptic cleft. Ach then binds to receptors located on the muscle cell, and this binding leads to muscle contraction.
Source: Adapted from Ref. 1.

25, VAMP (synaptobrevin), and syntaxin—forms, and this complex enables the synaptic vesicles containing ACh to dock with the plasma membrane. After membrane fusion, ACh is released into the synaptic cleft and then bound by receptors on the muscle cell. Once this binding occurs, the muscle contracts.

The foregoing describes normal muscle contraction; however, muscle contraction is greatly affected when BoNT enters animal cells. When this occurs, the toxin is first absorbed into the circulation from either a mucosal surface (lung or gut) or a wound. Once BoNT is absorbed, the bloodstream carries it to the peripheral cholinergic synapses, primarily the neuromuscular junction (figure 3.3). At the neuromuscular junction, the terminal H_c chain binds to gangliosides, located on the presynaptic membrane, and a protein receptor. The gangliosides are considered low-affinity receptors, and the ganglioside binding is believed to bring the toxin close to a high-affinity protein receptor or induce a conformational change at the protein receptor site to promote binding.

The second step in neuronal cell intoxication by BoNT is called productive internalization. It is distinguished by two major events: the internalization of the toxin through receptor-mediated endocytosis and the pH-de-

Neuronal cell intoxication

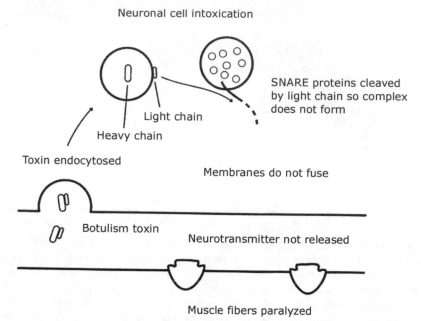

Light chain

Heavy chain

SNARE proteins cleaved by light chain so complex does not form

Toxin endocytosed

Membranes do not fuse

Botulism toxin

Neurotransmitter not released

Muscle fibers paralyzed

FIGURE 3.3. Transmitter release affected by botulinum neurotoxin exposure. Once BoNT has been absorbed into circulation from a mucosol surface or wound, the bloodstream carries it to peripheral cholinergic synapses located primarily in the neuromuscular junctions. At the neuromuscular junction, BoNT binds irreversibly to the presynaptic membrane through gangliosides and a protein receptor. The carboxy-terminal region of the H chain mediates this binding. Following binding, BoNT is internalized into the neuron through endocytosis, and its L chain is translocated across the endosomal membrane. The committed step to BoNT's toxic action is the enzymatic cleavage of SNARE proteins. Specifically, the L chain acts as a specific endopeptidase against synaptobrevin, syntaxin, or SNAP-25.[5]

pendent translocation of the L chain across the endosomal membrane. The acidity (low pH) of the endosome containing the BoNT is believed to induce conformational changes of the translocation domain (Hn). Although these processes of BoNTs are least understood, current research suggests that the structural changes induce a porelike or protein-transporter structure of the Hn domain, which allows the L chain to be translocated across the endosomal membrane. Consistent with these findings, productive internalization of BoNT may also result in the reductive cleavage of the disulfide bond between the H and L chains. Furthermore, separation of the H and L chains may facilitate the exposure of the zinc catalytic site that is required to cleave the SNARE proteins.

The last step of BoNT's toxic action is intracellular expression of catalytic activity and substrate cleavage. Specifically, the L chain acts as a specific endopeptidase against one of the three synaptic vesicle docking and fusion proteins, collectively referred to as SNARE proteins, that facilitate the exocytosis of neurotransmitters. The three SNARE substrate proteins include synaptobrevin, which is found on the synaptic vesicle, and syntaxin and SNAP-25, which are both found on the plasma membrane. The cleavage of the SNARE proteins occurs before they can form a complex, preventing the vesicles containing ACh from docking and fusing with the presynaptic membrane. This prevents the release of ACh from the synaptic terminal, leading to muscular paralysis.

BoNTs/B, D, F, and G cleave synaptobrevin; BoNTs/A, C, and E cleave SNAP-25; and BoNT/C cleaves syntaxin. The substrate specificity between serotypes is believed to be due to two substrate-recognition sites on the L chain—BoNT's catalytic domain. Furthermore, cleavage occurs only if the substrate binds to both sites. Substrates of a minimum length with a specific nine-amino-acid residue region, known as the SNARE secondary recognition (SSR) sequence, are recognized by different serotypes. Each BoNT substrate—SNARE protein—contains the SSR sequence in at least two locations, and each BoNT serotype is able to bind to all SSR sequences. However, substrate cleavage occurs only when the bound SSR sequence is the proper distance from the BoNT's scissile bond. Therefore, BoNT's substrate specificity is derived from the spatial relationship between a SSR sequence and the serotype's specific cleavage site on the different SNARE proteins.

CLINICAL DIAGNOSIS AND RESPONSE

There are four forms of botulism, classified according to their modes of BoNT transmission. Of the four, three are natural: foodborne, intestinal (subdivided into infant and adult), and wound botulism. The fourth, a synthetic form made by aerosolizing BoNT, is known as inhalation botulism, and it is the most probable candidate for a biological weapon aimed to cause botulism. All forms of botulism ultimately involve BoNT absorption and deliverance to peripheral cholinergic synapses, specifically the neuromuscular junction. As such, all forms of botulism present the same symptoms of afebrile (without fever), symmetric, descending flaccid paralysis.[1]

TYPES OF BOTULISM

Foodborne botulism is caused by ingesting food that contains the preformed BoNT. This occurs when foods that contain *C. botulinum* spores are improperly canned and/or processed. The incorporation of *C. botulinum* spores in foods is due to the bacteria's natural presence in soil. Faulty canning and preservation techniques that do not control for pressure, temperature, and pH create environments conducive to *C. botulinum* growth. As the spores germinate in the foods, the bacteria produce BoNT. After ingestion, BoNT is absorbed into the bloodstream from the mucosa of the stomach and the upper GI tract. It is then carried to peripheral neurons at myoneural junctions, where it initiates its toxic actions. Between 1950 and 1996, vegetables were the primary vectors of foodborne botulism. Other vectors included fish and marine animals and—to a lesser extent—beef, milk products, pork, and poultry.

Intestinal botulism occurs when *C. botulinum* spore–contaminated food is ingested. The spores then germinate in the victim's GI tract. The resulting vegetative cells multiply and produce BoNT intracellularly. Following autolysis of *C. botulinum* cells, BoNT is released into the gut. Via the same pathway described for foodborne botulism, BoNT carries out its toxic actions.

Although intestinal botulism can affect adults, it primarily affects infants. About 20% of infant intestinal botulism cases are attributed to honey containing type B *C. botulinum* spores. Other sources of intestinal botulism include corn syrup and environmental conditions susceptible to *C. botulinum* spore invasion.

Wound botulism occurs when *C. botulinum* spores germinate and produce BoNT in devitalized tissue. The toxin enters the victim's bloodstream and exerts its effects via the same pathway for foodborne and intestinal botulism.

Inhalation botulism—the manmade form of the disease and the most likely candidate for biowarfare involving botulism—occurs when aerosolized BoNT is inhaled.

SIGNS AND SYMPTOMS OF EXPOSURE

The ultimate clinical manifestation of botulism is essentially the same regardless of the type of botulism (foodborne, intestinal, wound, or inhalation) infection. These symptoms include symmetric descending flaccid paralysis with prominent bulbar palsies (paralysis of the muscles of the tongue, lips, palate, pharynx, and larynx), diplopia (double vision), dysarthria (labored speech), dysphonia (altered or difficult voice production), and dysphagia (difficulty in swallowing). However, the victim is afebrile and cognitively alert and aware.

Although the ultimate symptoms for all types of botulism are the same, the incubation period and initial symptoms differ. The average incubation period of foodborne botulism is between 18 and 36 hours. Probably because of the ingestion of "bad" food, abdominal pain, nausea, vomiting, and diarrhea tend to precede or accompany the neurological symptoms. The incubation period for intestinal botulism is between 8 and 22 days. As with foodborne botulism, GI symptoms may precede or accompany the neurological symptoms. The incubation period for wound botulism is between 4 and 21 days. Unlike foodborne and intestinal botulism, wound botulism is not accompanied by GI symptoms. The incubation period for inhalation botulism in humans is not known to any degree of certainty because of the few cases (three to date). The three known victims of inhalation botulism began to show neurological symptoms approximately 72 hours after exposure to an undetermined amount of aerosolized BoNT. In experimental studies, monkeys show signs of inhalation botulism 12 to 80 hours following exposure to a median lethal dose.

Botulism is often misdiagnosed as myasthenia gravis, stroke, Guillain-Barré syndrome, food poisoning, tick paralysis, chemical intoxication (from carbon monoxide, barium carbonate, methyl chloride, methyl alcohol, or atropine), mushroom poisoning, medication reactions, poliomyelitis, or diphtheria. Additionally, infant botulism is often misdiagnosed as meningitis, electrolyte–mineral imbalance, metabolic encephalopathy, Reye's syndrome, Werdnig-Hoffman disease, congenital myopathy, and Leigh's disease. Routine laboratory studies are usually not any more useful at distinguishing between botulism symptoms and symptoms of other diseases. However, a specialized laboratory test, known as a mouse bioassay, can determine the toxin type of a botulism infection. Furthermore, fecal and gastric specimens can be cultured to confirm the presence of *C. botulinum* spores, vegetative ells, and/or BoNT. Because these tests usually take days to complete (1–2 days for mouse bioassay and 7–12 days for cultures) and yield definitive results, physicians need to rely primarily on clinical signs when diagnosing botulism.

To avoid misdiagnosing patients, physicians should consider botulism, given the following indications: acute onset of GI symptoms, autonomic or cranial-nerve dysfunction (dry mouth, diplopia, dysarthria, and dysphagia), neck or peripheral muscle weakness, respiratory distress. Physicians should also be aware of the following signs in infants: poor feeding, diminished sucking, and reduced crying ability.

TREATMENT

Botulism does not discriminate—individuals of all ages, nationalities, and creeds are susceptible to the disease. (Note, however, that certain forms of botulism are more prevalent among specific groups of people. For instance, intestinal botulism usually occurs in infants, and injection-drug users are more susceptible to wound botulism.) Although botulism is considered a lethal disease because it can cause acute respiratory failure, improved treatments have substantially lowered the mortality rates associated with botulism. For instance, the mortality rate due to foodborne botulism was 25% between 1950 and 1959 but only 6% between 1990 and 1996. Similar reductions have been observed for intestinal and wound botulism.

The reduced mortality rate of botulism is attributed to advances in therapies and supportive care. The first line of treatment is to administer the equine trivalent botulinum antitoxin. The antitoxin contains antibodies for BoNTs/A, B, and E, which are the most common causes of human botulism. If botulism occurs from another BoNT serotype (C, D, F, or G), an investigational heptavalent antitoxin, from a stock maintained by the U.S. Army, can be administered. The antitoxin works to limit the neurological progress of botulism or shorten the duration of respiratory failure. It is most effective when administered early because the antitoxin can only neutralize BoNT molecules that have not yet bound to neuronal membranes. Another drawback is sensitivity associated with the antitoxin.

Secondary treatments for botulism include supportive care—tube feeding, mechanical ventilation, and treatment of secondary infections. Supportive care persists for the duration of paralysis, which can last for several months.

POTENTIAL DEFENSES

VACCINE

In addition to the trivalent antitoxin and supportive care for botulism treatment, prophylaxis can also be accomplished, with administration of a pentavalent vaccine (toxoid) for BoNT serotypes A–E. This toxoid, available only from the CDC, has been given to military personnel and laboratory workers who work with BoNTs, and it has proved to be both safe and effective.

PROPER FOOD PRESERVATION AND STORAGE

Other methods for preventing botulism relate directly to specific botulism forms. Because BoNT is inactivated by heating to a temperature of 85°C for 5 minutes, foodborne botulism can be prevented by briefly heating foods to this internal temperature. *C. botulinum* spores are considerably harder to destroy, and efforts to limit intestinal botulism may not always be effective. It is clear, however, that honey should not be fed to infants, because of the demonstrated presence of *C. botulinum* spores in many samples. If an aerosolized BoNT release is suspected, covering mucosal surfaces, such as the eyes, mouth, and nose, with clothing may hinder entry of the toxin.

DRUGS

Current research on the structure of BoNTs and the SNARE proteins (synaptobrevin, SNAP-25, and syntaxin) and on modes of neuronal cell intoxication may lead to future defenses against botulism. For instance, drugs may be developed to target specific active sites of BoNTs and SNARE proteins. The BoNT active site of membrane fusion to gangliosides is a potential target; specific gangliosides might also be targets. The findings of Yowler and colleagues show that BoNT/A activity is dependent on the presence of gangliosides. Interfering with BoNT binding to gangliosides may thus prevent internalization of the toxin and its effects.[11]

Another BoNT active site is its zinc-dependent protease site, which is responsible for cleavage of SNARE proteins. The prompt administration of chelators, such as EDTA and TPEN, may strip zinc from BoNTs and render them inactive. Although this treatment may be the most universal among all BoNT serotypes, the findings of Simpson and colleagues suggest that chelation therapy is ineffective because BoNT can regain zinc from endogenous stores.[8]

Other feasible defenses against botulism include developing drugs that target cleavage sites on SNARE proteins. Because all the SNARE proteins contain SSR sequences that are recognized by all BoNT serotypes, drugs that block BoNT binding to SSR sequences would prevent protein cleavage. These drugs would also act against multiple BoNT serotypes, making them extremely efficient defenses. Additionally, drugs that target cleavage sites of specific SNARE proteins by binding and protecting said sites from cleavage can also be developed. However, since each BoNT serotype cleaves specific proteins at type-specific sites (see table 3.2 for a partial list of type-specific cleavage sites on SNARE proteins), the development of these drugs may be time-consuming and inefficient.

In addition to structure-targeting drugs as defenses against botulism, the findings of Nowakowski and colleagues suggest that an immune response against BoNT can be generated from recombinant oligoclonal antibodies. Their data show that BoNT/A can be substantially neutralized both in vitro and in vivo by combining two or three monoclonal antibodies. Besides offering protection against botulism, oligoclonal antibodies have also been shown to be effective against HIV and anthrax. Furthermore, since the elimination half-life

TABLE 3.2 TARGET PROTEINS OF BONT SEROTYPES
AND CLEAVAGE SITES ON SNARE PROTEINS

NEUROTOXIN	INTRACELLULAR SUBTYPE TARGET
A	SNAP-25 (Gln^{197}–Arg^{198} bond)
B	Synaptobrevin (Gln^{77}–Phe^{78} bond)
C_1, C_2, D	SNAP-25, syntaxin, synaptobrevin (Lys^{59}–Leu^{60} bond)
E	SNAP-25 (Arg^{180}–Ile^{181} bond)
F	Synaptobrevin (Gln^{58}–Lys^{59} bond)
G	Synaptobrevin

Source: Adapted from Refs. 2 and 10.

Note: These sites may serve as targets for drugs against BoNT.

of these antibodies is approximately 4 weeks, oligoclonal antibodies can both confer protection as a prophylaxis regimen and be used for treatment.[7]

WEAPONIZATION

BoNT would most likely be used as a bioweapon in an aerosolized form to cause inhalation botulism or as a deliberate food contaminant to cause food-borne botulism.

AEROSOLIZATION

As mentioned earlier, BoNT is the most poisonous substance known. Lethal doses of BoNT—extrapolated from primate studies—are as follows: 0.09–0.15 μg intravenously or intramuscularly and 70 μg orally. Therefore, 1 g of BoNT, dispersed and inhaled evenly, would kill more than 1 million people. More realistic lethality estimates of BoNT are drawn from median lethal dose (LD_{50}) and specific dispersement-area studies. One such study calculates that 8 kg of concentrated BoNT dispersed under optimal meteorological conditions over an area of 100 km^2 would deliver an LD_{50} to the entire unprotected population of that area. The duration of aerosolized BoNT in a specific site depends on atmospheric conditions and the particle size of the toxin. Temperatures above 85°C and high humidity levels will degrade the toxin, and fine particles will dissipate into the atmosphere. It is estimated that BoNT decays between 1% and 4% per minute, and substantial inactivation of the toxin would be expected after 2 days. These extreme lethality properties of BoNT make the native toxin an attractive bioweapon in its aerosolized form.

GENE SPLICING

Recent advances in biotechnology and bioengineering techniques may passively augment the potential destructive nature of bioweapons. Specifically, gene splicing may allow for the creation of supergerms. These supergerms, such as Ebolapox, in which the Ebola genome (or part of it) is spliced into the smallpox genome, exploit the most lethal or transmittable features of each pathogen. The supergerm thus becomes significantly more potent and lethal than the single pathogen. Like Ebola, the gene for BoNT can be spliced into other pathogens, such as smallpox, which can confer a high rate of infectivity to the otherwise uncommunicative BoNT. Therefore, a supergerm consisting of a BoNT-producing highly contagious organism could be responsible for monumentally devastating effects of biowarfare.

REFERENCES

1. Arnon, S., R. Schechter, et al. 2001. Botulinum toxin as a biological weapon. *JAMA* 285(8): 1059–2081.
2. Binz, T., J. Blasi, et al. 1994. Proteolysis of SNAP-25 by types E and A botulinal neurotoxins. *J Biol Chem* 269(3): 1617–1620.
3. Center for Civilian Biodefense Strategies. 2002. *Botulinum Toxin.* Online: http://www.state.gov/s/ct/rls/pgtrpt/2000/2441.htm. Accessed Oct. 31, 2002.
4. Centers for Disease Control and Prevention. 1998. *Botulism in the United States, 1899–1996. Handbook for Epidemiologists, Clinicians, and Laboratory Workers.* Atlanta: Centers for Disease Control and Prevention.
5. Hanson, M. A. and R. C. Stevens. 2000. Cocrystal structure of synaptobrevin-II bound to botulinum neurotoxin type B at 2.0 A resolution. *Nature Struct Biol* 7(8): 687–691.
6. Lacy, D. B., W. Tepp, et al. 1998. Crystal structure of botulinum neurotoxin type A and implications for toxicity. *Nature Struct Biol* 5(10): 898–902.
7. Nowakowski, A., C. Wang, et al. 2002. Potent neutralization of botulinum neurotoxin by recombinant oligoclonal antibody. *PNAS* 99(17): 11346–11350.
8. Simpson, L. L., A. B. Maksymowych, et al. 2001. The role of zinc binding in the biological activity of botulinum toxins. *J Biol Chem* 276(29): 27034–27041.
9. Singh, B. R. 2000. Intimate details of the most poisonous poison. *Nature Struct Biol* 7(8): 617–619.

10. Yamasaki, S., A. Baumeister, et al. 1994. Cleavage of members of the synaptobrevin/ VAMP family by types D and F botulinal neurotoxins and tetanus toxin. *J Biol Chem* 269(17): 12764–12772.

11. Yowler, B. C., R. D. Kensinger, et al. 2002. Botulinum neurotoxin A activity is dependent upon the presence of specific gangliosides in neuroblastoma cells expressing synaptotagmin I. *J Biol Chem* 277(36): 32815–32819.

CHAPTER 4

■ ■

TULAREMIA

(FRANCISELLA TULARENSIS)

James Hudspeth

*F*rancisella tularensis, the bacterium causing the disease tularemia, has been a subject of biological weapons research since World War II, and was developed into a deployable weapon by both sides during the Cold War. It occurs infrequently in nature, infecting humans through contact with infected animals, bites from infected insects, inhalation of aerosolized substances such as rodent droppings, and the consumption of contaminated food and water. Although the disease can be readily treated with basic antibiotics, reducing the frequency of fatalities to around 2.5% from the preantibiotic mortality rate of 50%, its extreme infectivity and virulence mean that it remains a considerable threat. Because of these reasons, the U.S. Centers for Disease Control and Prevention rates *F. tularensis* as one of six Category A biological agents, agents with "potential for major public health impact."

HISTORY

Francisella tularensis, the causative agent of the zoonosis tularemia, was first described in the English-language literature as a bacterium infecting rodents in 1911 in Tulare County, California. Shortly thereafter, Lamb and Wherry identified the first case of human tularemia. Dr. Edward Francis, one of the premier researchers on tularemia, wrote several reviews on the disease after its discovery in humans, developing the hypothesis that the disease spread from rodents to humans through bloodsucking insects. In later commemoration of his work the bacterium was given its present name, Francis-*ella* tulare-*nsis*.

Reported cases of tularemia rose to high levels throughout the United States and world during the first half of the twentieth century, peaking in the United

States with 2291 cases in 1939 and worldwide during World War II among the German and Russian troops on the Eastern front. This was followed by an equally dramatic decline in cases over the past 50 years, to the present level of approximately 200 cases a year in the United States. Public health and clinical responses to tularemia include vaccination campaigns in the former Soviet Union and the development of effective antibiotics (the first being streptomycin, introduced in 1944). The concurrent decrease in hunting in the United States and other industrialized nations also reduced human contact with animals and arthropods infected with *F. tularensis*.

Tularemia remains endemic throughout the Northern Hemisphere, with recent outbreaks in Spain and Kosovo, but has largely ceased to be a serious public health threat. Recently two small outbreaks of pneumonic tularemia in Martha's Vineyard raised fears of a biological weapons attack, because of the scarcity of pneumonic disease in natural infections.

Tularemia was pegged as a possible biological weapon from the earliest such research, with Japan's Unit 731 using it in experiments on prisoners of war in Manchuria during the Japanese occupation of China. After World War II, both the Soviets and the North American–British biological weapon researchers worked to develop preparations of tularemia for battlefield usage, and in 1958 tularemia became the American weapon of choice for biological retaliation. Although American development ceased in 1966, Soviet research and development of tularemia weapons continued through the end of the USSR into the early 1990s, and Russian research may continue today.

Every major weapons program performed a number of open-air tests with both real and simulated biological weapons; typically, real weapons were tested on animals in remote locations, while supposedly harmless simulants were used over populated areas to measure the effective doses that would reach people at varying distances. One such test took place in the New York City subway system—interestingly, the British and the Russians also undertook tests in subway systems. Presumably the combination of numerous civilians, a lack of ultraviolet light, and pneumatic pressure caused by trains suggested to all three countries the possibility of a biological attack on their train systems.

The New York City test involved dropping lightbulbs filled with bacteria onto the tracks, then observing the number of bacteria seen at different points throughout the system. Combined with observations of the number of passengers per train and the average time they spent in the subway, this information allowed the team to predict the number of casualties that would have occurred if the bacterial simulant had been a virulent agent. Similar tests took place throughout the 1950s and 1960s, including some tests of live *F. tularensis* on animals.

MOLECULAR BIOLOGY

SUBSPECIES

F. tularensis is a small (0.2 × 0.2–0.7 μm), nonmotile, aerobic, gram-negative coccibacillus that infects mammals, birds, and arthropods. The species is divided into three subspecies, of which tularensis (often termed type A) and holartica (often termed type B) are of primary clinical importance. Type A is found almost exclusively in North America, although it has recently been isolated in Europe, while type B appears throughout the Northern Hemisphere. The two types can be distinguished through microbiological means, with type A producing acid when exposed to glycerol as well as demonstrating citrulline ureidase (a particular enzyme) activity.

The pathology and epidemiology of the two subspecies also differ; type A is considerably more virulent and has a main reservoir of ticks and host mammals, whereas the less potent type B is a waterborne pathogen with a primary reservoir of aquatic rodents. Developments in PCR techniques now permit genetic identification with a greater specificity than is possible with culture tests.[16,17] Recent PCR-based tests have even shown the ability to identify subspecies strains found in different regions and epidemics. Unless otherwise specified, the subspecies under discussion for the remainder of this chapter is the more virulent type A strain.

MECHANISM OF ACTION

Although *F. tularensis* is a facultative intracellular pathogen, capable of living either within cells or on medium in vitro, it acts as an obligate intracellular pathogen in vivo.[8] Regardless of its habitat, the bacterium requires an environment rich in iron and cystine, rendering standard laboratory media insufficient for the culturing of *F. tularensis*. It is a relatively hardy bacterium, capable of surviving in hay, water, soil, and meat for several days, but it does not form spores.

The cellular and molecular mechanisms of infection, proliferation, and the immune response in tularemia remain relatively obscure, although recent work by Gray et al. has discovered five genes necessary for the intracellular growth of *F. novicida*, a closely related bacterium.[11] Of particular interest are the similarities in the infectious process between *F. tularensis* and other intracellular bacteria such as *Listeria monocytogenes* (which infects immunocompromised patients) and *Mycobacterium tuberculosis* (the pathogen causing tuberculosis). Like many intracellular pathogens, *F. tularensis* infects macrophages and the parenchymal cells of the body through an unknown mechanism of entry not dependent on phagocytosis.[8] Once inside these cells *F. tularensis* proliferates within acidified vacuoles containing iron, with one to several bacteria in each compartment.[22] Note that these vacuoles are not lysosomes, nor do they fuse with lysosomes. Cytopathogenicity and apoptosis occur once a large number of viable bacteria have accumulated, with one experiment finding that the signs of incipient apoptosis appeared at 18 hours after bacterial entry. The same results also suggested that cell death in response to infection is apoptotic as opposed to necrotic.

Since the human inflammatory response to pathogen entry attracts macrophages to the afflicted area, *F. tularensis* essentially exploits the immune response to provide additional hosts (the responding macrophages) for further pathogen production. Moreover, in vitro experiments have shown that *F. tularensis* grows twice as fast inside macrophages arriving in response to inflammatory stimuli as inside macrophages local to the area of infection.[7] The accumulation of macrophages without the clearance of pathogen causes a positive feedback cycle resulting in the formation of granulomas, a process typical of intracellular pathogens in general. The sustained activation of the immune system presumably accounts for the lymphadenopathy, or lymph node swelling, characteristically observed in tularemia cases.

EXPERIMENTAL MODELS

Much of the research performed on tularemia uses mice as the model animal. Mice are typically infected with *F. tularensis* live vaccine strain (LVS) through intraperitoneal injections, resulting in death at very low doses.[2] Intradermal injections of LVS into mice, on the other hand, induce immunity to later LVS challenges, allowing survival of intraperitoneal doses 10^3 times the normal LD_{50}.

The level of confidence with which one can extrapolate from murine data to humans is somewhat unclear, for several reasons. Fortier demonstrated in 1994 several differences in the immune reaction to *F. tularensis* between humans and mice. Moreover, LVS is not normally virulent within humans but acts as a vaccine, as the name implies. Equally important from the perspective of biological weapons is a recent study from Conlan suggesting that the murine immune response to inhaled *F. tularensis* may differ significantly from the response to systemic infections.[2]

THE IMMUNE RESPONSE

The initial, nonspecific immune response in any infection comes primarily from neutrophils, which attack pathogens via the release of toxic factors such as active oxygen compounds, the activation of other immune cells, and the direct phagocytosis of foreign bodies. Further attacks on pathogens occur through the serum of the host organism, which contains bactericidal proteins collectively called complement. Virulent forms of *F. tularensis* seem relatively immune to both neutrophils and complement, an immunity that hinges in part on the bacterial capsule such strains possess. Mutants lacking the capsule can be phagocytosed and destroyed by neutrophils, and are susceptible to complement. *F. tularensis* also produces an acid phosphatase, AcpA, which plays a role in protecting the bacteria through inhibition of active oxygen compound production.

Although neutrophils are relatively ineffectual against the bacterium in vitro, they play a critical role in the initial containment of a mouse's first infection with *F. tularensis*. They presumably act to restrain bacterial proliferation while the specific immune response develops, although this may be specific to the murine response to LVS. Macrophages, the other phagocytic cells that contribute to the nonspecific immune response, fare even worse against *F. tularensis* while in an unactivated state, serving as bacterial hosts.

T-cell activation by intracellular pathogens occurs through an unknown mechanism, but presumably involves the presentation of antigenic proteins to nearby T cells by infected macrophages. Considerable research has been done on the contributions of the three primary subpopulations of T cells: CD4+ ("helper" T cells), CD8+ ("killer" T cells), and γδ T cells. Of these three types, γδ T cells seem to play little or no role in either immediate or long-term immune reactions, but the presence of either CD4+ or CD8+ T cells is necessary for effective resistance to *F. tularensis* infection in mice. The best protection results from the presence of both types of T cell.

T-cell-mediated protection could act through a variety of pathways, with one area of particular focus being macrophage activation by T cells. In response to infection, both CD4+ and CD8+ T cells are capable of releasing interferon gamma (IFN-γ), a cytokine that in turn causes macrophages to become active and release tumor necrosis factor alpha (TNF-α). Neutrophils can also release TNF-α when appropriately stimulated. IFN-γ, in conjunction with TNF-α, causes a number of cellular responses that combat *F. tularensis* infection, including increased phagocytosis by macrophages; sequestration of iron within activated macrophages, preventing *F. tularensis* and other intracellular bacteria from obtaining the necessary nutrients; and augmented release of nitric oxide from a variety of cells. The control and clearance of *F. tularensis* in mice have repeatedly been shown to rely on the presence of IFN-γ and TNF-α, and the efficiency of control correlates with the level of nitrogen oxides released by the immune system.[7,12]

B cells also play a part in the immune response of immunized mice to *F. tularensis*.[6] The role of B cells may hinge on their ability to suppress the mobilization of neutrophils, as accumulation of neutrophils in the spleen correlates with death in mice lacking B cells. The ability of antibodies to fight *F. tularensis* infection remains uncertain, with Elkins et al. finding no role,[6] while experiments by Fulop et al. demonstrate a limited protective effect conferred by antibodies against *F. tularensis* lipopolysaccharide (LPS).[10] Research does overwhelmingly indicate that IFN-γ, TNF-α, and either CD4+ or CD8+ T cells are necessary for a successful immune response and the development of long-term immunity to *F. tularensis*.[12,21] B cells appear to play a role in the development of immunity to secondary challenges.[6]

The recent paper by Conlan et al., discussed earlier, calls into question whether this general model of an immune response to *F. tularensis* holds true for both systemic and pulmonary infections. Specifically, mice depleted of neutrophils or IFN-γ and subsequently infected through injection show a greater bacterial load in the spleen, lungs, and liver, whereas similar animals infected by aerosol do not show such a marked increase.[2] This may be of import for research relating to the use of *F. tularensis* as a biological weapon, since the most probable avenue of infection with such a weapon would be through inhalation of an aerosol. If the immune response to a pulmonary infection is radically different from that to a systemic infection in mice, then extrapolations from experiments on mice using injected LVS would not provide an accurate experimental model for human infections caused by aerosols.

Unlike many potentially fatal bacteria, *F. tularensis* does not release or contain any toxin that causes host death. *F. tularensis* LPS is notably inert in its interactions with the immune system, unlike many other bacterial LPSs, which often stimulate LPS receptors present on several classes of immune cells. Death instead seems to be caused by complications related to pneumonia, a common symptom of tularemia, or by septic shock, a fatal drop in blood pressure resulting from the massive release of inflammatory cytokines in response to widespread infection.

CLINICAL DIAGNOSIS AND RESPONSE

LOCALIZED VERSUS SYSTEMIC INFECTION

As with mice, *F. tularensis* can infect humans through a number of different routes, including inhalation, ingestion, incisions, or abrasions, and the clinical presentation often depends on the route of infection. A system of seven ratings is sometimes used, distinguishing between ulceroglandular, glandular, oropharyngeal (throat), oculoglandular (eye), typhoidal, septic, and pneumonic tularemia. This list of ratings can be simplified by grouping the first four categories into "localized" infections.

Tularemia typically presents with an abrupt onset of fever, headache, chills, rigors, and sore throat regardless of the type of infection, and generally progresses with these symptoms as well as a subsequent loss of energy, appetite, and weight. Other frequent symptoms include coughing, chest tightness, nausea, vomiting, and diarrhea. These symptoms occur with severity sufficient to immobilize many people within the first two days of disease, regardless of the type of infection.

Localized presentations occur in response to a specific infection site such as a cut, tick bite, or entry into the conjunctiva. Adenopathy of the local lymph glands is common, as is the development of an ulcer at the site of inoculation. It is worth noting that localized *F. tularensis* infections can develop into systemic ones through hematogenous spread, and there is a documented instance in which draining of an enlarged lymph gland led to development of systemic disease.

In cases of systemic infection, *F. tularensis* exposure has usually occurred through aerosol inhalation or ingestion, and presents without any localized ulcers or lymph gland swelling.

GASTROINTESTINAL INFECTION

Infection through the gastrointestinal tract is most commonly seen with the type B subspecies, often after ingestion of contaminated water, although tularemia has occurred after people consumed meat contaminated with type A.

PNEUMONIC INFECTION

Pneumonia is common in the case of exposure to an aerosol; such aerosols have been naturally generated in grain inhabited by infected rodents, and by lawnmowers presumably sweeping over areas containing the bacterium. This is the type of tularemia most likely to result from a biological weapon attack using tularemia as an agent.

DIFFICULTY OF DIAGNOSIS

Several other types of bacteria cause initial symptoms similar to tularemia, and three genera in particular (*Brucella*, *Yersinia*, and *Pasteurella*) also have bacteria that respond similarly to microbiological tests. But even without considering these similar diseases, diagnosis of tularemia is difficult. Localized presentation of the disease is frequently recognized, but systemic infections are often not diagnosed—since tularemia is difficult to culture in a laboratory and often presents as a septicemia, it may be that some patients with systemic tularemia are cured via broad-spectrum antibiotics without a diagnosis having been made, resulting in an underreporting of cases. Definitive diagnosis is confirmed through either successful culturing of the bacterium or a significant rise in specific antibodies.[5]

Both of these methods have their limitations, however: culturing the bacteria is both difficult and somewhat dangerous to laboratory personnel, while the antibody response does not occur until several days after disease onset. Recently developed PCR-based techniques allow for a quicker identification of *F. tularensis* at a higher level of success than culturing.[16,17] Similar tests may soon provide routine identification of the specific strain infecting a patient, giving epidemiologists the ability to look for similarities to other outbreaks from different locations or times. This could provide clinicians with a better idea of the virulence and possible antibiotic resistances of the current infection. Strain identification would also play an important role in suspected bioterror attacks, giving the investigators information on the virulence and original location of the strain in question; this genotyping would parallel the efforts seen in 2001 to identify the strain of anthrax used in the postal attacks.

VACCINE

The best immunity to tularemia results from previous infection with a virulent strain, as shown in the experience of Dr. Francis. He contracted tularemia early in his career and never had another infection with systemic symptoms, despite four localized reinfections. The most effective prophylaxis for tularemia is currently immunization with the live vaccine strain (LVS) previously discussed. Although another vaccine that used killed bacteria, known as Foshay's vaccine, was previously available, it provided less immunity to the fatal aspects of the disease.[26] LVS does not provide 100% protection against either local or systemic infections, as seen in Saslaw's vaccine studies (figure 4–1).[26,27] However, since its introduction, LVS immunization is credited with reducing the number of typhoidal (systemic) laboratory cases of tularemia from 5.7 to 0.27 per 1000 employee-years.

The time to effective immunity from either vaccine is too long for preventive treatment after *F. tularensis* exposure, with pulmonary infection typically developing 3–5 days postexposure while signs of LVS-mediated immunity take about 2 weeks to appear.[27] Antibiotics, however, have been demonstrated as effective for immediate prophylaxis, with the current recommendation being for 14 days of prophylactic treatment.[5]

ANTIBIOTIC PROPHYLAXIS

Interestingly, Sawyer et al. have shown that prophylactic antibiotic treatment must begin several days postexposure to be effective.[28] Without this gap, sub-

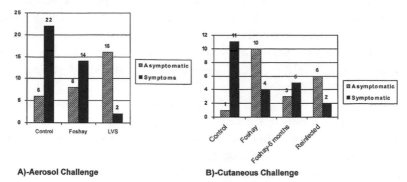

A)-Aerosol Challenge   **B)-Cutaneous Challenge**

FIGURE 4.1. Saslaw's seminal studies on efficacy of tularemia vaccine were performed in 1961 on prisoner volunteers using the virulent *F. tularensis* strain Schu S4. In (**A**), subjects were exposed to 10–52 organisms via an aerosol of particles with a diameter of 0.7 μm. In (**B**), subjects were injected with 7–12 organisms. In both cases, control patients r   no vaccinations.
cine approximately 1 month before the exposure. *Foshay–6 months* patients received the vaccine 6 months before exposure. *LVS* patients had received the live vaccine strain within the previous month. *Reinfected* patients had had earlier cases of virulent tularemia.

jects had an onset of tularemia shortly after they stopped taking antibiotics. Presumably, the suppressive action of the antibiotic prevented *F. tularensis* from fully activating the immune system, and the antibiotic and nonspecific immune system could not fully wipe out the bacteria. This adds further creditability to the idea that T-cell-independent mechanisms active initially can help contain the disease but T-cell-dependent mechanisms are necessary to wipe out the infection.

A sufficiently large dose of tularemia—in one case, 1000 infective doses—can sometimes overcome the immunity granted by the vaccine. While this may sound like a rather high threshold, it should be kept in mind that the infective aerosol dose for humans is at most 25 organisms.[27]

ANTIBIOTIC TREATMENT

Antibiotics currently provide the definitive treatment for tularemia. *F. tularensis* displays a large number of natural immunities, with the β-lactams, cephalosporins, and macrolides all being relatively ineffectual.[22] The first effective antibiotic against tularemia was streptomycin, which continues to be a drug of choice for tularemia treatment and is prescribed for a 10-day period.[5] However, streptomycin and the related aminoglycoside gentamicin have considerable

drawbacks as treatments: both require intramuscular or intravenous administration, both have relatively high toxicity profiles, and both are somewhat rare
in the U.S. medical system.[22] There are also occasional relapses of the disease
on aminoglycosides, although this remains by far the most effective class of
antibiotic. Additionally, there are known streptomycin-resistant strains of *F.
tularensis*.[28]

Fortunately, the tetracyclines and chloramphenicol have proven effective
against tularemia in extensive clinical use, and both of these classes have a low
toxicity and the capacity to be administered orally. These drugs also have a much
higher relapse rate than the aminoglycosides, presumably since they are merely
bacteriostatic whereas the aminoglycosides are bactericidal. Still, in a mass-casualty scenario with limited beds and a limited ability to provide intravenous
medication, tetracycline or chloramphenicol treatment would be suitable for
people with milder cases and for prophylaxis, as long as treatment is continued
for the recommended duration of 14–21 days.[5,28]

While the quinolones have not been used extensively to treat tularemia, the
clinical data thus far are largely positive, with treatment generally working well
and with a low relapse rate. Less positive experiences have also been recorded. In
vitro studies have shown ciproflaxin and the other quinolones to have considerable capacity against *F. tularensis*, perhaps because of their known ability to penetrate cells.[15,22] Studies in mice have shown both ciproflaxin and doxycycline
as effective against the virulent Schu S4 strain.[25] Since both of these antibiotics
can be given orally with a relapse rate equal to or lower than that of the tetracyclines, they remain a potent possible treatment, and have been endorsed by
the American Medical Association's Working Group on Civilian Biodefense as
suitable for treatment and prophylaxis of tularemia.[5]

ABSENCE OF RISK FACTORS

Tularemia infects all ages, sexes, and races equally. Other than compromised
immune systems, no known factors make one more susceptible to the disease.
Unsurprisingly, there is evidence that patients with serious underlying medical
conditions have a greater risk of developing severe tularemia, notably pneumonia. The death rate from tularemia in the modern world remains quite low,
around 2.3% since the introduction of streptomycin and current medical care.
The death rate from pneumonic tularemia, the form most likely to occur after
a biological weapons attack, is certainly higher: before antibiotics pneumonic
tularemia had a death rate of 50%, whereas localized cases killed approximately
5% of the time.[24]

Perhaps the most potent aspect of tularemia as a possible biological weapon
is its sheer infectivity. While it lacks the virulence of anthrax, smallpox, Ebola,

or botulinum toxin, the four most touted biological weapon threats, tularemia compensates via its incredibly high infectivity. It takes around 2500 spores to cause inhalational anthrax, whereas tularemia can be caused with fewer than 10 organisms under the skin, and fewer than 30 through an aerosol.[26,27] As discussed below, this has very important ramifications for the use of tularemia as a weapon.

WEAPONIZATION

The question of weaponizing *F. tularensis* is not whether it could be done but rather of the extent to which it has been done, and what further work might currently be underway. Both the USSR and the U.S.–UK–Canada programs produced *F. tularensis* in aerosol forms suitable for use in biological weapons, and both also developed specialized strains to serve as bioweapons.

CASUALTY ESTIMATES

Although the precise improvements made to tularemia will probably remain state secrets for the foreseeable future, a variety of sources have provided casualty estimates for attacks with weaponized tularemia based on expert speculation. The World Health Organization estimated in 1970 that the release of 50 kg of weaponized tularemia powder (with particles 1–7 μm in size, stabilized with additives) along a 2-km line above a developed city of 5 million would result in 250,000 incapacitated, among whom there would be approximately 5700 deaths. This estimate was probably conservative, as it assumed an infective dose of 250 organisms and estimated a death rate of 2.3% for treated pneumonic tularemia. If effectively disseminated in the admittedly difficult-to-generate aerosol form, tularemia can clearly cause many casualties. One report estimated the economic cost of a tularemia attack at $3.9 to $5.5 billion per 100,000 people exposed.[18]

The test in the New York City subway discussed earlier provided a striking example of how the high infectivity of *F. tularensis* increases its potential as a weapon. Given the travel of simulants through the subway system, the Army surmised that 12,000 cases of anthrax would have resulted, as opposed to 200,000 cases of tularemia. Pulmonary anthrax is frequently fatal without intensive support, which would be difficult with so many casualties, while pneumonic tularemia might be expected to kill 5% with widespread antibiotic treatment and limited medical support. So in this case, there might be 12,000 dead from anthrax but approximately 10,000 dead from tularemia, along with 190,000 heavily debilitated.

HIGH INFECTIVITY RATE

F. tularensis subspecies *tularensis* has a high natural infectivity, with respiratory exposure to 10–52 Schu S4 bacteria (in a liquid aerosol of 0.7-μm particles) giving an 80% infection rate in an experiment performed by Saslaw et al.[27] While no statistical information on the infectivity of aerosol tularemia has been released from the larger U.S. Army tests on Seventh Day Adventists, Saslaw's experiments clearly show an ID_{50} below 25 organisms for aerosol exposure, and suggest that tularemia can result from exposures as small as one or two organisms. Other experiments by Saslaw showed that intracutaneous exposures of fewer than 10 organisms can result in infection.[26] As numerous tests involving bioweapon simulants demonstrate, the importance of infectivity cannot be overemphasized when considering the use of *F. tularensis* as a biological weapon.

DECAY RATE

F. tularensis, although hardy in certain conditions, dies swiftly when exposed to sunlight and open air, as would occur during an aerosol release. One method to lower the decay rate would be an underground release, such as in a subway system, where the bacterial decay rate would approach the laboratory rate of 2% per minute.[28] One U.S. Army report estimated a decay rate of 5.3% per minute for *F. tularensis* release at night,[14] and the World Health Organization assumed a decay rate of 2% per minute for *F. tularensis* release at night.

USE BY THE UNITED STATES AND FORMER SOVIET UNION

Despite the release of several key documents, relatively little is known about the U.S. weaponization of tularemia. Production of tularemia began in 1955 at the Pine Bluff Arsenal, and by 1958 *F. tularensis* was considered the most probable "agent of retaliation employed by the United States" should biological warfare erupt.[13] The U.S. military later developed a more effective form of tularemia weapon, a dried powder as opposed to the earlier liquid slurry. The particles of liquid aerosols have a greater tendency to aggregate than powder aerosols, with such aggregation both increasing the chances of precipitation of the weapon out of the air and forming particles that may be larger than the optimal <5 μm for penetration of the lung. The U.S. military developed antibiotic-resistant strains around 1964.[14] Production of tularemia weapons continued through 1966, with production ceasing and all stockpiles of weapon being destroyed subsequent to President Nixon's ending of the U.S. biological warfare program in 1970.

Public knowledge of the USSR's work on biological weapons, including tularemia, has been considerably augmented by information from Dr. Kenneth Alibek, former deputy director of the USSR's biological weapons program.[1] He reveals a number of interesting, and frightening, aspects of the weaponization of *F. tularensis* in the Soviet Union. According to his book, the Soviets had developed a tularemia weapon by 1941. This weapon was allegedly used in the defense of Stalingrad during World War II, although Croddy and Krčálová contest that the epidemics of tularemia around Stalingrad were natural in cause.[4]

By 1982, Alibek himself had helped weaponize a vaccine-resistant *F. tularensis* strain, one capable of killing most immunized monkeys exposed to it. Another project that began in 1973, code-named Enzyme, focused on the use of genetic engineering to develop antibiotic-resistant strains of various diseases, including tularemia. In 1976, a goal was set to develop a triply antibiotic-resistant strain.

GENETIC ENGINEERING

The genetic manipulation of *F. tularensis* has been accomplished, by one group of scientists using a shuttle-vector plasmid to insert resistance genes for tetracycline and chloramphenicol.[23] The plasmid they generated was capable of replicating within both *F. tularensis* and *E. coli*, allowing for development of genes within the more collegial environment of *E. coli* before transferring them to *F. tularensis*. Among the genes of potential danger are those for resistance to antibiotics, radiation, and desiccation, genes coding for toxins from other bacteria, and genes that would decrease the incubation time of tularemia.

POTENTIAL DEFENSES

Tularemia remains poorly understood as a disease of biological warfare by the general populace precisely because it is readily treatable, as opposed to anthrax or smallpox. For the moment, the medical field retains the edge over *F. tularensis*, with multiple effective antibiotics and a reasonably useful vaccine. As far as what the future may hold, several areas of primary interest should be noted.

EXPERIMENTAL TREATMENTS

Several experimental treatments provide interesting possibilities for the treatment and prophylaxis of tularemia. One drug already in use for chronic granulomatous disease is IFN-γ, used to treat people with a congenital difficulty in clearing intracellular pathogens from their macrophages. In rare cases in which antibiotics are not suitable, IFN-γ might well be able to cure tularemia, or at least to reduce the severity of the disease.

A more experimental treatment consists of administering the CpG motif common in bacterial DNA as a method of activating macrophages to prevent infection. Klinman et al. demonstrated a considerable level of protection in mice against LVS challenge for up to 2 weeks per dose of CpG motifs, a protection that could be maintained indefinitely through repeated administrations.[20]

Also intriguing is the possibility of using antibiotics contained within liposomes and nebulized for treatment of pneumonic infection. The lipid membrane of the liposomes seems to inhibit clearance of the antibiotic, while nebulization allows the aerosol to penetrate deep into the lungs, paralleling the aerosolization of biological agents to increase their infective power. Conley et al. used this method of administering ciprofloxacin to effectively treat pneumonic tularemia in mice.[3]

All these approaches warrant further investigation, especially as none of them is applicable solely to tularemia.

VACCINES

Although the current live vaccine is generally considered to be sufficient, it does not provide complete immunity, with some vaccinated personnel still falling ill regardless of the method of exposure.[27] The vaccine also has problems typical of a live vaccine, such as varying immunogenicity between different batches and the remote possibility of a spontaneous return to virulence. A simpler vaccination consisting of several immunogenic *F. tularensis* proteins would not have these disadvantages, would be easier to prepare and store, and might provide more complete protection. Sadly, work on protein-based vaccinations has thus far failed to achieve notable success. Experiments in mice using lipopolysaccharide vaccination provided protection against LVS challenge, a protection that may be humoral in nature,[10] but failed to defend against challenge with the virulent Schu S4 strain.[9] Khlebnikov and his group reported success using a mixture of outer-membrane proteins as a vaccine in baboons, but this result has yet to be confirmed.[19]

ANTIBIOTICS

Finally, the development of new, more effective antibiotics is of great import both in preparation for biological attack and to deal with the far more threatening advent of antibiotic-resistant bacterial strains. This is especially necessary given the potential threat of engineered bacterial strains artificially given antibiotic resistance.

REFERENCES

1. Alibek, K. 1999. *Biohazard*. New York: Random House.
2. Conlan, J. 2002. Different host defenses are required to protect mice from primary systemic vs pulmonary infection with the facultative intracellular bacterial pathogen, *Francisella tularensis* LVS. *Microb Pathog* 32(3): 1.
3. Conley, J., et al. 1997. Aerosol delivery of liposome-encapsulated ciprofloxacin: Aerosol characterization and efficacy against *Francisella tularensis* in mice. *Antimicrob Agents Chemother* 41(6): 1288–1292.
4. Croddy, E. and S. Krčálová. 2001. Tularemia, biological warfare, and the battle for Stalingrad (1942–1943). *Mil Med* 166(10): 837–838.
5. Dennis, D., et al. 2001. Tularemia as a biological weapon: Medical and public health management. *JAMA* 284(21): 2763–2773.
6. Elkins, K., C. Bosio, and T. Rhinehart-Jones. 1999. Importance of B cells, but not specific antibodies, in primary and secondary protective immunity to the intracellular bacterium *Francisella tularensis* live vaccine strain. *Infect Immun* 67(11): 6002–6007.
7. Fortier, A., et al. 1992. Activation of macrophages for the destruction of *Francisella tularensis*: identification of cytokines, effector cells, and effector molecules. *Infect Immun* 60(3): 817–825.
8. Fortier, A., et al. 1994. Life and death of an intracellular pathogen: *Francisella tularensis* and the macrophage. *Immunol Ser* 60: 349–361.
9. Fulop, M., R. Manchee, and R. Titball. 1996. Role of two outer membrane antigens in the induction of protective immunity against *Francisella tularensis* strains of different virulence. *FEMS Immunol Med Microbiol* 13:245–247.
10. Fulop, M., et al. 2001. Role of antibody to lipopolysaccharide in protection against low- and high-virulence strains of *Francisella tularensis*. *Vaccine* 19(31): 4465–4472.
11. Gray, C., et al. 2002. The identification of five genetic loci of *Francisella novicida* associated with intracellular growth. *FEMS Microbiol Lett* 215(1): 53–56.
12. Green, S., et al. 1993. Neutralization of gamma interferon and tumor necrosis factor alpha blocks in vivo synthesis of nitrogen oxides from L-arginine and protection against *Francisella tularensis* infection in *Mycobacterium bovis* BCG-treated mice. *Infect Immun* 61(2): 689–698.
13. Hay, A. 1999. A magic sword or a big itch: An historical look at the United States biological weapons programme. *Med Confl Surviv* 14(3): 215–234.
14. Hay, A. 1999. Simulants, stimulants, and diseases: The evolution of the United States biological warfare programme, 1945–60. *Med Confl Surviv* 15(3): 214–234.
15. Ikaheimo, I., et al. 2000. In vitro antibiotic susceptibility of *Francisella tularensis* isolated from humans and animals. *J Antimicrob Chemother* 46(2): 287–290.
16 Johansson, A., et al. 2000. Comparative analysis of PCR versus culture for diagnosis of ulceroglandular tularemia. *J Clin Microbiol* 38(1): 22–26.
17. Junhui, Z., et al. 1996. Detection of *Francisella tularensis* by the polymerase chain reaction. *J Med Microbiol* 45(6): 477–482.
18. Kaufmann, A., M. Meltzer, and G. Schmid. 1997. The economic impact of a bioterrorist attack: are prevention and postattack intervention programs justifiable? *Emerg Infect Dis* 3(2): 83–94.
19. Khlebnikov, V., et al. 1996. Outer membranes of a lipopolysaccharide-protein complex (LPS-17 kDa protein) as chemical tularemia vaccines. *FEMS Immunol Med Microbiol* 13(3): 227–233.

20. Klinman, D., J. Conover, and C. Coban. 1999. Repeated administration of synthetic oligodeoxynucleotides expressing CpG motifs provides long-term protection against bacterial infection. *Infect Immun* 67(11): 5658–5663.

21. Leiby, D., et al. 1992. In vivo modulation of the murine immune response to *Francisella tularensis* LVS by administration of anticytokine antibodies. *Infect Immun* 60(1): 84–89.

22. Maurin, M., N. Mersali, and D. Raoult. 2000. Bacteriocidal activities of antibiotics against intracellular *Francisella tularensis*. *Antimicrob Agents Chemother* 44(12): 3428–3431.

23. Norqvist, A., K. Kuoppa, and G. Sandstrom. 1996. Construction of a shuttle vector for use in *Francisella tularensis*. *FEMS Immunol Med Microbiol* 13: 257–260.

24. Pullen, R. and B. Stuart. 1945. Tularemia: analysis of 225 cases. *JAMA* 129(7): 495–500.

25. Russell, P. 1998. The efficacy of ciprofloxacin and doxycycline against experimental tularemia. *J Antimicrob Chemother* 41(4): 461–465.

26. Saslaw, S., H. T. Eigelsbach, H. E. Wilson, J. A. Prior, and S. Carhart. 1961. Tularemia vaccine study I: Intracutaneous challenge. *Arch Intern Med* 107: 689–1701.

27. Saslaw, S., H. T. Eigelsbach, J. A. Prior, H. E. Wilson, and S. Carhart. 1961. Tularemia vaccine study II: Respiratory challenge. *Arch Intern Med* 107: 702–714.

28. Sawyer W. 1966. Antibiotic prophylaxis and therapy of airborne tularemia. *Bacteriol Rev* 30: 542–550.

CHAPTER 5

■ ■

EBOLA VIRUSES

Rohit Puskoor and Geoffrey Zubay

The viral family *Filoviridiae* represents one of the most serious threats to public health in the twenty-first century. Filoviruses are the causative agent for viral hemorrhagic fevers, so named because of the high fevers and profuse bleeding caused by the viruses. This family of viruses encompasses the Ebola viruses and the Marburg virus. The most well known of the viruses is Ebola Zaire, an extremely devastating strain of Ebola that causes the most fatal of the fevers, with a fatality rate upward of 90%.

There is currently no prophylactic treatment against the Ebola virus. The concern is compounded by our current lack of knowledge about its origins and its pathogenesis in a human host. The natural hosts for filoviruses have not yet been identified. Owing to the ability of the virus to spread by aerosolization, its high mortality rate and transmissibility, and the lack of an effective prophylactic therapy, Ebola is characterized as a Biosafety Level 4 pathogen; this makes research work with Ebola extremely tedious, because of the compliance measures that must be observed.

HISTORY

The virus family *Filoviridiae* comprises the Marburg and Ebola viruses. The Ebola virus is further divided into four major strains: Ebola Zaire, Ebola Sudan, Ebola Ivory Coast, and Ebola Reston. Ebola is endemic in Africa.

The first strain of Ebola virus to be identified was Ebola Sudan. The first known outbreak of this strain infected 284 people and was characterized by a mortality rate of 53%. Just a few months after this devastating outbreak was controlled, another strain appeared in Yambuku, Zaire. This strain, called Ebola Zaire (EBOZ), proved to be the more fatal strain by far. Of the 318 people it infected, 200 died. These nearly simultaneous outbreaks in the Sudan and Zaire in 1976 led to the arrival of international scientific teams to the region to investigate the outbreaks.[13] They concluded that the closing of medical facilities due to staff deaths had actually helped end the outbreaks, by eliminating the primary location of virus dissemination. Unlike those in the hospitals, patients in the villages were immediately quarantined, helping to stop virus spread in those locations. Following the two major outbreaks, Ebola appeared again in a single case in Tandala, Zaire, in 1977 and in a small outbreak in Sudan in 1979.

Ebola Reston (EBOR) was first identified in 1989 during an outbreak in Reston, Virginia,[13] of hemorrhagic fever among monkeys imported from the Philippines. The effects on the monkey population were devastating, and discovery of viral particles in the alveoli of the dead monkeys indicated that this strain could be aerosolized. The outbreak and subsequent identification of the virus as a strain of Ebola worried some that Ebola Reston would cause a highly lethal U.S. Ebola epidemic in humans. Although two Reston monkey handlers were infected by the virus, both infections were asymptomatic and neither experienced any complications, leading researchers to theorize that the Reston strain was avirulent in humans. Reston outbreaks recurred in 1994 and 1996.

The fourth and final known strain of Ebola, Ebola Ivory Coast (also known as Ebola Côte d'Ivoire, or EBO-CI), was identified in 1994.[5] A female ethologist performing a necroscopy on a chimpanzee from the Tai Forest accidentally infected herself. She was transported to Basel, Switzerland, for therapeutic care and made a full recovery. This is the only known case of Ebola Ivory Coast.

Between 1994 and 1995, 49 patients were hospitalized at Makokou General Hospital in northeastern Gabon, presenting Ebola-like symptoms.[6] The original diagnosis, based on polymerase chain reaction (PCR) of serum samples, was for yellow fever. After the outbreak ended 6 weeks later, some clinicians noted that the symptomatology was not characteristic of yellow fever, so additional analyses were performed. They quickly realized that Ebola Zaire, concomitantly with yellow fever virus, had been the causative agent of the outbreak.

This outbreak occurred in two waves, with a first wave of patients presenting in December of 1994 and a second wave during January and February of 1995. The first wave of patients originated from small gold-panning encampments on the outskirts of the forests, known as Mekouka, Andock, and Mikembe. Twenty-three patients from Mekouka, four from Andock, and five from Mikembe arrived at Makokou Hospital presenting symptoms including fever, diarrhea, and vomiting. The second wave of patients comprised cases of secondary and tertiary infection. This outbreak had a 59% fatality rate.

The second outbreak (February 1996) occurred in Mayibout 2, a small village on the Ivindo River, about 40 km south of Mekouka and Andock. Eighteen individuals who had skinned and chopped a dead chimpanzee became ill with fever, bloody diarrhea, and headache. These patients were also sent to Makokou

General Hospital, where four of them died. A fifth patient died after being sent back to Mayibout 2. Eleven secondary and tertiary cases appeared, as well as two additional primary cases that were unconnected with the handling of the chimpanzee. Of these two later cases, one died. A total of 31 individuals were affected by this outbreak, which manifested a mortality rate of 68%.

The third and final Gabon-based outbreak probably started July 13, 1996, with the death of a 39-year-old hunter in a logging camp in Booue, over 200 km from Mekouka. Six weeks later, in the middle of August, a second hunter at the same camp died from an illness with similar symptoms. The disease spread throughout Booue—24 cases, with 17 fatalities, were seen by November 13, 1996. Two cases were also seen in Johannesburg. A doctor who had fallen ill after treating a Booue Ebola case, not knowing he was carrying Ebola virus, traveled to Johannesburg for treatment, where he fell ill and infected a nurse who cared for him. The nurse was infected on November 2 and died November 24. At the end of November, a second wave of this outbreak spread through three distinct locations in Booue. The last official patient was seen January 18, 1997. The epidemic was declared ended in March 1997.

The Kikwit outbreak of Ebola Zaire attracted the attention of both the mainstream and the tabloid press, especially during its last weeks, because of the size of the affected population and the recently released novel *The Hot Zone*. Large donations were sent to the World Health Organization and directly to the Democratic Republic of the Congo (DRC) to help end the epidemic. Public interest was high at this point, as many people in other countries worried that the virus would continue to spread and pondered whether it could reach Europe, Asia, and the Americas.

Kikwit is a rural town on the banks of the Kwilu River in DRC, with an estimated population of 400,000. The city is located centrally, between the western city Kinshasa (the capital of DRC), the eastern territory of DRC, and the southern city Angola. Kikwit General Hospital is the main medical care facility in the city, with a total patient capacity of 375 patients and comprising 12 one-floor pavilions dedicated mostly to inpatient care. Monsango General Hospital, a nearby hospital with better facilities than Kikwit General Hospital, also served as a point-of-care facility during the outbreak, which affected Monsango residents to a slight extent. Kikwit Hospital also sent some patients to the Monsango hospital when they became overburdened.

The outbreak began during January of 1995, in the middle of the rainy season. A charcoal worker in the city of Kikwit was identified as the first case. The 1995 outbreak was the largest ever observed in an urban center. Because of the highly concentrated population, urban outbreaks usually present a greater risk for viral transmission than do rural outbreaks. Cases were observed as early as January 1995, but the disease was not positively identified as Ebola hemorrhagic fever (EHF) until May 10, 1995, when patient samples were analyzed at the Centers for Disease Control and Prevention in Atlanta, Georgia. Starting on May 13, international research and rescue teams descended upon the area and were able to monitor the progress of the Ebola outbreak, take note of the symptomatology, and develop appropriate control and containment procedures through observation of patients.

On April 7, 1995, the first case of Ebola hemorrhagic fever appeared at the Monsango hospital, a student nurse from the Kikwit hospital who died seven days after being admitted. Many subsequent cases occurred, usually of family, colleagues, or health care providers of infected cases. The mode of transmission for Monsango victims was usually by contact with infected blood. However, the outbreak in Monsango was smaller than the outbreak in Kikwit, possibly owing to better sanitation and sterilization practices. Success in dealing with the Monsango outbreak showed that Ebola could be easily controlled through the use of two inexpensive techniques: good hygiene and barrier nursing.[10]

MOLECULAR BIOLOGY

Filoviruses possess a nonsegmented, negative-stranded RNA genome approximately 19,000 bp long. The genomic RNA is antisense to the gene. All genes in the RNA genome are transcribed into monocistronic (one protein product per transcript), polyadenylated mRNAs. Seven genes encode eight distinct proteins, in the following order from the 3′ end of the viral genome to the 5′ end: nucleoprotein (NP), viral protein 35 (VP35), viral protein 40 (VP40), envelope glycoprotein/secreted glycoprotein (GP/sGP), viral protein 30 (VP30), viral protein 24 (VP24), and an RNA polymerase (L) (figure 5.1). The viral proteins are numbered according to their approximate molecular weight in kilodaltons (kd). The fourth gene from the 3′ end of the viral genome encodes both envelope glycoprotein (GP) and secreted glycoprotein.

THE FILOVIRUS VIRION

The filovirus virion comprises the nucleocapsid, a viral envelope, and an intracellular matrix between the nucleocapsid and the viral envelope. The virion particles have a molecular weight of 3–6×10^8. The nucleocapsid comprises a helical core of genomic RNA and RNA-associated proteins. The membrane envelope, derived from the host-cell lipid membrane, surrounds the nucleocapsid.[22] The envelope surface is studded with GP peplomers, one of the gene products of the Ebola. This protein is believed to mediate viral binding to the host cell, virion entry, and the immunosuppression that accompanies Ebola infection.

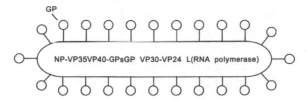

FIGURE 5.1. The structure of a filoviral virion. The RNA genome is encapsulated by a viral membrane studded with GP peplomers. The RNA genome is presented in the order it is transcribed, with NP at the extreme 3′ and L RNA polymerase at the 5′ end.

Electron microscopy of Ebola virus reveals that this virus can form either of two general structures. The primary form is that of long, filamentous structures that sometimes branch with one another. Otherwise, they are seen as shorter U-shaped or circular particles.[15] The string-like structures can be as long as 14,000 nm. The structure length associated with peak infectivity of the filamentous Ebola virus is 970 nm; by comparison, the average Ebola filamental virion is 1200 nm long. All virions have uniform diameters of 80 nm.

The following sections briefly describe the structures and functions of the proteins encoded by the Ebola genome starting from the 3′ end.

NUCLEOPROTEIN

NP is one of four proteins that form the ribonucleoprotein (RNP) core of the virion particle (the other three are L, VP35, and VP30). NP is the primary structural protein of the RNP complex. The N-terminal domain of NP is hydrophobic and contains all the cysteine residues. The C-terminal domain is hydrophilic and acidic, and contains most of the proline residues.

The N-terminal likely binds genomic RNA, while the variable C-terminal domain interacts with other virus proteins, such as the VP40 matrix protein. NP functions much like histones in human cells, tightly packing the RNA genome of filoviruses into a compact core for packaging into viral particles.

VIRAL PROTEIN 35

VP35 is an essential cofactor in transcription and replication. Ebola mutants that are defective in VP35 lack the ability to express proteins or to replicate their genomes. Ebola VP35 is functionally linked to the Ebola viral RNA polymerase.

Synthesis of IFN-β plays an important role in the initiation of the type I IFN response, because of its critical role in the IFN signaling cascade. In general, virus-infected cells display strong induction of the IFN-β promoter. However, in the presence of VP35, the promoter is completely blocked.[4] By inhibiting production of IFN, VP35 averts the antiviral state in both infected and noninfected cells, thus creating a favorable environment in as-yet-uninfected cells for viral infection.

VIRAL PROTEIN 40

Membrane association of VP40 induces the formation of hexamers necessary for virus assembly and budding. After entry of the virion into the host cell, filoviruses replicate in the *cytoplasm*.

THE ENVELOPE GLYCOPROTEIN

GP is responsible for cell-surface-receptor binding and fusion of the viral and host cell membranes. It comprises two disulfide-linked subunits, GP1 and GP2. GP1 comprises the N-terminal domain of preGP, while GP2 comprises the C-terminal domain (figure 5.2).

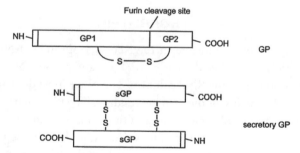

FIGURE 5.2. Structure of glycoprotein and secretory glycoprotein. The Ebola glycoprotein (GP) contains two disulfide-linked subunits, GP1 and GP2. GP is cleaved by furin (cleavage site indicated) in the Golgi following carbohydrate modification of the protein. The GP2 contains a transmembrane-anchoring region at its C-terminus. It also contains an immunosuppressive domain. The GP1 domain contains most of the carbohydrate modifications and a signal peptide at its N-terminus that directs it to the extracellular surface of the viral membrane. Ebola secretory GP (sGP) is a double disulfide–linked, antiparallel homodimer. It is believed to inhibit neutrophil activation.

Source: Adapted from Takada, A. and Y. Kawoka. 2001. The pathogenesis of Ebola hemorrhagic fever: *Trends Microbiol* 9(10): 506–511.

GP, the membrane-bound product of the GP/sGP gene in Ebola virus, is produced by transcriptional editing, which results in the addition of an extra adenosine within a run of seven adenosines in the mRNA transcript of the gene.[16] Transcriptional editing is seen only in the Ebola GP/sGP gene. Ebola GP–encoding mRNA synthesis is the result of a reading-frame shift during transcription in the –1 direction, owing to stuttering on the polyU template (figure 5.3). Only 20% of the GP/sGP transcripts are modified, so the GP is produced from only 20% of all GP/sGP transcripts. Consequently, sGP is the primary product of this gene in the Ebola virus.

Filovirus GP is modified by glycosidic residues as it is transported through the exocytotic pathway. Mature GP features both N-linked and O-linked carbohydrate modifications that contribute about half of the protein's 130–170-kd molecular weight.

GP is the only surface protein of the filoviral virion. GP oligomerizes to form the surface spikes seen on the membrane surface of filoviral virions. These spikes, also known as peplomers, or GP peplomers, mediate virus entry through binding to host-cell receptors (figure 5.4). These peplomeric spikes, which are

1. Virus enters the body via mucosal surfaces, skin abrasions, or parenteral introduction.
2. Macrophage infection and virus amplification.
3. The mononuclear phagocytic cell system is a primary target of filovirus replication.
4. Dissemination via lymphatic and vascular systems. Endothelial cells are targets for replication.
5. Infection of endothelial cells, as well as cytokine production, contributes to vascular injury and increased permeability, which leads to hemorrhage and other circulatory dysfunctions.

FIGURE 5.3. Pathway followed in filovirus infection. The filoviral virion enters the body through a permeable surface, such as via an open wound, injection site, or mucosal surface, or by inhalation. Mononuclear phagocytic cells, such as macrophages, are a primary target of filoviruses. Infection of macrophages serves two purposes: it suppresses the immune system and it allows the virus to spread to distant tissues and organs in the body as the macrophages and monocytes circulate. The destruction of the endothelial monolayer leads to the violent hemorrhaging characteristic of filovirus infections.

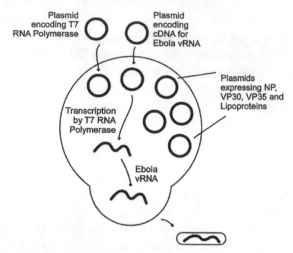

FIGURE 5.4. Cells previously transfected with plasmids expressing NP, VP30, VP35, and Ebola RNA polymerase are cotransfected with plasmids containing a cDNA clone of the Ebola RNA genome and T7 RNA polymerase. T7 RNA polymerase is capable of regenerating the viral RNA genome from a cDNA template. Thus, after transfection, T7 RNA polymerase transcribes the cDNA clone to recreate Ebola genomic RNA, which is then transcribed and replicated by the transcription and replication complex made up of NP, VP30, VP35, and Ebola RNA polymerase. This process yields functional Ebola virions from cDNA. This system is most useful for introducing mutations into the negative-stranded RNA genome, which was heretofore impossible because of the instability of RNA and the lack of a means for genetic manipulation of RNA. Once the genome has been retrotranscribed into cDNA using reverse transcriptase, the cDNA can be manipulated using established techniques of molecular genetics. The mutations are maintained in the genomic RNA when the cDNA is utilized to synthesize functional Ebola virions using the reverse genetics system, which thus provides a way to study the various genes of Ebola virus and mutate the virus into a more or less lethal form for purposes of weaponization or vaccination.

composed of trimers of the GP1-GP2 heterodimer, probably assemble in the endoplasmic reticulum.

GLYCOPROTEIN 2

Near the N-terminal region of GP2 there is a highly conserved fusion domain that is able to insert itself into membranes containing phosphatidylinositol (a phospholipid). This fusion domain is believed to mediate the fusion of host and viral membranes. GP2 also contains the transmembrane-anchor domain of the GP protein.

Another highly conserved domain of GP2 comprises the 26 residues in the C-terminal third of the protein, which functions as an immunosuppressive domain. In vitro, this domain has shown great ability to inhibit T-cell proliferation and antigen-stimulated proliferation of lymphocytes.

FURIN

After all carbohydrate modifications are complete, preGP is cleaved by the pro-protein convertase furin to form mature GP, made up of two subunits, GP1 and GP2, linked by a single disulfide bond. Proprotein convertases are eukaryotic proteins that recognize specific cleavage sites in proteins. Proprotein convertases activate many host-cell proteins and viral surface proteins through their cleavage action. Cleavage often exposes and activates protein domains necessary for fusion activity, such as the GP2 fusion domain.

Furin, which is expressed in most human cells, is a processing enzyme–produced constitutive secretory pathway. Furin is localized to the trans-Golgi network, but cells have been observed to secrete a truncated version. Furin recognizes the sequence R-R-X-K/R-R, a sequence observed at positions 497–501 in the immature GP of most Ebola subtypes.

SECRETED GLYCOPROTEIN

Unlike the GP, the C-terminal domain of sGP does not contain a hydrophobic transmembrane-anchor domain, explaining why it is secreted rather than inserted into the viral membrane. Membrane-bound GP and sGP share approximately 300 N-terminal residues, but GP differs from sGP by its 380 C-terminal residues (compared with sGP's 70 C-terminal residues).[17] The sGP's 70 C-terminal residues are mostly hydrophilic and positively charged, further explaining its preference for the extracellular environment over the bilipid layer.

The sGP is observed at high concentrations in the blood of patients in later stages of the disease. While sGP plays a role at the humoral effector level, helping the virus evade the host immune response.[25] Thus, sGP binds to human neutrophils, which inhibits early activation of the neutrophils by preventing the physical interaction of the neutrophil receptor and CR3.[26] The inactivation of the neutrophil phagocyte system by sGP prevents the host from generating an immune response.

VIRAL PROTEIN 30

VP30 is a minor nucleoprotein that is tightly associated with the RNP core. In Ebola VP30 deletion mutants, a basal level of transcription was observed, indicating that transcription did occur, albeit at an extraordinarily low rate. In addition, Ebola VP30 deletion mutants produced mRNA molecules that lacked poly-A tails).

VIRAL PROTEIN 24

The structure and function of VP24 are not homologous in any way to other known viral proteins. It is termed a matrix protein because of its interaction with the viral membrane. It has been shown that VP24 does not completely dissociate from the RNP complex under isotonic conditions. It may serve to link the VP40 submembrane lattice to the RNP complex.

RNA POLYMERASE (L)

Each virus particle is released with L associated with the RNP core. These RNA-dependent RNA polymerases initiate transcription and replication upon viral infection of the host cell. The L (or large) protein is the largest protein produced by filoviruses. It is also the least abundant protein produced by the filoviruses. The precise function of the L protein in transcription and replication is unknown.

The replication complex of Ebola virus is made up of NP, L, VP35, and VP30, all associated with the nucleocapsid core. These replication complexes are highly specific.[20]

CLINICAL DIAGNOSIS AND RESPONSE

Filoviruses infect several cell types. However, their primary targets are hepatocytes and endothelial cells, both of which, interestingly, produce more furin than any other cell type in the human body.

Filovirus budding from infected endothelial cells occurs initially from the apical surface. At later stages, budding also occurs from the basal surface, encouraging viral spread to the underlying tissue and leading to the destruction of endothelial cells. Endothelial cell infection and destruction is associated with the violent hemorrhages seen in the later stages of a viral hemorrhagic fever. Ebola causes high fever, hemorrhaging, and shock. The virus kills a large percentage of its victims with 5–7 days of symptom onset.

MAJOR ROUTES OF INFECTION

The most common routes of filoviral entry are through the mouth, an open wound, and the conjunctiva. Contact with late-stage patients is most risky, because of the increased concentration of viral particles in their blood and secretions.[11] Contact between a permeable, unprotected surface (such as a mucous membrane or broken skin) and the blood or fluids of an infected patient places an individual at high risk for contracting the virus.[4] Thus, the virus can enter by viral exposure to small skin breaks or a mucous membrane. It can also enter by the ingestion, inhalation, or injection of infectious material, or by viral exposure to the pharyngeal surface during ingestion or inhalation.

MACROPHAGES AND MONOCYTES

Circulating monocytes and macrophages, as well as tissue macrophages, are believed to be major sites of viral replication in Ebola infection.[18] Monocytes and macrophages belong to the mononuclear phagocytic system, which is involved in the host innate immune response. Circulating phagocytes help spread the virus to distant locations in the body.

Electron microscopy of infected phagocytic cells reveals numerous viral particles encapsulated in vacuoles in the cell cytoplasm. Vacuole-encapsulated viral particles provide a way for the virus to evade the host immune system—virus particles can be passed on to other cells by exocytosis of the vacuoles or secretion into the extracellular environment following cell lysis.

B- and T-lymphocyte depletion and spleen destruction have been noted in Ebola fatalities.[2] Lymphocyte apoptosis may be caused by interactions with altered antigen-presenting cells. Macrophages, a leading antigen-presenting cell, are primary targets of the Ebola virus, and lymphoid damage is concurrent with viral replication in local macrophages. The depletion of B lymphocytes may explain the lack of a humoral immune response generated in Ebola fatalities. Overstimulation of T cells by IFN-γ may lead to downregulation of Bcl-2 and T-cell receptors. Lymphocytes expressing lower levels of Bcl-2 are more susceptible to apoptosis. The uncontrolled production of factors such as TNF-α or IFN-γ in the wake of lymphocyte destruction may increase the permeability of the endothelium and cause coagulation failure and liver disorders.

THE PATHOGENESIS OF FILOVIRUSES

The transmission of Ebola virus involves patient contact with blood or body fluids from an infected individual. Risk of successful transmission increases in situations where contact is made with an individual in the later stages of disease.[14]

At this point, it is unknown how infective body fluids are, or how exactly the disease enters the body. It is believed that the virus enters the body by contact between contaminated fingers and the mouth or the eyes (the latter explains the conjunctivitis seen with Ebola infection). Researchers also believe that the virus can be aerosolized and enter through the inhalation pathway.[11] Ebola virus was isolated in human lung specimens from the Kikwit outbreak. During the same outbreak, five patients died who had had no contact with any other Ebola victims. Also, as Ebola virus has been identified in sexual fluids, it is possible for Ebola to be transmitted during sexual activity.[14]

Ebola viruses cause focal necrosis of liver, lymphoid organs, kidneys, testes, and ovaries. Endothelial cells, macrophages, monocytes, and hepatocytes are the primary cellular targets of the virus. A marked elevation of the cytokines, interleukin-2, interleukin-10, TNF-α, interferon-α, and interferon-γ were noted in fatal Ebola hemorrhagic fever cases.[4]

THE POSTINFECTION PERIOD

The virus incubates for 4–10 days in the host after infection, after which flu-like symptoms including fever, headache, fatigue, arthralgia or myalgia, sore throat, nausea, and cough appear. Symptoms specific to filoviruses manifest as well, including dysphagia or odynophagia, anorexia, abdominal pain, and diarrhea.[3] Conjunctivitis is an early clinical sign and should alert the clinician to suspect filoviral infection if the patient may have come into contact with an infected person or animal. Conjunctival symptoms were seen in around 48% of Kikwit patients and 58% of Yambuku patients.[7] A skin rash is also seen sometimes, but it is difficult to identify in dark-skinned patients.

Overall patient condition rapidly deteriorates. Later-stage symptoms include hiccoughs, oliguria, and shock. Tachypnea and hemorrhaging, such as epistaxis, hematemesis, melena, petechiae (pinpoint hemorrhages under the skin), ecchymosis (large, bruise-like bleeding under the skin), bleeding at needle puncture sites, and hemorrhagia, also manifest. Hemorrhage and shock develop in 10–20 days. Some patients also develop neurological symptoms, including seizures and delirium, often resulting in a coma. Death usually occurs within 3 weeks of infection.[4]

In surviving patients, recovery is a long process. Recovering individuals are still a reservoir of the disease: infectious viral particles have been detected in the fluids of recovering patients up to 82 days after the initial presentation. Patients often complain of arthralgia and a feeling of malaise. Some patients experience permanent hearing loss, uveitis (inflammation of the uvea), orchitis (inflammation of the testes), parotitis (inflammation of the parotid glands), and/or tinnitus (buzzing or ringing sound in one or both ears). A small number of survivors of the Kikwit outbreak complained of pain in the eyes, hyperlacrimation, abnormally high sensitivity to light, and lack of visual sharpness, all of which are characteristics of uveitis.[3] All these patients benefited from topical treatment of 1% atropine and steroids. The ethologist infected with Ebola Ivory Coast experienced hair drying, loss of hair elasticity, and eventual massive hair loss one month after the onset of her symptoms. The hair loss persisted for about 3 months.[5]

BREAKDOWN OF THE IMMUNE RESPONSE

The inability of most patients to develop an antibody immune response can be traced to the infection of the mononuclear phagocytes and fibroblastic reticular systems. Fibroblastic reticular systems, associated with lymph nodes, usually amplify the immune response. Thus, infection of lymph nodes disrupts antigen trafficking and the production of cytokines. Infection of circulating macrophages and monocytes transmits the early stages of the disease. In general, these cytokines have been implicated in causing shock and increased vascular permeability during filovirus infection.

LABORATORY TESTS

Ebola is easy to isolate and culture. However, tests with live virus must be performed in Biosafety Level 4 containment laboratories, because of the extreme virulence and contagiousness of the disease. After it has been isolated, the virus can be detected via immunofluorescent techniques, ELISA, immunohistochemistry, or PCR.[4] Western blot and radioimmunoprecipitation methods are also commonly used as confirmatory tools, but they are too cumbersome and time-consuming to be used as diagnostic tools in an epidemic setting.

ELISA tests for IgG and IgM antibodies directed against Ebola antibodies have been developed for evaluation of the sera of humans and animals potentially infected with the Ebola virus.[8] IgM-capture ELISA is a very effective method for diagnosing acute infection and appears to be quite useful as a diagnostic tool. Anti-Ebola antigen IgM antibodies can be detected as early as 6 days after infection and persist for up to 90 days postinfection. The IgG antibody response is less rapid, but it has been shown to persist in survivors for up to 10 years, suggesting its potential as a confirmation tool for months or even years after infection. While IgG assays are not recommended for diagnosis, because of the superiority of the IgM-capture assay for detecting the virus soon after infection, it could be useful in field investigations of the virus.

Large quantities of the Ebola virus are present in the epidermis, owing to its high affinity for endothelial cells. Because of this, skin biopsies present a way of confirming Ebola infection postmortem.[19] Formalin-fixed biopsy samples are not infectious, and can be shipped anywhere without refrigeration or other special precautions. Skin biopsies are also less invasive than liver biopsies or autopsies, and can be easily analyzed using immunohistochemical methods.

ISOLATION AND CONTROL RESPONSES

During the Kikwit outbreak, medical care facilities were central in amplifying further transmission of the Ebola virus.[21] Lack of proper isolation supplies and training in barrier nursing techniques, as well as poor hygiene practices, caused health care workers to serve as vectors for the virus. More money and training are needed: the procedures set in place during the Kikwit outbreak were quickly abandoned in area hospitals, because of lack of supplies and funding.

Better surveillance is also necessary. Case diagnosis and isolation alone are effective in the middle of an outbreak, but it is difficult to properly identify filovirus infection in sporadic cases or at the start of epidemics. It is preferable that investigations for the natural reservoir of the viruses begin during an outbreak, rather than afterward, when there is no guarantee that the virus will still be present in its natural habitat. In general, investigators have started searching for the reservoir months after the onset of an epidemic, owing to late identification and confirmation of the disease and problems with weather.

While it would be convenient to have Ebola diagnostic tools on site in endemic areas, it is highly unlikely that such measures will be taken because of the cost. It is critical that barrier nursing be put into place immediately after filovirus infection is suspected, and that samples be taken for confirmatory analysis. Formalin-fixed skin biopsies enable clinicians in endemic areas to send samples to laboratories for testing.

SUPPORTIVE STRATEGIES

Treatment is supportive and focused on alleviating symptoms. Adequate hydration and nutritional support is necessary. Treatment with antibacterial and antimalarial drugs to prevent secondary infections is also highly recommended. Intravenous infusions were shown to slightly moderate the fatality rate in the 1996 outbreak in Gabon, compared to previous Ebola Zaire outbreaks.

Finding the natural reservoir of the virus is a key step in developing a successful therapy.[4] By studying how the virus is able to perpetuate in nature without eliminating its natural host population, one can develop a therapy for treating it in non-natural hosts (i.e., humans and nonhuman primates). Researchers at the National Institute of Virology in South Africa have injected many plants and animals with the virus to examine whether they are able to internalize the virus without developing disease. So far, only bats have been found to support replication and growth of high levels of viral particles without developing the associated disease.

WEAPONIZATION

Few of the pathogens covered in this book equal or surpass filoviruses in transmissibility; none surpasses them in virulence. This explains why the CDC has classified Ebola virus as a Category A biological weapon. Ebola was among the most often analyzed pathogens at Biopreparat, the Russian think tank for biological weaponization. Among the more virulent products of this institution is Ebolapox, which could be used by a rogue group to infect and kill large segments of a population. As a weapon of mass destruction, Ebola has the drawback of degrading quickly upon contact in the air. This might make Ebola virus difficult to weaponize.

AEROSOLIZATION

Formal experiments have shown that all filoviruses are stable in small aerosol particles. In addition, research done with lab animals has confirmed aerosolization as a mode of transmission in these animals. Indeed, during the Ebola Reston outbreak in 1989, presence of virus in nasal and oropharyngeal secretions and the alveoli of infected monkeys and the pattern of transmission within and between rooms where monkeys were housed indicated that the virus was transmitted by aerosol in this outbreak. However, human patients who had not had prior direct contact with blood or fluid secretions from an infected human are rarely found, indicating that even if Ebola was spread by aerosolization, it is not a major mode of transmission.

EBOLAPOX

Ken Alibek has frequently stated his belief that Soviet scientists also developed a recombinant chimera of the Ebola and smallpox genomes.[1] Using reverse transcriptase to create a double-stranded DNA copy of the disease-causing genes of Ebola (most likely its GP/sGP gene), scientists inserted these genes into the smallpox virus to create a chimera virus capable of producing a combination of Ebola and smallpox's cytopathic effects. Alibek maintains that such a hybrid virus would be stable.

The hybrid would cause pruritic smallpox, or "blackpox." Blackpox is the most severe form of smallpox, characterized by hemorrhaging in place of pustule formation. Rather than develop blisters, the skin of the patient becomes black all over. Blood vessels leak, causing internal hemorrhaging. A weapon composed of Ebolapox would possess the violent hemorrhaging and high fatality rate characteristic of the Ebola virus and the contagiousness of the smallpox virus. Naturally occurring blackpox is fatal within 7 days due to toxemia. Because of blackpox-induced toxemia and Ebola-induced hemorrhaging, an Ebolapox hybrid virus would likely cause near-100% fatality rates if released.

Because of a lack of understanding of how such a virus would work on the molecular level, no drug compound or antibody synthesized could effectively combat a hybrid of both smallpox and Ebola, two pathogens that are already very difficult to combat individually.

REVERSE GENETICS SYSTEMS

It has recently been proven that functional negative-strand RNA viruses can be obtained from cDNA clones of the virus. As DNA is much more stable and researchers have a vast array of tools with which to genetically manipulate DNA, the development of a reverse genetics system for negative-strand RNA viruses represents a true breakthrough. Once in DNA form, site-directed mutagenesis, restriction enzyme cleavage, and recombination can be performed to mutate the genetic structure of the virus. These changes will be contained in the RNA genome when the cDNA (contained in a plasmid) is transcribed to create the RNA genome of the virus.[12]

Reverse transcriptase–PCR (RT-PCR) is performed on the filovirus RNA genome to produce double-stranded cDNA molecules, each of which contains one strand that is antisense to the RNA genome and one strand that is sense to the genome. The antisense strand will later serve as the template for recreating the RNA genome. Once this cDNA has been synthesized, it is cloned in a plasmid, after which the viral genome can be altered in a variety of ways. The plasmid containing the mutated cDNA clone is then cotransfected with a plasmid encoding T7 RNA polymerase (which is capable of generating viral genomic RNA from a DNA template) into cells that have already been transfected with plasmids expressing the NP, L, VP30, and VP35 proteins (which are required for replication and transcription of filoviral genomic RNA).

Posttransfection, T7 RNA polymerase recognizes the template strand of the cDNA clone and transcribes the filoviral RNA genome. NP, L, VP30, and VP35 already present in the cell are then able to replicate and transcribe the filoviral genome, causing the formation of functional virions from a DNA copy of the filoviral genome. Volchkov and colleagues at the Institut für Virologie in Marburg, Germany, used the reverse genetics system for Ebola to create a mutated virus in which the editing site in the GP/sGP gene was eliminated.[23,24]

The mutant virus no longer produced sGP, as compared with wild-type Ebola virus, which produces 80% sGP and 20% GP. The mutant virus was significantly more cytotoxic than the wild-type virus, indicating that production of sGP downregulates the virulence of Ebola virus. However, a mutant form of Ebola that overexpressed GP or expressed a toxic non-Ebola protein could be a dangerous tool for bioterrorism. Thus, the reverse genetics system provides a way to produce highly virulent mutated viruses for the purpose of biological warfare or biological terrorism.

DEFENSES

Research on viral proteins has been aided by the development of recombinant DNA technology, which allows researchers to perform biochemical and other experiments with these proteins without the risks of working with infectious viral particles. With this and other research systems in hand, we may ultimately be able to explain the extreme virulence that characterizes Ebola hemorrhagic fever. Current research will enhance our knowledge of the mechanisms of Ebola virus infection and could lead to the development of effective vaccines and antiviral compounds.

VACCINES

A significant number of patients suffering from Ebola infections recover spontaneously from acute disease, indicating that their recovery is mediated by a more efficient immune response that is able to overcome the immunosuppression caused by the virus. This suggests a route for vaccine development. The crucial role of T-cell responses in the control of Ebola hemorrhagic fever has recently been recognized. In Ebola patients, recovery was associated with the introduction of cytotoxic T cells in the patient.

Neurological problems were seen in a significant percentage of patients treated by transfusion, indicating some issues with this type of treatment. Based on the success of this limited study, more research should be done on the effectiveness of passive immunization during the next outbreak.

ANTIVIRAL DRUGS

The antiviral drug ribavarin has proven to be ineffective against Ebola infections. S-adenosylhomocysteine (SAH) hydrolase is a host-cell enzyme that is a popular target for antiviral drug compounds. Inhibitors of the enzyme block transmethylation, thus preventing any steps dependent on methylation, such as the formation of the 5′ cap on viral mRNA transcripts, making viral mRNA less stable and virtually unreadable by the host-cell translation machinery.

The enzyme is believed to recognize qualitative differences in feedback inhibition from viral transmethylation and host-cell transmethylation, allowing inhibitors to specifically block viral transmethylation without interfering with host-cell transmethylation.

Compounds designed to inhibit SAH hydrolase activity have been shown to shut down Ebola viral replication in vitro, indicating that SAH hydrolase is a drug target that must be further explored as a general antiviral target.[9] Carbocyclic 3-deazaadenosine (Ca-c^{3}Ado) in particular has proven to be of significant efficacy against filoviruses. When the drug was administered on day 3 after infection in mice, the treated mice were symptomatic but slowly recovered over the course of 2 weeks. The next tests for the drug will include observation of the drug's effects in nonhuman primates, as well as an examination of whether the drug can be effective if administered later after infection.

HYPERIMMUNE SERUM

In order to develop hyperimmune serum for the treatment of filovirus infections, an immunization schedule was developed to induce an immune response to Ebola antigen within sheep and goats, in which the virus is not pathogenic. Goat serum was administered to lab guinea pigs, and proved to be effective for protection against Ebola virus if administered up to 48 hours postinfection.[9] It is the belief of the researchers involved in this study that hyperimmune serum developed in either goats or horses would be effective in protecting humans from Ebola virus provided the serum was administered within 72 hours following infection. However, this prophylaxis tool would probably be used most often by medical care providers, who are most likely to know exactly when they came into contact with the virus.

REFERENCES

1. Alibek, K. W. 1999. *Biohazard: The Chilling True Story of the Largest Covert Biological Weapons Program in the World.* London: Hutchinson Publishers.
2. Baize, S., et al. 2000. Apoptosis in fatal Ebola infection: Does the virus toll the bell for immune system? *Apoptosis* 5: 5–7.
3. Basler, C. F., et al. 2000. The Ebola virus VP35 protein functions as a type I IFN antagonist. *Proc Natl Acad Sci USA* 97(22): 12289–12294.
4. Baxter, A. G. 2000. Symptomless infection with Ebola virus. *Lancet* 355: 2178–2179.

5. Bukreyev, A. A., et al. 1993. The GP-protein of Marburg virus contains the region similar to the 'immunosuppressive domain' of oncogenic retrovirus P15E proteins. *FEBS* 323(1,2): 183–187.

6. Bwaka, M. A., et al. 1999. Ebola hemorrhagic fever in Kikwit, Democrat Republic of the Congo: Clinical observations in 103 patients. *J Infect Dis* 179(suppl 1):S1–7.

7. Colebunders, R. and M. Borchert. 2000. Ebola haemorrhagic fever: A review. *J Infect* 40: 16–20.

8. Dessen, A., et al. 2000. Crystallization and preliminary X-ray analysis of the matrix protein from Ebola virus. *Acta Crystallog* D56: 758–760.

9. Dessen, A., et al. 2000. Crystal structure of the matrix protein VP40 from Ebola virus. *EMBO J* 19(16): 4228–4236.

10. Feldmann, H., et al. 1994. Characterization of filoviruses based on differences in structure and antigenicity of the virion glycoprotein. *Virology* 199: 469–473.

11. Feldmann, H., et al. 1996. Filovirus-induced endothelial leakage triggered by infected monocytes/macrophages. *J Virol* 70(4): 2208–2214.

12. Ferron, F., et al. 2002. Viral RNA-polymerases-a predicated 2′-O-ribose-methyltransferase domain shared by all mononegavirales. *Trends Biochem Sci* 27(5): 222–224.

13. Floyd-Smith, G., E. Slattery, and Lengyel P. 1981. Interferon action: RNA cleavage pattern of a (2′–5′)oligoadenylate-dependent endonuclease. *Science* 212(4498): 1030–1032.

14. Formenty, P., et al. 1999. Human infection due to Ebola virus, subtype Côte d'Ivoire: Clinical and biologic presentation. *J Infect Dis* 179(suppl 1): S48–53.

15. Georges, A.-J., et al. 1999. Ebola hemorrhagic fever outbreaks in Gabon, 1994–1997: Epidemiologic and health control issues. *J Infect Dis* 179(suppl 1): S65–75.

16. Guimard, Y., et al. 1999. Organization of patient care during the Ebola hemorrhagic fever epidemic in Kikwit, Democratic Republic of the Congo, 1995. *J Infect Dis* 179(suppl 1): S268-S273.

17. Haller, O., M. Frese, and G. Kochs. 1998. Mx proteins: mediators of innate resistance to RNA viruses. *Rev Sci Tech* 17(1): 220–230.

18. Huggins, J., Z. X. Zhang, and M. Bray. 1999. Antiviral drug therapy of filovirus infections: S-adenosylhomocystein hydrolase inhibitors inhibit Ebola virus in vitro and in a lethal mouse model. *J Infect Dis* 179(suppl 1): S240–247.

19. Kudoyarova-Zubavichene, N. M., et al. 1999. Preparation and use of hyperimmune serum for prophylaxis and therapy of Ebola virus infections. *J Infect Dis* 179(suppl 1): S218–223.

20. Muhlberger, E., et al. 1999. Comparison of the transcription and replication strategies of Marburg virus and Ebola virus by using artificial replication systems. *J Virol* 73(3): 2333–2342.

21. Peters, C. and J. LeDuc. 1999. An introduction to Ebola: The virus and the disease. *J Infect Dis* 179(suppl 1): ix–xvi.

22. Sanchez, A., et al. 2001. *Filoviridiae*: Marburg and Ebola viruses. In D. M. Knipe and P. M. Howley, eds. *Fields Virology*, 4th ed. Philadelphia: Lippincott Williams and Wilkins.

23. Volchkov, V., et al. 2001. Recovery of infectious Ebola virus from complementary DNA: RNA editing of the GP gene and viral toxicity. *Science* 291(5510): 1965–1969.

24. Volchkov, V. E., et al. 1995. GP mRNA of Ebola virus is edited by the Ebola virus polymerase and by T7 and vaccinia virus polymerases. *Virology* 214: 421–430.

25. Wickelgren, I. 1998. A method in Ebola's madness. *Science* 279(5353): 983–984.

26. Yang, Z.-Y., et al. 1998. Distinct cellular interactions of secreted and transmembrane Ebola virus glycoproteins. *Science* 279: 1034–1037.

CHAPTER 6

■ ■

INFLUENZA VIRUS

Rian Balfour

Influenza virus (IV) causes the respiratory disease known as influenza ("the flu"). The virus infects 10 to 20% of the U.S. population yearly. Of those infected, approximately 114,000 are hospitalized and 20,000 die due to influenza and/or complications caused by the virus. Despite these overwhelming statistics of infection and mortality, the Centers for Disease Control and Prevention (CDC) does not classify IV as a bioterrorism threat. However, the characteristics of the virus do indeed render it a likely, but highly ignored, germ weapon threat. A pathogen that kills more people than the human immunodeficiency virus (HIV), claimed the lives of 40 million people in less than two years, and routinely mutates into highly lethal forms should be considered a potential bioweapon.

HISTORY

Following Livy's *History of Rome*, which described an influenza-like disease that affected the Roman and Carthaginian armies in 212 B.C., no record of influenza or influenza-like symptoms appeared until 1781. Between 1781 and 1782, an influenza pandemic infected two-thirds of Rome's population and three-fourths of Britain's population. The disease continued to spread in North America, the West Indies, and South America. The spread of this pandemic culminated with an influenza epidemic in New England, New York, and Nova Scotia in 1789. 1781 marked the inception of the average 10- to 40-year cycle of emergent influenza epidemics and pandemics.

Like clockwork, the first half of the nineteenth century witnessed isolated epidemics in several regions of the world. Most notable were the outbreaks in Asia, which started in 1829, and Russia, which commenced during the winter of 1830. The Asian outbreak of influenza spread to Indonesia by January

1831, and the Russian outbreak spread throughout Russia and westward between 1830 and 1831. By November 1831, the influenza outbreak had reached America, and epidemics remained prevalent until 1851.

Between 1889 and 1890, a Russian Flu pandemic occurred, and it was the most deadly influenza pandemic to that date. It began in Central Asia during the summer of 1889 and spread to spread to Russia, China, North America, parts of Africa, and major Pacific rim countries. A conservative estimate of mortalities is 250,000 in Europe and between 500,000 and 750,000 worldwide.

In 1900, the *Journal of the American Medical Association* published an article stressing the severity of influenza, the virulence of the virus, and the importance for physicians to take note of symptoms early in order to avoid the complications associated with the infection. The author wrote, "People have long since ceased to regard influenza as a joke; and the profession is coming to realize that it ranks among the more serious maladies with which we have to deal." The article also describes the diversity of symptoms related to influenza and implies that the more virulent virus causes more severe symptoms:[5]

> Influenza is so variable in its forms…. Various divisions have been suggested, based mainly on the organs affected. To my mind, the most natural and rational division is into the two types of the disease so constantly observed—the catarrhal and the neurotic…. The catarrhal type of the disease divides itself into the respiratory and the gastro-intestinal or abdominal [and spreads to] … the respiratory or alimentary muscosoe. The neurotic type may be divided into the cerebral, the neuralgic, the cardiac and the rheumatic forms. And a blending of these types and forms produces a common and serious variety, the typhoid form.

Although the article spotlighted influenza as a serious health threat, the Spanish Flu pandemic of 1918 is responsible for the threat of influenza reaching the agenda of public health officials and claiming its place as a true public health threat. Considered the most lethal and infectious pandemic ever, it resulted in the death of more than 20 million people in North America, Europe, the Alaskan wilderness, and remote Pacific Islands between 1918 and 1919, contemporaneous with the end of World War I. While the Spanish Flu derives its name from an outbreak in Spain, the virus first appeared at the military Camp Funston in Kansas on March 11, 1918. On that morning, a cook presented at the infirmary with typical flu-like symptoms of a low-grade fever, mild sore throat, slight headache, and muscle aches, and the doctor recommended that he get bed rest. By noon that same day, 107 soldiers were sick with the same symptoms. By March 13, 1918, 522 people were sick and many were ill with pneumonia. Within days, military camps throughout the United States reported outbreaks of the flu, and thousands of sailors stationed off the East Coast fell sick. On March 18—only one week after the first report—every state in the United States reported cases of influenza. By April, the virus had crossed the Atlantic and proceeded to infect individuals in Europe, China, Japan, Africa, and South America. This wave of the influenza pandemic, now referred to as the "First Wave," continued until the summer of 1918 and was characterized by the virus' high communicability but relatively low lethality. Yet, despite the low lethality, by the end of the First Wave over 800,000 people (of whom 28% were Americans) had died.

In late August 1918, a second, more virulent, form of the virus emerged and infected individuals worldwide in 6 months. This form dominated the virus of the "Main Wave" of the pandemic. Between September and November, the virus killed over 10,000 people per week in some U.S. cities—158 of 1000, 148 of 1000, and 109 of 1000 people died in Philadelphia, Baltimore, and Washington, D.C., respectively. As noted earlier, it spread throughout Europe, the Alaskan wilderness, and remote islands of the Pacific. The U.S. death toll was approximately 850,000 individuals, which—surprisingly—made it the region of the world least affected by the virus. In Alaska, 60% of the Eskimo population died. In Samoa, 80–90% of the population was infected, and many of the survivors died from starvation because they were too weak to feed themselves. Ocean liners that set sail from Europe would arrive in New York with 7% fewer passengers than they had embarked with; those who perished fell ill to influenza, and the confined area of the ship made it a productive breeding ground for its survival and transmittal. By October 1919, this virulent flu strain seemed to have vanished. During its 18-month world rampage, it claimed the lives of at least 20 million people (more than World War I, which had a death toll of only 9 million).

The mortality figure of 20 million is remarkable for several reasons. It confirms its status as the most deadly pandemic ever—for comparison, its death toll was greater than the total mortality for the 4-year "Black Death" bubonic plague, between 1347 and 1351. In addition, the mortality rate associated with the Spanish flu was 2.5% among infected individuals, which is exceptional because the mortality rate for other influenza epidemics is less than 0.1%. Another unique feature of this pandemic was the demographics of the deaths. Most deaths occurred in adults between the ages of 18 and 30 whereas about 90% of deaths usually associated with influenza occur in individuals 65 years and older. Infected individuals first developed a rapid onset of flu-like symptoms of high fever, chills, headaches, muscle ache in the back and legs, and dry cough. While most individuals recovered from the infection within a week, others experienced a "depression" in which symptoms of malaise and aches persisted for several months. Still others died within 2 to 3 days of showing symptoms from the pulmonary edema and hemorrhage into the lungs. However, most deaths were due to pneumonia caused by secondary bacterial infections.[19]

The unusual features of the 1918 pandemic stimulated research to determine the strain and its characteristics responsible for the extraordinary number of deaths. While the IV was not isolated until 1933 by Sir Christopher Andrewes, Wilson Smith, and Sir Patrick Laidlaw, a considerable amount of research had been carried out prior to 1918 and during the pandemic to discover the cause of influenza. In 1892, a microbiologist named Pfeiffer isolated the bacterium *Bacillus influenzae* following an influenza outbreak in 1890. This bacterium was proposed to be the cause of influenza. The pandemic of 1918 allowed researchers to gain extensive bacterial cultures from lung specimens. They isolated a wide selection of bacteria, not merely *Bacillus influenzae*, which led them to believe that influenza was caused by a then-undiscovered virus while bacteria were responsible for the severe secondary complications of pneumonia.[19]

Researchers continued to investigate the cause of influenza and the strain of the virus. In late 1930s, antibody titers of 1918 flu survivors were analyzed, and researchers inferred from phylogenic studies of past influenza outbreaks that the 1918 strain was an H1N1 IV that was closely related to a swine virus. Since wild waterfowl are believed to be the natural reservoir for IV, another inference was made that the H1N1 strain emerged from an avian virus that is also capable of infecting swine, and subsequently was transmitted from swine to humans. Subsequent studies conducted by Taubenberger, in which he analyzed the RNA of a 1918 Spanish Flu victim isolated from a formalin-fixed paraffin-embedded lung tissue sample, confirmed that the virus was an H1NI strain with a common avian ancestor for both the human and the swine strains. (These processes are discussed in more detail later in this chapter.)

Between 1957 and 1958, another influenza pandemic—the Asian flu of strain H2N2—began in China and spread throughout the Pacific. It is estimated to have infected between 10% and 35% of the population; however, its mortality rate of 0.25% was more in line with other influenza epidemics and pandemics besides the Spanish flu. The Hong Kong flu of 1968 of strain H3N2 was responsible for 34,000 deaths in the United States and 700,000 deaths worldwide between 1968 and 1969.[19] Another noteworthy outbreak of IV (strain H5N1) occurred in Hong Kong in 1997. A priori, the scientific community believed that only H1, H2, or H3 strains could infect humans; however, this avian IV infected 18 humans between the ages of 1 and 60 years old and claimed six lives. The significance of this outbreak is that an avian IV directly infected humans, whereas all previous epidemics and pandemics caused by avian IVs required an intermediary host.

Another noteworthy event in the evolution of IVs is an instance of an IV crossing a species barrier. In 1998, an avian IV of strain H9N2 was isolated from pigs in China, and in 1999 the same strain was isolated from individuals in China presenting with mild flu symptoms. The relevance of animal-specific IVs will be discussed in further detail in relation to weaponizing IVs.

MOLECULAR BIOLOGY

Influenza virus causes a highly contagious infection of the respiratory tract that belongs to the family *Orthomyxoviridae*. The three types of IV are designated A, B, and C and classified according to antigenic differences in the their nucleoprotein (NP) and matrix (M1) proteins. Influenza A viruses are further distinguished by the antigenic differences in their hemagglutinin (HA) and neuraminidase (NA) proteins (table 6.1). The three types of IVs (A, B, and C) infect humans, but influenza A viruses can also infect other mammals and birds. Influenza A viruses are also responsible for influenza pandemics and epidemics partly because of their ability to infect nonhuman hosts. Influenza B and C viruses infect only humans, and influenza C viruses cause only mild infections and do not cause epidemics. Influenza A and B viruses are the focus of this chapter.

TABLE 6.1 FUNCTIONS OF INFLUENZA A PROTEINS

RNA SEGMENT	ENCODED POLYPEPTIDE	FUNCTION
1	PB2	Component of RNA transcriptase complex
2	PB1	Catalyzes nucleotide addition to RNA
3	PA	Component of transcriptase and replicase complex
4	HA	Major surface glycoprotein
5	NP	Binds to RNA to form coiled ribonucleoprotein
6	NA	Surface glycoprotein
7	M1	Interacts with ribonucleoproteins and NS2
8	NS1 and NS2	Reduces IFN activity; contains trafficking signals

Source: Adapted from Knipe, D. M. and Howley, P. M., eds. 2001. *Fields Virology*, 4th ed. Philadelphia: Lippincott Williams and Wilkins.

The IVs are composed of 1% RNA, 21–24% lipids, 70% protein, and 5–8% carbohydrate. The virions measure between 80 and 120 nm in diameter. They are enveloped by the membrane derived from the host cell's plasma membrane. Morphologically, influenza A and B viruses have approximately 500 protein spikes, which measure 10–14 nm long and span the lipid envelope. The spikes correspond to the HA and NA proteins in a ratio between 5:1 and 4:1 HA: NA.

The genomes of influenza A and B are composed of eight negative-sense single-stranded RNA segments whereas the genomes of influenza C virus are composed of seven segments. The eight segments of influenza A and B viruses encode 10 genes and consist of 890 to 2341 nucleotides per segment—a total of approximately 13,588 nucleotides per genome, depending on subtype. The 10 genes of influenza A viruses encode 11 proteins: three integral membrane proteins (HA, NA, and M2), three subunits of the IV RNA polymerase complex (PB1, PB2, and PA), matrix protein (M1), nucleoprotein (NP), nuclear export protein (NEP), and two proteins that compromise the antiviral host mechanisms of the cell (NS1 and PB1-F2). Influenza B viruses contain the same proteins as influenza A viruses except that they encode the integral membrane protein NB instead of M2, which acts as an ion channel.[22] Influenza C viruses contain different integral membrane proteins. The integral membrane proteins of influenza C are HEF (hemagglutinin-esterase-fusion protein), which is structurally similar to HA, and CM2, which is structurally similar to M2 and NB.

THE INTEGRAL MEMBRANE PROTEINS

Hemagglutinin Hemagglutinin is located on the IV's surface and comprises about 30% of the virion's total proteins. It was named for its ability to agglutinate erythrocytes by mediating the attachment of sialic-acid-containing receptors. HA is a glycoprotein essential to the IV life cycle. It allows the virion to bind to the host cell and permits the fusion of the viral envelope with the endosomal membrane, which facilitates the virion's entry into the host cell's cytoplasm. Like all glycoproteins, HA is synthesized in the rough endoplasmic reticulum, modified in the Golgi, and finally transported to the cell surface. The fourth RNA segment encodes the protein, which is translated as a single polypeptide chain.[9]

HA is cleaved into two chains joined by a disulfide bond. The cleavage site is dependent on the presence of basic amino acids. This site is usually conserved between viruses that infect specific species. For instance, human IVs contain the amino acid arginine (R) at the cleavage site (HA_1-PSIQVR-GL-HA_2) that is necessary for proteolysis. The protease responsible for the cleavage is believed to be tryptase Clara, which is released from Clara cells located only in the epithelial lining of the respiratory tract. A consequence—for the virus—of the specific cleavage site and the finite tissue distribution of the protease is that the influenza infection is confined to the upper respiratory tract, where Clara cells are found. Noteworthy are the cleavage sites of avian IVs, which rely on the presence of at least one basic amino acid. The ones that rely on the single basic amino acid in the cleavage site (HA_1-PEKQTR-GL-HA_2) are characterized by nonvirulent, low pathogenic, avian IVs whereas the viruses that rely on the presence of more than one amino acid in the cleavage site (HA_1-KKREKR-GL-HA_2) have the versatility of being cleaved by ubiquitous proteases that are present in the Golgi apparatus of all cells. As a result, these latter viruses are able to produce systemic infections and are typically considered highly pathogenic strains.[20]

Once HA is cleaved, one of the resulting chains (HA_1) contains 328 amino acids while the other chain (HA_2) contains 221 amino acids. HA_1 contains a receptor-binding moiety that functions as a sialic-acid-binding site. While human IVs recognize moieties containing 5-N-acetyl-neuraminic acid, IVs that infect swine and equine species can also recognize and bind to N-glycolylneuraminic moieties. Different viruses distinguish between the linkage of sialic acid to galactose and can recognize only a specific complex that is dependent on the amino acid residues in the HA receptor-binding pocket. For instance, human IVs only recognize and bind to sialic acid linked to galactose by 2,6 linkages whereas avian strains recognize and bind to 2,3 linkages.

Furthermore, the binding is dependent on and corresponds to a specific amino acid of HA at the 226 position. For instance, HAs that have leucine at the 226 position selectively bind to 2,6 linkages, which occur in human influenza strains, while HAs that have glutamine at the 226 position selectively bind to 2,3 linkages, which occur in avian IVs. The presence of the specific linkages in the host cells determine which IV strain they are susceptible to. Since humans cells contain only 2,6 linkages, they cannot be directly infected by an avian strain. Some species, such as swine, contain cells that have both 2,6 and 2,3 linkages, making them susceptible to infection by human and avian strains, and this renders them vectors for the transmittal of avian-based IVs, which are generally considered the IV strains most pathogenic to humans.[17] This receptor specificity of HA creates a barrier that usually prevents avian IVs from directly infecting humans. An intermediary host, such as the pig, is usually required for humans to become infected with an avian-derived IV.

Whereas HA_1 is responsible for the binding of IVs to host cells, HA_2 is essential for permitting viral entry into the host cells' cytoplasm. The functional regions of HA_2 include a transmembrane region and an amino acid terminus that facilitates the fusion of viral envelope with the endosomal membrane and permits viral entry into the cell. When HA is properly inserted into IVs, at the normal cellular pH (7.4) it assembles into a trimer.[8]

Neuraminidase Like HA, NA is a glycoprotein located on the surfaces of influenza A and B viruses that is also important for the viruses' life cycles by facilitating the release of progeny viruses from host cells. NAs of influenza A and B viruses are both encoded by the sixth RNA segment, and they consist of 453 and 466 amino acid residues, respectively. The main domain of NA is a single hydrophobic domain located at the N-terminus of the protein. It spans the lipid bilayer and functions as a signal, which targets NA to the ER membrane, and as an anchor, which stabilizes the protein's attachment to the membrane.[1]

NA catalyzes the degradation of sialic acid. It cleaves N-acetylneuraminic acid (sialic acid) residues from IVs as well as the host cell's glycoprotein receptors.[21] Since HA normally binds to these sialic acid residues, the cleavage allows new budding copies of IVs to emerge from the host cell without binding to the cell's receptors or other progeny viruses.[11]

Influenza A viruses are subdivided into subtypes according to antigenic differences of HA and NA proteins. To date, there are 15 HA (H1–H15) and nine NA (NA1–NA9) subtypes among influenza A viruses that convey antigenic variation of each HA and NA subtype. The emergence of the various subtypes is a result of evolution caused by antigenic shifts and drifts. (The processes involved with antigenic shifts and drifts as well as their consequence to the virulence of the virus are discussed later in the chapter.)

M2 and NB Proteins The M2 and NB proteins are the ion-channel proteins of influenza A and B viruses, respectively. Like HA and NA, M2 and NB are integral membrane proteins. However, they are smaller: M2 contains 97 amino acids and NB contains 100 amino acids. While M2 is encoded by the seventh RNA segment of influenza A viruses, NB is encoded by the sixth RNA segment of influenza B viruses, which contains a second open reading frame that overlaps the NA reading frame by 292 nucleotides.[14] The two proteins have small extracellular domains (24 amino acids for M2) and larger cytoplasmic domains (54 amino acids for M2). Structurally, M2 is composed of two dimers linked by a disulfide bond, and the protein assembles itself into a homotetramer.

The transmembrane domain of M2 and NB functions as an active ion channel during the early stages of the viral infection, between viral penetration and uncoating. Since the virion's RNA is introduced into the host cell in the form of ribonucleoproteins (RNPs), M2 and NB proteins—functioning as ion channels—allow for effective uncoating of the virion's RNA.[21,22]

THE RNA POLYMERASE COMPLEX

The proteins PB1 (basic polymerase subunit 1), PB2 (basic polymerase subunit 2), and PA (acidic polymerase subunit) form the IV RNA polymerase complex. The three proteins are encoded by first, second, and third RNA segments, respectively. The first and second RNA segments contain 2341 nucleotides and encode PB1, which contains 759 amino acids, and PB2, which contains 757 amino acids, respectively. The third RNA segment contains 2233 nucleotides and encodes PA, which consists of 716 amino acids. The proteins are named because of their reactions on isoelectric gels; PB1 and PB2 are basic and PA is acidic. These proteins are collectively referred to as P proteins and, along with M2 and NB, are essential for viral mRNA synthesis. They are all synthesized in the cytoplasm of the host cell and then migrate (either as a complex or individually) into the nucleus to facilitate RNA synthesis.[2]

The P proteins form a heterotrimeric polymerase complex, and the complex is responsible for transcribing and replicating the virion's genome. Transcription involves copying virion RNA (vRNA) into messenger RNA (mRNA), and replication involves converting vRNA into cRNA and cRNA into vRNA. The IV polymerase complex is stabilized by the interaction of the complex with the vRNA promoter, which consists of 13 nucleotides on the 5′ end and 12 nucleotides on the 3′ end of every RNA segment of the virion's genome. (The processes on viral transcription and replication are discussed in detail later in this chapter.)

NUCLEOPROTEIN, NUCLEAR EXPORT PROTEIN, AND MATRIX PROTEIN

Nucleoprotein, NEP, and M1 mediate the trafficking of RNPs into and out of the host cell's nucleus, where transcription and replication occur. NP is encoded by the fifth RNA segment, which is 1565 nucleotides in length; the protein contains 498 amino acids. The protein is synthesized in cytoplasm, and it is intrinsically targeted to the nucleus by one of its two nuclear localization signals, which occur between residues 198–216 and residues 1–38. NP coats viral RNA, forming RNPs, and contains the nuclear localization signal, which facilitates the import of RNA into the host cell's nucleus. Because of NP's function in coating viral RNA, it is one of the specific IV antigens. To this end, it is a target of cytotoxic T lymphocytes in the host immune response.

NEP is a 121-amino-acid protein encoded by eighth RNA segment. It contains the nuclear export signal and facilitates the export of RNPs from the nucleus to the cytoplasm late in the infectious cycle.

M1 is a matrix protein encoded by the seventh RNA segment; it contains 252 amino acids. Located in the lipid envelope of the virion, M1 works in conjunction with the NEP to mediate the export of RNPs into the host cell's cytosol. Furthermore, M1 coats RNPs and facilitates exocytosis of viral progeny.

This process of RNP trafficking is believed to require an active mitogen-activated protein, namely, the MAP kinase pathway.[16]

NS1 AND PB1-F2

The proteins NS1 (nonstructural protein) and PB1-F2 (basic polymerase subunit 1–frame 2) combat the host cell's defense mechanisms. For instance, the secretion of interferons (IFNs) is an important cellular response to viruses because they promote the transcription of antiviral genes that hinder viral replication. To survive, IVs must counter the actions of IFNs; they do this using NS1, which is encoded by the eighth RNA segment and consists of 230 amino acids. The NS1 protein prevents the activation of IFN transcription factors. Additionally, it inhibits the activation of protein kinase R (PKR), which catalyzes the production of IFNs. As a consequence, the actions of the NS1 protein allow IVs the opportunity to survive and replicate in host cells.[22]

PB1-F2—a newly discovered IV protein derived from an overlapping reading frame of PB1 located in the second RNA segment—appears to also create havoc with the host cell's defense mechanisms. The protein associates with the mitochondria of the host cell, and induces monocyte apoptosis and degrades proteosomes and proteases. All of this promotes IV survival by destroying the host's natural defense cells.[12]

THE EARLY STAGE OF THE INFECTIOUS CYCLE

When an individual inhales IV, it enters the host cells of the nose, throat, or lungs, where the virus replicates. Influenza viruses are attracted to cells that possess polarized epithelium, which are cells that contain apical and basolateral surfaces (figure 6.1). These cells are primarily epithelial cells, and IVs specifically target the epithelial lining of the respiratory mucosa. Therefore, when the virus is inhaled, it first encounters the apical, or outer, face of the epithelial cells. Following replication the virus is also released from the apical surface into the respiratory tract. In this manner, the host can exhale IVs so that they can infect new hosts. Furthermore, the absence of basolateral release prevents the systemic spread of IVs within the host. However, systemic spread may still occur due to secondary bacterial infections and/or a highly pathogenic virus. (These conditions are discussed later in this chapter.)

The IV infection is initiated when HA binds to sialic-acid-containing molecules, such as cell-surface glycoproteins and/or glycolipids on the host cell's plasma membrane. The site of HA responsible for binding is characterized by a shallow depression, known as the HA-receptor-binding pocket, at the distal tip of the protein. As described earlier, different IVs specifically bind to sialic acid

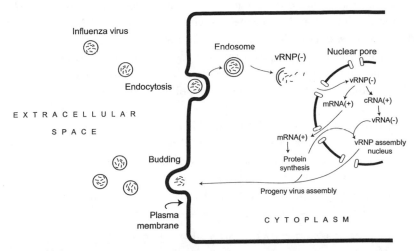

FIGURE 6.1. The replication cycle for influenza virus. The steps involved include the following: (a) The virus binds to receptors on the surface of the host cell. (b) The virus is internalized into an endosome. (c) Fusion and uncoating events, result in (d) the release of the viral genome in the form of viral ribonucleoprotens: vRNAs into the cytoplasm. The vRNPs are then imported into the nucleus for (e) replication. (f) Positive-sense viral messenger RNAs (mRNAs) are exported out of the nucleus into the cytoplasm for (g) protein synthesis.

linked to galactose by either 2,3 or 2,6 linkages, and these linkages are specific to certain hosts. Therefore, binding is the initial determinant of viral pathogenicitiy, and this specific binding is ultimately dependent on the sequence of amino acids contained in the HA receptor–binding pocket.[17]

Following HA binding, IVs enter the host cell via endocytosis. The viruses are endocytosed into clathrin-cooated vesicles, with membranes derived from the host cell, and the vesicles are also released from the apical surface into the respiratory tract. In this manner, the host can exhale influenza virues so that they can proceed to infect new hosts. Furthermore, the absence of basolateral release prevents the systemic spread of IVs within the host. However, systemic spread may still occur due to secondary bacterial infections and/or a highly pathogenic virus.[22] (These conditions are discussed later in the chapter.)

In addition to viral fusion, the release of viral components, such as M1 and vRNPs, into the host cell cytoplasm is also initiated and mediated by the low endosomal pH. This stage is referred to as virus uncoating, in which RNPs (which contain the NP, PB1, PB2, and PA proteins) are delivered to the cytoplasm. In addition to the acidic requirements, uncoating is also dependent on the presence of an ion-channel protein, such as M2 and NB in influenza A and B viruses, respectively.[22] The presence of such a protein allows the viral components to become exposed to the acidic endosomes. This low-pH environment is necessary to disrupt interactions between vRNPs and M1 proteins. The disruption of these interactions is required to transport vRNPs into the nucleus because the vRNP-M1 complex is too large to pass through the nuclear pore, while the "uncoated" vRNPs are able to traverse the pore complex. Furthermore, nuclear import of vRNPs is facilitated by NP, which contains the nuclear localization signal.[2] Once dissociated from the vRNP, M1, as well as newly synthesized M1 proteins, can enter the nucleus by diffusion or targeted to the nucleus by its punitive NLS that has recently been identified.

THE LATE PHASE OF THE VIRUS LIFE CYCLE

Following the early stage of the influenza viral cycles, the late stage is initiated by the transcription of vRNA. At this stage, the IV RNA polymerase complex assembles from three protein subunits PB1, PB2, and PA. Following assembly, transcription and replication of the virion's genome are carried out in the host cell's nucleus.

Replication and transcription of IVs differ from those of most RNA viruses in that these processes are carried out in the nucleus of the host cell rather than the cytoplasm. Furthermore, they are unlike any other RNA viruses because of their unique dependence on the host cell's nucleus and selective functions. For instance, the influenza polymerase molecules (PB1, PB2, and PA) are incapable of initiating viral mRNA synthesis without a primer in vitro. In vivo, the polymerase requires the capping of the viral RNA, and the capped RNA serves as the primer for initiating transcription. Through a process known as cap snatching, the IV selectively cleaves 10 to 13 nucleotides from the 5′ ends of the cellular transcripts and directs them to serve as the primers for its own viral mRNA synthesis.

As noted earlier, the P proteins (PB1, PB2, and PA) catalyze viral mRNA synthesis and function as a complex. Following the capping of viral RNA, the PB2 protein recognizes and binds to the capped primer. Furthermore, studies have shown that PB1 catalyzes the addition of new nucleotides. To date, the specific role of PA is not known.

The viral mRNAs are then transported to the host cell's cytoplasm to be translated into the IV's 11 proteins. The integral membrane proteins—HA, NA, and M2 (NB for influenza B viruses)—are exported to the endoplasmic reticulum for translation. Following translation, they are transported to the host cell's plasma membrane.

PROTEIN AND VIRAL EXPORT

Newly synthesized NP, PB1, PB2, and PA are transported back to the nucleus to augment RNA replication. Later in the infectious cycle, M1 and NEP move back into the nucleus and mediate RNPs export into the cytosol.

Viral budding occurs at the host cell's plasma membrane. The M1-coated RNPs interact with the viral plasma membrane proteins—HA, NA, and M2 (or NB). Furthermore, NA desialylates sialic acid from viral proteins and the plasma membrane.[21] This allows new IVs to emerge from the host cell without sticking to the cell's membrane or other IVs.

Once IV invades the host cell, a complex struggle of defense and counter-defense actions begins. The HA proteins, which are inserted into the host cell's plasma membrane during the middle and late stages of the infectious cycle, bind to NKp46 and NKp44 proteins. The binding of these proteins activates natural killer cells that mobilize the immune system against IV. Simultaneously, NS1 disarms the immune-system response by inhibiting interferon production and compromising cellular apoptotic pathways, which are also mediated by PB1-F2.[22]

CLINICAL DIAGNOSIS AND RESPONSE

Although many consider influenza to be a mild, benign disease, it is the sixth most common cause of death in the United States, and it is responsible for the deaths of about 20,000 Americans each year. The flu season in the Northern hemisphere extends from November to March, and an IV infects 10–20% of all U.S. residents each year.[1]

Although influenza A viruses are primarily responsible for influenza pandemics and epidemics, the characteristics of an influenza infection due to types A and B are quite similar. These are discussed below.

COMMUNICABILITY

Influenza virus is an airborne virus, and it is transmitted from person to person when an infected individual releases virus particles into the air by sneezing, coughing, or speaking. Epidemiological studies, measuring the human infectious dose–50 (HID_{50}) of influenza A wild-type virus, have shown that the aerosol route is the most effective mode of transmission. The HID_{50} by intranasal administration by drops is 127 to 320 tissue culture infectious dose–50 ($TCID_{50}$) while the HID_{50} by aerosol administration is 0.6 to 3 $TCID_{50}$. When someone inhales the virus, it enters the nose, throat, or lungs, where it multiplies and causes flu-like symptoms.[17] IVs can also be spread if a person touches an object that contains virus particles and subsequently touches his or her mouth or nose.

As stated previously, flu is very contagious. An individual can spread the flu one day before symptoms arise, even if he or she has no symptoms at all. Once the symptoms start, adults continue to be contagious for an additional 3 to 7 days, and children continue to be contagious for over 7 days. In most instances, the incubation period for flu is about 1 to 2 days, and symptoms generally arise 1 to 4 days after infection.[4]

Once the host is infected, the IV replicates within the respiratory tract, and the virus can be recovered from the upper and lower respiratory tracts of infected individuals. Studies in which volunteers were inoculated with the virus indicate that viral replication peaks 48 hours following inoculation and its rate of replication declines thereafter. Most viral shedding occurs 1 to 6 days following inoculation, and little shedding is observed 6 to 8 days following inoculation. There is a correlation between the viral shedding and the severity of symptoms, and peak viral titers in symptomatic volunteers ranged from 10^3 to 10^7 $TCID_{50}$/mL of nasopharyngeal wash. Furthermore, viral antigen can still be detected for several days after the virus can no longer be recovered from cells and secretions of infected individuals. In children naturally infected with IV, viral recovery most frequently occurs 1 to 2 days following the onset of symptoms and can be recovered up to 13 days thereafter. The peak titer occurs on the first day of symptoms and averages 10^4 $TCID_{50}$.[17]

EFFECTS ON THE RESPIRATORY TRACT

Influenza viruses induce the most significant pathological effects in the lower respiratory tract of infected hosts. Individuals who do not have any complications due to the virus show inflammation of the larynx, trachea, and bronchi, along with mucosal inflammation and edema. Ciliated and mucus-producing epithelial cells are stripped to their basal layer 1 day after the onset of symptoms. In other areas of the lower respiratory tract, the basement membrane is exposed. These cells are then infiltrated with neutrophils and mononuclear cells, which trigger submucosal edema and hyperemia (an accumulation of blood) within this area.[22] As mentioned above, IVs and their antigens are usually localized within epithelial cells and mononuclear cells; however, their presence within the basal cell layer is an indicator of a highly virulent IV strain and systemic spread of the virus. At the cellular level, IVs are responsible for cell destruction by terminating cell protein synthesis and inducing apoptosis, primarily through the NS1 and PB1-F2 proteins described earlier. Three to five days following the onset of symptoms, mitosis is observed within the basal cell layer, and this signals the initiation of epithelial regeneration. While the virus is still active and replicating within the host, regenerative (via host cellular mechanisms) and destructive (via viral mechanisms) processes can proceed simultaneously. Complete regeneration of necrotic epithelium usually takes up to 1 month.

THE FIRST LINE OF DEFENSE:
THE INNATE IMMUNE RESPONSE

The first protective measure against IVs may be assumed to be the apical surface of epithelial cells, which limit the systemic spread of the virus. Ironically, however, this surface also facilitates the release and transmission of virus particles from person to person. Fortunately, the respiratory tract contains more explicit protective features against IV infections; these include the mucin layer, ciliary action, protease inhibitors—which may impede effective viral entry into the host and virus uncoating—and cytokines. The cytokines are most notable in the innate immune response. Once the virus invades the epithelial cell, proinflammatory cytokines, such as interleukin-6 (IL-6) and IFN-⊠ are summoned and released from the cells. The concentration of these cytokines peak by the second day following infection, and their release corresponds contemporaneously to the onset of symptoms, mucus production, fever, and viral load.[13] IFN also serves to impede the replication of IVs, which is supported by the observation that the NS1 protein is an IFN antagonist. Furthermore, deletion of NS1 in an IV mutant leads to an extremely attenuated virus. In summary, IFN facilitates recovery from an IV infection while cytokines direct the early immune responses of the host.

THE SECOND LINE OF DEFENSE:
THE COGNATE IMMUNE RESPONSE

Humoral Immunity Humoral immunity toward influenza is conferred by antibodies against the virus. This is the main response that human bodies have against infections—80% of naturally contracted infections initiate an antibody response. In short, when a foreign agent, such as a bacterium or a virus, is introduced into the body, the cognate immune response to this offender includes a humoral response, which directs antibodies that target specific antigens of a foreign agent, and a cellular response, which directs T cells to destroy body cells affected by the foreign agent. The humoral response of the immune system includes a primary and secondary response. The primary response is to direct B cells to divide, produce, and secrete antibodies specific to a primary infection. The first set of antibodies released in this phase are specific for prior infections; B cells have a memory for past antigens and are able to quickly release and produce such antibodies specific to these antigens. If the foreign agent is a new invader, specific B cells do not hold in their memory the antibodies specific for the new antigen. However, B cells are eventually able to produce specific antibodies after a delayed period of time. The secondary response is characterized by the production and secretion of antibodies in response to an infection caused by an agent against which B cells have memory for antibodies.

When an individual is infected with an IV, the humoral immune response produces antibodies specific to the HA, NA, NP, and M proteins, antigens that are specific to the IV strain. Antibodies specific for the HA and NA proteins are responsible for resistance to IVs while antibodies specific for M1 and NP proteins, which are found within the virion, do not contribute to immunity. Note that the M2 protein, which has an extracellular domain, is being investigated as a target site for monoclonal antibodies and a universal vaccine. Because this protein has an extracellular domain, antibodies recognize this domain as an antigen and subsequently bind to it and target for destruction the cell that contains the virus. (This is discussed in further detail under "Potential Defenses" later in the chapter.) As is common with any infection, the serum level of antibodies specific to an antigen or a foreign agent directly correlates with the resistance and immunity an individual has against an invader. In influenza, specifically, the serum level of antibodies specific to HA and NA proteins corresponds to how well the body is able to fight the infection. Therefore, after an influenza infection, the antibodies in the blood serum are able to prevent a subsequent reinfection of the same type and strain of IV, and this immunity usually persists for a long period of time. However, research and data suggest that individuals can be reinfected by antigenically similar IVs. The implication of this is that highly pathogenic strains may induce only an incomplete immunity and a subsequent infection is required to confer complete immunity and allow the antibodies to be specific for several antigenic sites on these strains.

Cellular Immunity As explained above, the cellular immune response involves lymphocytes that target cells affected by the foreign agent for destruction, thereby contributing to the removal of the virus. The role of cellular immunity in response to an IV infection is known through the work by Karzon using the murine model. The key cells in this response consist of two classes of cytotoxic lymphocytes (CTLs), which have either the CD8+ or CD4+ phenotype. The CTLs of the CD8+ type are measured in the blood of infected or vaccinated individuals between the sixth and 14th day after infection (or vaccination) and are undetected after the 21st day. In vitro studies have revealed a subtype of CTLs known as memory CTLs.[6] This subtype has been detected in the blood after the antigen is introduced into the bloodstream, and they destroy infected cells. Specifically, they lyse cells infected with any subtype of a specific IV.

Therefore, they target cells infected by any strain of influenza A virus, but they do not target cells infected with any type of influenza B or C virus. However, the more commonly known and better studied subtype of CTLs is specific for IV subtypes. Therefore, they are able to target and lyse only cells infected with a specific virus subtype (e.g., they target and lyse a cell that is infected by influenza A H1N1 but cannot target or lyse a cell infected with influenza A H2N1).

The second class of lymphocytes involved in the cellular immune response consists of cells that have the CD4+ phenotype. These cells belong to the T-helper class of T cells, known as CD4+ T cells, and they contribute to cellular immunity in two ways. First, as their name implies, they facilitate production, secretion, and proliferation of B-cell and CD8+-cell antibodies by secreting growth factors specific for these cells. Second, they are also cytotoxic cells, thereby targeting and lysing cells that are affected by the foreign agent. The specific functions of CD4+ T cells are being further investigated for their role in human immunity. A defense system involving CD8+ and CD4+ T cells functions to destroy infected cells, which subsequently clears the virus from an individual's system.[6]

MUCOSAL ANTIBODY RESPONSE

In addition to the humoral and cellular response to an influenza infection, the human body also combats the IV through nasal secretions. Francis et al. first published their findings that these secretions have a neutralizing activity toward IVs in the journal *Science* in 1950. Subsequent studies by Feery et al. attributed this activity to antibodies that were present in bronchial secretions in mice. Furthermore, more recent studies on humans classify the antibodies as immunoglobulin A and M (IgA and IgM) isotypes, which target both the HA and NA proteins and are produced locally within the upper respiratory system. The presence of these antibodies contributes to resistance against IVs.[17]

SIGNS AND SYMPTOMS

An IV infection can produce a wide spectrum of symptoms, which may range from none at all to severe symptoms related to secondary infections, such as those associated with viral pneumonia. While there are subtle differences between the symptoms caused by specific strains, the general symptoms caused by influenza A and B viruses are similar across types and strains.

The typical symptoms of influenza include fever, headache, malaise, dry cough, sore throat, nasal congestion, and body aches. Unlike those of a common cold, these symptoms appear suddenly and persist for 1 to 2 weeks. The incubation period, which depends on the viral load encountered and the immune response in the host, ranges from 24 hours to 5 days. The fever, which is usually between 38°C and 40°C but can rise to as high as 41°C, is a main sign of an influenza infection; it peaks within 24 hours of the onset of symptoms. The duration of fever and body aches is 3 to 5 days, and the cough and malaise can last for 2 or more weeks.

Although the clinical features of an influenza infection in children and adults are similar, there are some notable differences in symptoms. A higher fever, often accompanied by convulsions, is present in juvenile influenza. Furthermore, children are more susceptible to secondary infections during an influenza infection, including otitis media, croup, pneumonia, myositis, and lower-respiratory-tract disease.

COMPLICATIONS

Although everyone is susceptible to IV, 90% of the 20,000 deaths each year in the United States occur in people 65 years and older. These deaths are usually attributed to complications of the flu, including pneumonia, bronchitis, sinusitis, and ear infections. The flu can also exacerbate health problems such as asthma and congestive heart failure. Very young children are also more susceptible to complications due to influenza.

Some of the major complications affect the lower respiratory tract: syndromes of pneumonia (known as primary viral pneumonia), combined viral-bacterial pneumonia, and bacterial pneumonia. Primary viral pneumonia was first recorded as an IV complication during the 1957 Asian Influenza Pandemic (H2N2), when it was linked to fatal cases. However, it is apparent that primary viral pneumonia was also a complication of earlier influenza epidemics and pandemics. The syndromes of pneumonia usually occur in individuals at high risk for complications, such as the elderly. The symptoms that accompany primary viral pneumonia include a rapid respiration rate (30–60 breaths per minute), tachycardia (pulse greater than 120 per minute), cyanosis (occurs in 80% of cases), high fever (average of 39°C), and hypotension. The symptoms usually appear very soon after the onset of an influenza infection—typically within 6 to 24 hours. Furthermore, the symptoms can progress to hypoxemia and death within 1 to 4 days. The signs that precede the fatal outcome include frothy hemoptysis, tachypnea, and cyanosis.[17]

The second pneumonia syndrome, known as combined viral-bacterial pneumonia, is essentially clinically identical to primary viral pneumonia. However, it is about three times more common than the latter; involves accompanying bacterial infections, such as *Streptococcus pneumoniae*, *Staphylococcus aureus*, and *Haemophilus influenzae*; and symptoms appear later after the onset of influenza than in primary viral pneumonia. It is noteworthy that specific strains of *S. aureus* secrete proteases that are capable of cleaving the HA polyprotein, thereby activating the virus and contributing to its virulence. While the normal fatality rate associated with combined viral-bacterial pneumonia is between 10 and 12%, the fatality rate due to a *S. aureus* coinfection is much higher—approximately 42%.[17]

Bacterial pneumonia differs from viral-bacterial pneumonia in that the latter is a coinfection whereas the former is pneumonia following an influenza infection. Typical in bacterial pneumonia is an individual with an influenza infection who initially seems to be improving but then develops chills, pleuritic chest pain, and augmented cough production that is rich in blood and virus. Unlike in primary viral pneumonia, cyanosis and increased respiration rate are not as common, and of the three pneumonia syndromes this is the mildest, with a fatality of only 7%.[17]

Other complications associated with IV infections include myositis (muscle inflammation), Reye's syndrome, and central nervous system (CNS) disorders. Myositis in adults often permeates the whole body, but it can also be associated with myocarditis. Noteworthy symptoms include diffuse, numb pain throughout the body, tenderness, and weakness of muscles. During such an episode, the body produces increased levels of muscle enzymes, known as myoglobinemia. Acute renal failure can accompany myositis, which may be fatal. Reye's syndrome, generally seen only in children and adolescents, is characterized as a neurological and metabolic disorder. Specific symptoms include noninflammatory encephalopathy and hepatic dysfunction. Reye's syndrome most commonly follows infections caused by respiratory viruses, such as the IV, and it is commonly associated with IV complications in children. The incidence rate of Reye's syndrome in children under the age of 18 is between 0.37 and 0.87 per 100,000 children, and the fatality rate has been reported to be between 22% and 42%. Other complications related to influenza viral infections affect the CNS of humans. Symptoms associated with CNS disorders include irritability, drowsiness, psychosis, delirium, and coma. The pathological manifestations of these symptoms are currently unknown.

Additionally, two specific neurological disorders, known as influenzal encephalopathy and postinfluenzal encephalitis, have been linked to influenza infections. Influenza encephalopathy usually appears during the viral infection, and it may progress to cause death. Postinfluenzal encephalitis, as indicated by the name, appears approximately 2 to 3 weeks following the viral infection. This form of encephalitis is rare and usually not fatal.[4]

LABORATORY TESTS

Influenza is difficult to diagnose because the initial symptoms are similar to those caused by many other infections. Laboratory tests can be done to confirm an IV infection. These tests usually target specific influenza viral antigens, allowing medical professionals and/or researchers to identify the type and strain responsible for the infection.

Viral cultures can be performed from nasopharyngeal or throat swab, nasal wash, or nasal aspirates. The samples should be collected from the first 4 days of infection. While the viral culture provides results within 3 to 10 days, a rapid influenza test provides results within 24 hours. The rapid tests, which can be performed in physicians' offices, are more than 70% accurate for detecting influenza and over 90% accurate for determining IV type. Serum samples can also be tested for influenza antibodies to diagnose a recent infection. Unfortunately, results for this test take over 2 weeks to obtain.[4]

ANTIVIRAL DRUGS

The general treatment of flu for most victims is to get plenty of rest, drink liquids, and take medication to relieve the symptoms. In addition an assortment of antiviral drugs, such as amantadine, zanamivir, rimantadine, and oseltamivir, can be used to reduce the severity and duration of the symptoms.

The antiviral drugs used to treat influenza are prescription drugs and are effective only at reducing the duration of an IV infection. Additionally, they must be administered within the first 2 days of illness. The drugs approved for the treatment of IVs are: amantadine and rimantadine, which are effective only against influenza A viruses, and zanamivir and oseltamivir, which are effective against both influenza A and B viruses.[18] All the antiviral drugs inhibit viral replication, but they differ in their mode of action.

Zanamivir and Oseltamivir Zanamivir and oseltamivir are neuraminidase inhibitors. They prevent the IV NA protein from removing sialic acid from sialic acid–containing receptors. Viral budding, and subsequent downstream replication of IVs, are inhibited when sialic acid remains on the host cell and virion membranes. Since sialic acid is found on both surfaces of the cell and the virion, the emerging IVs tend to stick to the cell plasma membrane or other viruses in the presence of neuraminidase inhibitors.[21]

Zanamivir must be administered intranasally for optimal efficacy whereas oseltamivir is effective if taken orally. Both drugs have produced only minimal side effects in studies using animal models and human clinical trials, and they are approved by the U.S. Food and Drug Administration (FDA) for use in humans for prophylaxis and treatment of influenza. (Oseltamivir is also considered safe for use in children as young as 1 year old.) The use of zanamivir or oseltamivir reduces the duration of major influenza symptoms such as headache, malaise, cough, and sore throat by 1.5 to 3 days when the drug is first administered within 36 to 48 hours following the onset of influenza symptoms.

Under drug pressure of neuraminidase inhibitors, IVs are capable of mutating into variants resistant to NA inhibitors; this has been shown through in vitro studies. The first sign that resistance against these drugs is being developed is the appearance of mutations surrounding and affecting the receptor-binding site for HA. This lowers the binding affinity, which compensates for the loss or reduction of NA function. The second step of emerging resistance is the appearance of mutations surrounding the NA inhibitor-binding site, which eventually renders the NA inhibitor ineffective. The emergence of resistant variants has not been observed in vivo with human clinical trials; however, large-scale trials as well as wide use of NA inhibitors will be the ultimate determinant of whether these mutations also occur in vivo.

Amantadine and Rimantadine The antiviral drugs amantadine and rimantadine inhibit influenza A virus replication by blocking the ion-channel M2 protein. The blockage of the M2 protein inhibits the delivery of IV RNPs from the endosomes to the cytosol. This specifically prevents the influx of H^+ ions from the endosomes into the virion, which is necessary to disrupt the vRNP-M1 interactions required to import the vRNPs into the host cell's nucleus for transcription and replication.[10]

In vitro, 0.2 to 0.4. µg/ml of amantadine and rimantadine are required to inhibit the plague formation of influenza A whereas >100 µg/ml are required to inhibit the plague formation of influenza B viruses. The recommended adult dose of these drugs is 200 mg. At this dosage, side effects are minimal and the level of the drugs is about the level required to inhibit influenza A viruses, which further explains why these drugs specifically target influenza A viruses.

Epidemics involving H1N1, H2N2, and H3N2 influenza A viruses have revealed that these drugs can be used for prophylaxis against these virus strains by contributing to a high percentage of protection against infection caused by them. Furthermore, the drugs have been shown to be effective for treating the aforementioned strains. As with neuraminidase inhibitors, the inhibitory effects of amantadine and rimantadine are not always long-lasting because the gene that codes for M2 can mutate and confer resistance from the drugs quickly. This can occur within 24 to 48 hours following treatment.[7]

VACCINES

The best way to protect against influenza is to get vaccinated once a year. Because IVs undergo frequent mutations, especially with reference to their HA and NA proteins, new vaccine cocktails, containing different influenza strains, need to be produced yearly. Each year's vaccine is trivalent—conferring resistance against three strains of IV. The vaccine contains two influenza A viruses and one influenza B virus.

The determination of which strains of IVs will comprise the vaccine is done via a global surveillance network consisting of approximately 200 WHO laboratories in 79 countries. Four WHO Influenza Collaboratory Centers (located in Atlanta, London, Melbourne, and Tokyo) coordinate the work of the laboratories. Throughout the year, IVs isolated from patients are sent to these centers. The strains are then analyzed, and the network makes recommendations as to the strains they expect to be circulating the following year. The FDA Vaccines and Related Biological Products Advisory Committee also make recommendations about upcoming IV strains. After agreement has been reached, the vaccine manufacturers make the trivalent vaccine from inactivated viruses.

Inactivated IV vaccines have been FDA-approved for use in humans for approximately 50 years. Currently the only influenza vaccine approved for human use is an inactivated virus vaccine grown in chicken eggs. The seed viruses used for the vaccines are naturally occurring viruses. They are able to replicate productively and to high titer levels in the allantoic cavity of chicken eggs. These viruses are then purified (although some residual egg proteins may remain) and chemically inactivated. One egg is capable of yielding enough virus particles for one to three vaccine doses. Two types of vaccines are currently used: the whole-virus (WV) vaccine and the subvirion virus (SV) vaccine, which is composed of split or purified IV antigens.[15]

The vaccines consist of either a monovalent influenza A H1N1 virus or a trivalent mixture consisting of H1N1, H3N2, and an influenza B virus. They are standardized according to the amount of HA antigen included. The Advisory Committee on Immunization Practices has mandated that the immunoreactive HA quantity should be 15 µg per component for adults and children 3 years and older and 7.5 µg for children under 3 years of age. The immunoreactive NA quantity, however, is not standardized because the protein is unstable, especially when subjected to purification and storage. The vaccine dose is 0.5 mL, and it contains about 10 billion IV particles, as well as small, inert quantities of endotoxin, egg-derived protein (which may be reactive in individuals allergic to egg products), free formaldehyde, and preservative.[6] The vaccine is administered intramuscularly.

The effectiveness of the vaccines depends on whether the IV strains used are good matches to the circulating strains, the individual's age, and his or her prior exposure to an influenza A strain that closely resembles one used in the vaccine cocktail. Generally, the vaccines in current use prevent an IV infection in 70 to 80% of healthy individuals under the age of 65 while they prevent infection in only 30 to 40% of individuals 65 years and older. If efficacy is defined as the vaccines' ability to prevent death, the effectiveness rate is about 80%, even in high-risk individuals such as the elderly. Since vaccines work by inducing an immune response, thereby generating antibodies for influenza HA and NA antigens contained within the vaccine cocktail, an individual who had previously been infected with an IV antigenically similar to an antigen used in the vaccine will generate a better immune response because his or her immune system was primed by the primary infection. Therefore, greater protection may be conferred to this individual than to one with no prior exposure to an antigenically similar virus. Accordingly, each individual generates a different immune response to antigens, and this affects the amount of antibodies generated. In fact, in the case of a new pandemic IV strain, research suggests that individuals will require two vaccine doses. Also, if a new influenza strain emerges that is not in the vaccine, the vaccine will be unable to protect against it because the antigens the vaccine contains, and the antibodies generated in response to those antigens, are not the antigens of the new strain and therefore the generated antibodies are not effective against the new strain.[13]

Medical professionals recommend that people over the age of 50 and high-risk individuals get the flu vaccine. The risks associated with vaccination are minimal, and, since the vaccine includes only inactivated flu virus particles, there is no risk of developing an influenza infection from the vaccine.

POTENTIAL DEFENSES: NEW VACCINES

Most of the research targeted to developing defenses against IVs focuses on improving IV vaccines. There is also some work investigating rational structure-based drug design. The work on vaccines focuses on developing cold-adapted IV vaccines, vaccines containing live IVs, adjuvant vaccines, MDCK cell–derived vaccines, and a universal vaccine. These approaches seek to create more effective vaccines that confer greater protection and remain effective longer.

Since the currently used vaccines contains inactivated, killed IVs, it does not offer complete protection because the immune response may not be as great as it is in response to a live virus. Live-virus vaccines would probably be more effective at generating a longer-lasting and stronger immune response than killed-virus vaccines; hence, research is being directed toward their development. The current live-vaccine research seeks to develop cold-adapted, genetically engineered, and replication-defective live-IV vaccines.

Cold-adapted live-IV vaccine has actually been administered to millions of children in Russia without signs of adverse reactions and spread of virulent IV strains. Research has been directed toward such vaccines for over 20 years in the United States, but they have yet to gain FDA approval. In one effort, temperature-sensitive IV master strains were adapted to grow in chicken kidney cells and embryonated eggs at 25°C.[15] These master strains are presumably good candidates for vaccines because of their retained pathogenicity and high attenuation. The proposal for vaccines includes a 6-to-2 reassortment ratio, in which six genes are taken directly from the cold-adapted master strain and 2 genes correspond to the surface antigens (HA and NA) of current circulating IV strains. The advantages of this vaccine include its ability to induce local neutralizing immunity and cell-mediated immune responses, the increased protection against infection in children aged 6 months to 9 years, the ability to administer it as a nasal spray as opposed to the intramuscular injection required for the current inactivated virus vaccine, and the reduction of secondary bacterial infections.[15]

Genetic engineering techniques and reverse genetics technology are also being utilized to develop other live-IV vaccines. One of these efforts aims to engineer site-specific mutations in the genomes of negative-strand RNA viruses such as IVs. The ability to accomplish this opens up new avenues for the creation of live-attenuated-virus vaccines. In one study, the promoter region of the NA gene of an influenza A virus was exchanged with the same region from an influenza B virus, resulting in an attenuated virus. The exchange led to an attenuated virus. Other studies focus on genetically engineering mutations in the PB2 gene that have been demonstrated to be deleterious to the replication of IV.[3,15]

Reverse genetics technology has afforded researchers the ability to investigate the effectiveness of live-IV vaccine candidates that express mutated NS1 genes. Specifically, the plasmid-only rescue system of reverse genetics allows IVs to be grown with deletions in their genomes. For instance, another method of creating a vaccine with live attenuated IV is to include a virus whose genome includes a truncated NS1 gene or deleterious mutations in the gene. This vaccine might be effective since the NS1 protein has IFN-antagonist activity, and a defective or missing NS1 protein would leave the IV exposed to attack by the host's immune system. Therefore, a strong immune response would be generated by the individual and the live virus would be highly attenuated.

Another approach to creating live-attenuated-virus vaccines would be to use replication-defective IVs. The main aim here would be to develop viruses that can undergo only one replication cycle, which would render them attenuated within the vaccinated individual but still robust enough to induce an immune response. Attempts at creating these viruses include the generation of IVs that lack the gene for the NEP protein or the gene for the M2 protein. These mutants retain the ability to express viral proteins, which induce antibody- and cell-mediated immune responses, while losing the ability to replicate.[15,17]

Work is also progressing on developing a DNA vaccination, which involves the injection of plasmid DNAs to induce an immune response.[18] A vaccine of this type would include at least one IV protein. The use of HA proteins in this system has shown promising results in small-mammal-model studies.

Many questions have been posed as to whether a universal vaccine for IVs can be created. The vaccines are currently based on the IV protein HA as an antigen. An attempt at creating a universal vaccine uses M2 as the antigen. Unlike the variant HA and NA proteins between strains of influenza A viruses, the extracellular domain of the M2 protein is highly conserved between strains. The findings of Neirynck et al. suggest that a vaccine based on the M2 protein could protect against infections caused by influenza A viruses.[14]

The possibility exists of using a fusion protein vaccine as a potential universal vaccine for IVs. The extracellular domain of M2, from the 23rd amino acid has been fused to the hepatitis B virus core (HB_C) protein to create a hybrid gene for $M2HB_C$. The resulting $M2HB_C$ protein structurally resembles the M2 protein; however, the HB_C portion of the protein stimulates the antibody response. Therefore, an $M2HB_C$-based vaccine amplifies the immunogenic response in host cells, protects against all influenza A virus strains, and provides passive immunization with sustaining effects.

WEAPONIZATION

VIABILITY

Influenza viruses continuously evolve with a high rate of antigenic shifts and antigenic drifts. Antigenic shifts produce small changes of the HA and NA proteins, which lead to yearly influenza epidemics. Antigenic drifts involve big changes in the viruses' proteins due to recombinations and replacements of gene segments, which lead to influenza pandemics. The strains for which humans have no immunity are likely to be the causative agents of influenza pandemics.

To some extent, the high rate of mutation of IVs is an inherent property of viruses with RNA chromosomes. RNA chromosomes do not have the exquisite corrective enzymes available to most DNA chromosomes. Amino acid substitutions of HA and NA occur at a rate of 0.5–2 % per year and 0.45–1.01% per year, respectively.[17] The amino acid substitutions eventually change the protein's conformation so that the original antibody can no longer bind to the original antigenic site, which renders the host's immune system unable to neutralize a modified IV strain. Influenza A and B viruses both undergo antigenic shifts, but these shifts in influenza B viruses are mostly linked to isolated outbreaks of influenza.

Influenza A viruses also undergo genetic reassortment in which genes for influenza proteins are swapped between strains. These antigenic shifts, are facilitated by the segmented structure of the IV genome into right chromosomes as well as the ability of influenza A virus to infect nonhuman hosts. Influenza A viruses are divided into subtypes based on antigenic differences of the viruses' HA and NA proteins. To date, there are 15 HA types and 9 NA, which have evolved due to antigenic drifts. Any particular influenza virion can contain only one type of HA and one type of NA protein. However, different IVs, carrying different HA and NA protein types from each other, can simultaneously invade and infect a cell. If this occurs, the viruses are replicated together, and the chromosome bearing HA and NA proteins can form new combinations.

Another consequence of antigenic drifts that has led to pandemic strains of IVs, and could also probably lead to extremely virulent IV strains, is that the genetics for HA and NA proteins can undergo genetic reassortment with IVs that infect other mammals and birds. As a rule, humans cannot be directly infected by avian IV strains because these viruses recognize a specific receptor that humans do not possess; however, other mammals, such as swine, possess receptors that both mammalian and avian IVs recognize. Therefore, swine, if infected simultaneously by mammalian and avian IVs, could serve as an intermediary host of IV so that humans could become infected by strains of IVs that have avian origins. Because humans typically do not have antibodies against these new strains, the outcome might be new influenza pandemics.

The Spanish Flu of 1918 is an example of an exceptionally virulent pandemic strain of IV (H1N1), which upon genetic analysis has its origin in an avian wild waterfowl reservoir.[19] However, because humans cannot typically be infected by an avian IV strain because of its receptor specificity, an intermediary host—the pig, which can be infected by both human and avian IVs—was required to spread the avian IVs. Another noteworthy outbreak of an avian IV occurred in Hong Kong in 1997. The culprit in this outbreak was avian H5N1 virus that was directly transmitted from live chickens to 18 humans. Of the 18 infected individuals, ranging from 1 to 60 years old, six died. The significance of this outbreak is that humans were directly infected by avian IVs, whereas all previous epidemics and pandemics caused by avian IVs required an intermediary host.[9]

GENETIC ENGINEERING AND REVERSE GENETICS

Key characteristics of good bioterrorist weapons are ease of production and ease of obtaining the pathogens. Influenza viruses meet these criteria as they are easily grown in embryonated hens' eggs. Furthermore, many strains of human influenza A viruses can be isolated directly from the allantoic or amniotic cavities of embryonated eggs. Subsequently, the viruses can be grown in these eggs either directly or after adaptation.

Although most attention has been given to the HA and NA proteins as the main virulence factors, the host range and pathogenesis of these viruses appear to be polygenic. This likelihood is an outgrowth of the research aimed at producing attenuated IVs for vaccines. Such work implicates the roles of NP, PB1, PB2, PA, M1, M2, and NS1, in addition to HA and NA, as determinants for either host specificity or virulence. Thus, ironically, one way to create a more virulent IV strain uses the same techniques being utilized to develop treatments and vaccines for these viruses.

For example, it is well known that HA is the main protein conferring pathogenicity for influenza A viruses. As mentioned above in "Molecular Biology," the HA protein must be cleaved into two subunits in order to be active. This requires a protease that recognizes specific cleavage sites. HA-possessing cleavage sites that contain multiple basic amino acids are associated with more virulent IV strains because the proteases that cleave these sites are found in all body cells. Because of their wider range, these IVs are able to replicate and spread to nonrespiratory cells and induce more serious infections. Currently, influenza A viruses containing the H5 or H7 proteins contain a polybasic cleavage site and are associated with highly pathogenic IVs. In terms of weaponizing IVs, mutations can be engineered at the cleavage sites of HA proteins so that they mirror the sites found in the H5 and H7 proteins.

MANIPULATING THE GENETICS OF INFLUENZA VIRUS

The genetic makeup of IVs can be affected by natural mutations, genetic reassortment, and complementation, which can be used to create a more virulent strain. Natural mutations are explained at the beginning of this section. The latter events can be manipulated by genetic engineering and reverse genetics techniques. Genetic reassortment involves the exchange of RNA segments between two viruses, resulting in a mutual exchange. Complementation occurs when a cell is infected with two defective viruses with defections at different loci. These viruses can then recombine to complement each other and restore function in defective genes.

In addition to mutating the genome of IVs, new techniques can be exploited to splice genes into the virus and/or induce reassortment to include genes that would increase the pathogenicity of the virus. For instance, the high rate of transmission, which is an innate characteristic of the IV, and the deleterious pathogenic effects of another pathogen, such as Ebola, can be combined to create a supergerm. A bioterrorist attack could also be a two-stage assault, in which the IV is the initial pathogen followed by a more virulent pathogen. The primary influenza infection could make individuals more susceptible to secondary infections.

REFERENCES

1. Barbey-Martin, C., B. Gigant, et al. 2002. An antibody that prevents the hemagglutinin low pH fusogenic transition. *Virology* 294: 70–74.

2. Brownlee, G. and J. L. Sharps. 2002. The RNA polymerase for influenza A virus is stabilized by interaction with its viral RNA promoter. *J Virol* 76(14): 7103–7113.

3. Centers for Disease Control and Prevention. 2001. *Pandemic Influenza*. Online: http://www.cdc.gov/od/nvpo/pandemics/fluprint.htm. Accessed October 18, 2002.

4. Centers for Disease Control and Prevention. 2002. Antiviral drug information. Online: http://www.cdc.gov/ncidod/diseases/flu/fluviral.htm. Accessed July 15, 2002.

5. Crawford, G. E. 2000. Influenza. *JAMA* 283: 449.

6. Das, P. 2002. New influenza strain and subtype identified. *Lancet Infect Dis* 2: 128.

7. Dowdle, W. R. 1997. The 1976 experience. *J Infect Dis* 176(suppl 1): S69–72.

8. Feery, B., M. G. Evered, and E. I. Morrison. 1979. Different protection rates in various groups of volunteers given subunit influenza virus vaccine in 1976. *J Infect Dis* 139: 237–241.

9. Hatta, M. and Y. Kawaoka. 2002. The continued pandemic threat posed by avian influenza viruses in Hong Kong. *Trends Microbiol* 10(7): 340–344.

10. Hay, A. 1992. The action of adamantanamines against influenza A viruses: Inhibition of the M2 ion channel protein. *Semin Virol* 3: 21–30.

11. Klenk, H.-D., R. Wagner, et al. 2002. Importance of hemagglutinin glycosylation for the biological functions of influenza virus. *Virus Res* 82: 73–75.

12. Lamb, R. A. and M. Takeda. 2001. Death by influenza virus protein. *Nature Med* 7(12): 1286–1288.

13. Monto, A. 1997. Prospects for pandemic influenza control with currently available antivirals. *J Infect Dis* 176(suppl 1): S32–37.

14. Neirynck, S., T. Deroo, et al. 1999. A universal influenza A vaccine based on the extracellular domain of the M2 protein. *Nature* 5(10): 1157–1163.

15. Palese, P. and A. Garcia-Sastre. 2002. Influenza vaccines: Present and future. *J Clin Invest* 110: 9–13.

16. Pleschka, S., T. Wolff, et al. 2001. Influenza virus propagation is impaired by inhibition of the Raf/MEK/ERK signalling cascade. *Nature Cell Biol* 3: 301–305.

17. *Straits Times*. 2002. Flu virus that tricks body's defences threatens pandemic. London.

18. Sugaya, N. 2000. Influenza vaccine, anti-influenza drugs, and rapid diagnosis in Japan. *J Infect Chemother* 6: 77–80.

19. Taubenberger, J. K., A. H. Reid, et al. 1997. Initial genetic characterization of the 1918 Spanish influenza virus. *Science* 275: 1793–1795.

20. Whittaker, G. 2001. Intracellular trafficking of influenza virus: Clinical implications for molecular medicine. *Expert Rev Mol Med*: Feb. 8:1–13.

21. Wutzler, P. and G. Vogel. 2000. Neuraminidase inhibitors in the treatment of influenza A and B: Overview and case reports. *Infection* 28(5): 261–266.

22. Yewdell, J. and A. Sastre-Garcia. 2002. Influenza virus still surprises. *Curr Opin Microbiol* 5: 414–418.

CHAPTER 7

■ ■

HANTAVIRUS

Payal Shah

The genus *Hantavirus*, of the viral family *Bunyaviridae,* currently encompasses more than 20 genotypic variants; it is thought that other, as yet unidentified strains also exist. Hantaviruses are transmitted by rodents in the family *Muridae*. Hantavirus strains can be categorized into those that cause hemorrhagic fever with renal syndrome (HFRS) and those that are the etiological agents of hantavirus pulmonary syndrome (HPS).

HFRS and HPS differ primarily in terms of target organ and disease severity. Occurrences of HFRS have been localized to Europe and Asia, where hundreds of thousands of cases each year yield a case fatality rate of up to 15%. HPS is localized to the Americas, where hundreds of cases are reported annually with a 50–60% case fatality rate. The dominant feature of both HFRS and HPS is vascular dysfunction, thought to be causally associated with the immune response rather than with the direct cytopathic effect of the virus.

Much remains to be learned regarding hantavirus pathogenesis in humans. Largely because of this lack of knowledge, no vaccines are currently available in the United States for HPS-producing viruses. A vaccine for one type of HFRS-causing hantavirus has been offered to laboratory workers at the U.S. Army Medical Research Institute of Infectious Diseases. In addition, an inactivated (noninfectious) derivative of an HFRS-causing virus has been used to make a vaccine that is widely used in Asia; however, no vaccines for other HFRS-producing agents are currently available in the United States.

As causative agents of hemorrhagic fever, viral strains that produce HFRS are classified as Category A potential bioweapons, genotypes that produce HPS are categorized as Category C emerging pathogens. Although nonpathogenic hantaviruses exist, they are not the prime focus of this discussion.

HISTORY

HFRS may have been recognized in China as early as 1000 years ago, and Russian clinical records from 1913 describe a disease with symptoms characteristic of HFRS. Other outbreaks attributed to HFRS occurred again in Russia (1932), as well as among Japanese troops in Manchuria (1934) and in Sweden (1934). A 1951 outbreak in Korea that affected approximately 3200 United Nations soldiers resulted in a 7% mortality rate;[5] as a result, Korean hemorrhagic fever (KHF) became a widely recognized health threat. KHF, the clinically similar epidemic hemorrhagic fever (identified in Japan), and nephropathia epidemica (recognized in Scandinavia and Russia) were first collectively designated as HFRS in 1983 by the World Health Organization. Subsequent studies have implicated HFRS as a causative agent for illnesses during both the first and the second World Wars.[6]

HPS was not identified until a 1993 outbreak in the United States' Four Corners region (comprising Colorado, New Mexico, Arizona, and Utah). Isolated cases of HPS subsequently occurred in Louisiana, Florida, and New York. Cases have since been seen in Canada as well as in Central and South America. A 1996 Argentina outbreak captured public attention, as it highlighted potential evidence for person-to-person HPS transmission.

JAPANESE STUDIES OF HFRS

By 1940, Japanese physicians had compiled a clinical and pathological description of the 1934 outbreak[5] of what was called epidemic hemorrhagic fever. They reported an estimated incubation period of 2–3 weeks and a mortality rate of over 10%. They identified the following elements of the syndrome: hemoconcentration (decreased blood fluid composition, increased blood concentration), an abnormal white blood cell response, hemorrhage, hypotension at the time of febrile resolution, and an ensuing phase of renal insufficiency. During this time, in attempts to isolate the causative agent of the observed illness, Japanese investigators conducted studies using human subjects. They claimed to induce disease by injecting, into humans, filtrates of tissues from *Apodemus agrarius*, a species of rodent that in 1978 was indeed discovered as the host of the HFRS-causing Hantaan hantavirus (HTNV). They also claimed to produce similar results by injecting patients with tissue filtrates of mites who fed on *Apodemus* mice; however, mite transmission has never been confirmed.

RUSSIAN STUDIES

Beginning in 1932, annual outbreaks of HFRS were seen in the Amur River Valley of Siberia.[5] In a 1939 epidemiological study, workers in two camps 4 km apart acquired infection; interestingly, 31 workers in one camp, all of whom lived near a refuse dump, were afflicted with the disease, whereas only one individual from the other camp acquired infection. A 1959 report documents rodent contact to have been very common in and near the camp in which the 31 infected individuals resided.

A 1961 outbreak in Moscow was reported to have affected 113 of 186 workers and visitors to a scientific laboratory; the lab had just received a shipment of rodents, including *Clethrionomys glareolus*, from Kirov county, Russia. No fatalities were recorded. *C. glareolus* is now known to host Puumala virus (PUUV) virus, the least lethal (<1% case fatality rate) of the HFRS-causing hantaviruses.

In 1967, two Russian scientists, Yankovski and Povalishina, provided more insight into the disease.[5] They noted that the incubation period could be up to 6 weeks, that disease occurred mainly in forested areas and along rivers, and that the cycles of virus activity paralleled observed population cycles of field rodents, particularly of *C. glareolus*, which had been implicated in the 1961 outbreak. Finally, they suggested that transmission of the disease occurred via inhalation of dust contaminated by rodent excretions.

A 1951 OUTBREAK IN KOREA

Between 1951 and 1954, over 3000 cases of an acute febrile illness were seen among UN soldiers engaged in the Korean conflict.[6] Approximately one-third of these cases demonstrated hemorrhagic symptoms; the mortality rate from HFRS (then called Korean hemorrhagic fever, or KHF) was approximately 7%. The outbreak drew international attention, and it became evident that KHF was clinically and epidemiologically similar to previously identified viral hemorrhagic diseases identified in Eastern Europe. The rodent reservoir was not identified until almost 25 years later, when it was found to be *Apodemus agrarius*, the host of HTNV.

THE 1993 FOUR CORNERS OUTBREAK

In May 1993, a previously unknown disease afflicted two formerly healthy young adults in Gallup, New Mexico; they displayed pulmonary and respiratory dysfunction and passed away soon after. Autopsy and investigation results perplexed physicians, who called on public health officials in Arizona, Utah, and Colorado. Similar cases were reported in these states, while more such cases appeared in New Mexico. Within 1 month of investigation, antibodies to hantavirus were detected in patient sera; within 2 more weeks, the murid rodent *Peromyscus maniculatus* was identified as the carrier of the causative agent. The manifestation of the disease was different from that of the previously identified hantavirus disease HFRS; here, renal and hemorrhagic presentations were not prominent. Genetic and serological testing resulted in the identification of a new hantavirus: Sin Nombre (literally, "no name") virus (SNV). The new form of hantavirus disease was subsequently termed HPS. In 1993, 48 patients were diagnosed with HPS; 29 individuals died, yielding a mortality rate of approximately 60%. In this year, 14 HPS patients were treated at University of New Mexico Hospital, with a mortality rate of 36%.[1]

It is now widely believed that the 1993 outbreak did not mark the emergence of HPS. Retrospective studies, instigated by that outbreak, suggested that HPS had caused the deaths of several individuals who had passed away from previously unexplained pulmonary and respiratory complications. The oldest such case was that of a 38-year old man from Utah who passed away in 1959.

Scientists also hypothesized that the 1991–1992 El Nino–southern oscillation (ENSO) resulted in elevated precipitation levels that allowed an increase in rodent population densities, and consequently, an increased likelihood of transmission to humans.[4] They suggested that this was an essential factor contributing to the 1993 outbreak in the southwestern United States. In accordance with this hypothesis, in 1997–1998, a second strong ENSO occurred; subsequently, the number of HPS cases due to SNV increased dramatically in 1998–1999 with respect to average annual caseload. Furthermore, a large majority of the 1998–1999 case patients reported indoor exposure to deer mice.

A 1996 OUTBREAK IN ARGENTINA

In the spring of 1996, a significant increase in the number of HPS cases around El Bolson, Argentina became evident.[7] During the same period, two physicians from Buenos Aires, one with no history of having been near El Bolson, were also diagnosed with HPS. Comparative sequence and phylogenetic analyses were performed on viral genomes within infected individuals; polymerase chain reaction (PCR) products of partial sequences of viral genomes were studied to assess homology. Results indicated that all hantavirus variants in the 1996 Argentina outbreak were genetically similar.

Because of the geographic distance between the first 20 cases detected in El Bolson, and the two subsequent cases (130 and 1400 km from El Bolson), it was deemed unlikely that the disease had been transmitted to all these individuals by a local rodent population hosting the same variant of the virus. Interestingly, the infected individuals were linked: they were all friends, coworkers, physicians, patients, or acquaintances of one another. One individual, referred to as Case M, had never visited El Bolson but had traveled in an automobile with Case L for 20 hours and had stayed with her infected parents, Cases J and H. Person-to-person transmission thus seemed possible.

The factors that suggest person-to-person transmission include both similar viral genetic sequence in the patients and hospital contacts and the lack of other HPS cases in cities where infectees resided. However, it should be noted that virus genotypes in the rodent populations near El Bolson were not analyzed, so the possibility that the dominant virus genotype existed in all mice that transmitted the hantavirus cannot be eliminated from discussion. Furthermore, one review of hantavirus epidemiology concluded that there was no evidence of person-to-person spread there.[6]

DIFFERENCES BETWEEN HFRS-
AND HPS-CAUSING STRAINS

The main HFRS-causing hantaviruses include HTNV, Dobrava-Belgrade (DOBV), Seoul (SEOV), and PUUV. Corresponding mortality rates are 5–15% (both HTNV and DOBV), 1–5%, and <1%, respectively. Distribution of HFRS-causing hantaviruses is localized predominantly in Asia and Europe. The prototypic HPS-causing hantavirus is SNV, with a mortality rate of approximately 50–60%. Distribution of HPS-causing hantaviruses is localized largely in the Americas.

Approximately 100,000–200,000 cases of hantavirus disease (HVD) are reported throughout the world each year. In the United States, HPS cases have been reported in 31 states. The average U.S. HPS patient is 38 years old. A substantial majority of HPS cases occur in males (59%), Caucasians (77%), and individuals living in rural areas (75%). HPS in children is rare in the United States but accounts for a larger percentage of cases in South America. Because the major risk factor is contact with infected rodents, which often occurs through field labor in rural areas, the above statistics likely reflect that HPS cases occur in those who the greatest occupational risk.

TRANSMISSION

Rodents in the family *Muridae* are the natural hosts of hantaviruses. Strains of hantavirus are species-specific, with a single species of murid rodent hosting a single strain of hantavirus. Such a strong parallel between host and virus phylogeny is indicative of a coevolutionary relationship, perhaps over thousands of years. It appears that the presence of hantavirus in an individual rodent does not result in any disadvantages in terms of its survival or reproduction. Once persistently infected, the rodent will secrete *virions* for an undetermined period of time. These virions may be shed in the rodent's

Viral Strain   Rodent Host   Subfamily

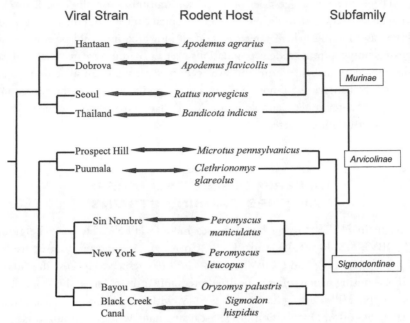

FIGURE 7.1. Phylogenetic tree depicting parallel phylogenies of rodent reservoirs and hantavirus strains. The diagram is based on pairwise comparison of S-segment nucleotides encoding the nucleocapsid protein. The high degree of concordance between phylogenies implies a coevolutionary trend that has probably existed for thousands of years.

Source: Adapted from Centers for Disease Control and Prevention. Online: http://www.cdc.gov.

urine, feces, and saliva; a human usually becomes infected via inhalation of aerosolized virions.

Infection via rodent bites is rare but has been cited. In addition, contact with infected rodent excreta through skin lesions or the eyes may occur. Infections contracted through all modes of transmission are regarded as equally pernicious. Vertical transmission of hantaviruses in pregnant, HPS-afflicted women has not been demonstrated, although only two such live births were examined.[2] Person-to-person transmission has been examined, as detailed above, but remains unverified. Reinfection with homologous hantavirus types is not believed to occur in humans. The infective dose is unknown, but it is assumed to be low, as persistently infected rodents do not excrete large amounts of virus.[3]

MOLECULAR BIOLOGY

Like all bunyaviruses, hantaviruses are composed of enveloped virions that have a negative-sense RNA genome. The genome exists in the form of three circular RNA-containing ribonucleocapsids: the L (large) segment codes for an L protein; the M (medium) segment codes for two envelope glycoproteins; and the S (small) segment codes for the nucleocapsid protein. Following RNA and protein synthesis, virion components assemble and bud to form new virions.

Pathogenesis is thought to occur via an immune-mediated mechanism rather than through direct cytopathic effects of the virus. Beta-3 integrins, receptors normally present on endothelial cells, macrophages, and platelets, are thought to mediate viral entry into host cells. Correspondingly, the prime cell types targeted by hantaviruses are endothelial cells, monocytes, and platelets.

VIRION ORGANIZATION

When examined by electron microscopy, the virions are roughly spherical with a diameter of approximately 100 nm, although oval particles with lengths of up to 210 nm may also be seen. Virions are enveloped by a bilayered phospholipid membrane 5 nm thick; the interior of this envelope appears to have a highly organized, gridlike pattern of subunits, differentiating hantaviruses from other viruses in the *Bunyaviridae* family. Glycoproteins 1 and 2 (G1 and G2) project (approximately 6 nm) from the phospholipid bilayer. The lipid bilayer surrounds a granulofilamentous interior, which contains ribonucleoprotein structures that include the viral RNA genome. Virion composition is >50% protein, 20–30% lipid, and 2–7% carbohydrate;[6] the remainder of a virion particle is composed of RNA.

Hantaviruses, like other enveloped viruses, are sensitive to and inactivated by detergents, organic solvents, heat, and low temperatures. Nevertheless, HTNV infectivity is maintained at 0–4°C for up to 12 hours. After exposure to such low-temperature conditions for 12 hours, hantaviruses can be inactivated by high salt concentrations or nonphysiological pH extremes (<7 or >8).

THE HANTAVIRUS GENOME

The hantavirus genome is composed of three segments of negative-sense RNA; each segment is contained individually within a noncovalently closed (ribonucleocapsid intraprotein interactions are noncovalent) circular ribonucleocapsid. Each such structure is associated with a single nucleocapsid (N) protein of approximately 48 kd. The combined length of the segments is approximately 13 kb Each segment

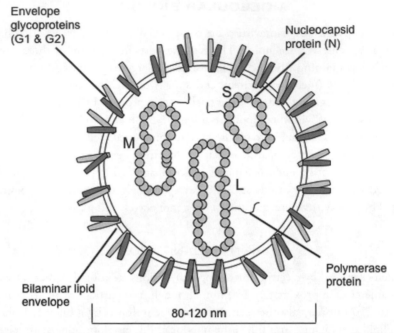

FIGURE 7.2. The virion envelope is composed of a phospholipid bilayer of host origin embedded with G1 and G2 glycoproteins. The virion contains three circular ribonucleocapsids—L, M, and S—which contain RNA encoding L polymerase, G1 and G2, and nucleocapsid protein, respectively.   *Source*: Adapted from Ref. 8.

has a conserved 3′-terminal nucleotide sequence (AUCAUCAUC), complementary to a 5′-terminal nucleotide sequence; such sequences are capable of base-pairing to form panhandle structures. Other viruses that contain such sequences utilize pan-handle structures in initiation and termination of viral transcription; specifically, in initiation, the loop of the panhandle contains a polymerase-binding site. It is likely that the structures serve a similar function in hantaviruses.

CODING STRATEGY

The three genomic segments are designated L (large, approximately 6.5 kb), M (medium, approximately 3.6–3.7 kb), and S (small, approximately 1.7–2.1 kb). These three segments code for a total of four translation products.

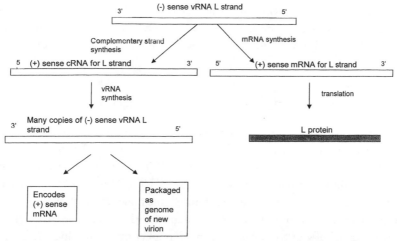

FIGURE 7.3. Generalized replication and transcription pathway. The negative-sense vRNA is replicated by an RNA-dependent RNA polymerase (RdRp), via a positive-sense cRNA intermediate. The negative-sense vRNAmay then either encode positive-sense mRNA, or it be packaged as the genome of a new virion. Messenger RNA will be translated to form the protein products that assist in the production of, or become components of, new virions. The pathway is shown for the L vRNA strand; there is an identical sequence of events for M and S vRNA strands as well.

Segment L encodes a 247-kd protein with polymerase, replicase, and endonuclease activity. As a polymerase, the L protein is responsible for copying the three negative-sense genomic vRNA strands into their respective mRNA strands. As a replicase, the polymerase will generate entire, complementary copies of the genome (cRNA) that are then used as templates for vRNA synthesis. As an endonuclease, during replication, the L protein cleaves host mRNAs 10–18 nucleotides beyond the 5′ guanosine cap; the capped oligonucleotides are then utilized as primers during the transcription of mRNA molecules coding for viral proteins. The L protein may also function as a helicase to unwind the encapsidated vRNAs during synthesis.[8]

Viral RNA segment M encodes a polypeptide that is cotranslationally cleaved into envelope glycoproteins G1 (64–67 kd) and G2 (54 kd). The protease involved is thought to be host cellular signal peptidase, located on the inner surface of the endoplasmic reticulum. G1 and G2 are integral membrane proteins; they have asparagine-linked sugars, and induce neutralizing antibody responses in animals. G1 and/or G2 are thought to play roles in viral entry and exiting.

Segment S encodes the nucleocapsid protein, N (48 kd). The nucleocapsid protein is associated with vRNA and cRNA strands, but not with mRNA strands. N is also thought to play a role in assembly and budding of new virions. Unlike other bunyaviruses, hantaviruses are not known to produce an NSs (nonstructural, S-segment-derived) protein.

SYNTHESIS OF vRNA, cRNA, AND mRNA

The 5′ terminal residue of HTNV vRNA is thought to be a monophosphorylated U residue rather than a G. The "prime and realign" model[8] proposes that a pppG (guanosine triphosphate) aligns at the third nucleotide position (C) of the 3′ repeating terminal sequence (AUCAUCAUC). This initiates RNA synthesis; after several nucleotides are synthesized, the polymerase (L protein) slips, and the nascent RNA moves so that it is aligned with the very first base of the repeating terminal sequence, while the initial pppG priming residue is left overhanging. The L protein then cleaves the overhanging G residue. The 5′ end of the nascent RNA becomes a monophosphorylated U. This model is proposed for the synthesis of vRNA and cRNA.

The synthesis of mRNA may also follow a similar model; however, instead of the pppG priming residue utilized during vRNA/cRNA synthesis, the primer used during mRNA synthesis is a 5′-guanosine-capped residue that is cleaved by the L endonuclease from the host mRNA. The G residue may align at the C of the second or third triplet and, after synthesis of a few oligonucleotides, the nascent RNA may slip backward two nucleotides. The G at the 3′ end *of the G residue* would then be aligned with a C of either the first or second triplet. This type of mechanism is supported by the common observation that one or two of the triplet repeats are deleted in the mRNAs.

Although termination sequences for RNA synthesis have not been elucidated for hantaviruses, sequence analysis shows that mRNAs appear to terminate shortly after the open reading frames (ORFs) present in each gene segment; consequently, mRNAs do not contain noncoding information, but do contain the host-derived guanosine-capped 5′ residue. It has been suggested that accumulation of N might affect the ability of the L polymerase to selectively recognize termination signals,[8] allowing for termination of mRNA synthesis shortly after the ORFs, and ignore these signals, allowing for the later termination of vRNA/cRNA synthesis.

TRANSLATION AND POSTTRANSLATIONAL

Following synthesis, L and S mRNA segments, which code for the hantaviral L and N proteins, are translated on free ribosomes; the M mRNA segment, which codes for G1 and G2 proteins, is translated on membrane-bound ribosomes. No precursor polypeptide of a length that corresponds to the M segment ORF is detectable in the cell at any time; this indicates cotranslational cleavage of the G1/G2 precursor protein, most likely by host-cell signal peptidase. G1 and G2 undergo glycosylation with asparagine-linked carbohydrates.

ASSEMBLY AND BUDDING

Hantaviruses usually bud into the Golgi cisternae; two hantaviruses, including SNV, also bud from cell membranes. In order to be encapsidated, vRNA and cRNA molecules must associate with the N protein. It is presumed that vRNA and cRNA molecules interact with the N protein immediately following synthesis. Furthermore, it is believed that encapsidation requires a specific interaction between N and the viral nucleic acid, which may involve a signal at the 5′ end of the RNA, as is the case for other viruses in the *Bunyaviridae* family. Messenger RNA is distinct from the other RNA molecules primarily in its 5′-capped structure and the deletion of 3′ terminal complementary sequences. It is likely that these features prevent its association with the N protein, thereby inhibiting mRNA encapsidation and subsequent inclusion in new virions.

For all hantaviruses, G1 and G2 are synthesized in the endoplasmic reticulum (ER) and are then transported to the Golgi. HTNV virion formation requires the dimerization of G1 and G2 in the ER with a specific conformation. If either glycoprotein were individually expressed, it would remain mainly in the ER. When SNV glycoproteins are individually expressed, conversely, they are able to migrate from the ER to the Golgi; however, migration proceeds more slowly, and G1 must form homotrimers to migrate at all. Because the hantavirus genome does not encode any matrix proteins, it is believed that the N protein might interact directly with G1 and/or G2 to induce budding.[8]

IMMUNE SYSTEM RESPONSE

Current evidence suggests that disease produced by hantavirus is not directly mediated by the virus. There is no evidence of a cytopathic effect caused by hantavirus; that is, infection generally results in minimal disturbance of normal cellular function.

The damage caused by hantavirus infection appears to result indirectly from the immune response. Thus, elevated levels of CD8+ T and B lymphocytes are seen in the blood and in kidney biopsies from certain HFRS infections. Increased levels of immune mediators, including TNF-α, IL-6, and IFN-γ, are observed in both the kidney and the blood of individuals with hantavirus infections; it is suspected that these mediators cause the endothelial cell permeability that leads to pulmonary edema. Furthermore, an increased number of cytokine-producing cells are observed in the lungs of individuals with fatal HPS but not in the lungs of patients with non-HPS acute respiratory syndrome. Finally, postmortem tissue from patients who had HPS, as well as those who had HFRS, reveal an increase in blood-cell concentration, suggesting that this might be the cause of edema, and subsequently of organ failure.

β-3 INTEGRINS

Hantaviruses enter cells via interaction with β-3 integrins. β-3 integrins are present on endothelial cell, macrophage, and platelet surfaces; accordingly, endothelial cells, monocytes, and platelets are generally regarded as the principal cell types targeted by the virus. β-3 integrins are normally involved in the regulation of vascular permeability and platelet function. The hantavirus–β-3 integrin interaction is thought to be responsible for the vascular dysfunctionality that is the predominant pathophysiological feature of both types of hantavirus disease. Hypotension due to hantavirally induced vascular dysfunction is frequently the cause of death in infected individuals.

CLINICAL PRESENTATION AND FINDINGS

HFRS progresses through five phases: a febrile phase that is well-defined in terms of clinical findings; a hypotensive phase, associated with hemorrhage, edema, and clinical shock; an oliguric phase during which many HFRS fatalities occur; a diuretic phase; and a convalescent phase, with a gradual (2–3-month) recovery period. This disease course is generalized, and variations exist, depending on the strain of hantavirus responsible for the disease.

HPS is generally associated with three phases: an early, prodromal phase, during which symptoms are nonspecific and diagnosis of hantavirus is difficult; a cardiopulmonary phase, during which respiratory and circulatory abnormalities are present; and a convalescent phase. Clinical variations of this triphase syndrome exist; in particular, some hantaviruses, including Black Creek Canal, Bayou, and Andes viruses, are accompanied by renal insufficiency, while renal symptoms are virtually absent in SNV.

HEMORRHAGE AND RENAL DYSFUNCTION IN HFRS

Hemorrhagic fever renal syndrome is a term that broadly describes the clinical manifestations of the disease. Hemorrhage is the escape of blood from vessels; it is thus associated with the most prominent feature of both types of hantaviral disease: vascular dysfunction. Furthermore, the designation "hemorrhagic" reflects many of the features of the disease, including low blood volume leading to low blood pressure, visible internal bleeding, and leakage of not only fluid but other blood components through the damaged vessels into extravascular regions of the body. The term "renal" refers to the kidney, whose main function involves filtering wastes from the blood and excreting them into the urine. It is readily conceivable, then, that renal malfunction might result in the presence of metabolic wastes in the blood.

THE FIVE PHASES OF HFRS

The incubation period for HFRS is approximately 1–2 weeks, with extremes ranging from 4 to over 40 days. This period is followed by a febrile phase, which usually lasts for approximately 3–6 days. The febrile phase is marked by abrupt onset of fever, chills, malaise, and myalgia, and often headache and backache. Orbital (eye) pain is a characteristic feature, although it may be absent; conjunctival hemorrhage may also be observed. An affected individual's face, neck, and anterior chest may appear flushed. Petechiae (small purple/red flat spots due to hemorrhages in or under the skin) appear on the soft palate, face, neck and axillary folds. The clinical findings during this phase are relatively well-defined; albuminuria (abnormally high amounts of the plasma protein albumin in the urine) and microhematuria (microscopic amounts of blood in the urine, as opposed to gross hematuria, in which blood urine is visible to the eye) are observed; tubular cells may also be present in the urine. Intense proteinuria generally develops during either the febrile phase or the hypotensive stage of disease that follows.

Within the first week, the hypotensive phase of the disease develops abruptly. This phase may last for hours or days; severe hemorrhage, if it occurs, begins at this point. This hemorrhage, coupled with edema, may lead to hypovolemia (decreased blood volume), clinical shock, and the hallmark feature of this phase, hypotension. Clinical shock, which results from inadequate blood supply to vital organs, is marked by such signs as pale or bluish complexion; weak, rapid pulse; shallow, rapid breathing; thirst; nausea; vomiting; blurring of vision; and, in severe cases, loss of consciousness. Approximately 33% of all HFRS-related deaths are linked to multiorgan hypoperfusion (decreased blood flow, circulatory shock) at this stage. Changes in blood composition are marked; an increase in hematocrit (red blood cell concentration in the blood) accompanies thrombocytopenia (a low blood platelet count). Leukocytosis is also observed. Levels of urea and creatinine, which are normally excreted via the kidney, increase in the blood, indicating renal dysfunction. Electrolyte imbalance is uncommon.

Elevation of blood pressure to normal levels is indicative of the onset of the oliguric phase, which lasts for 3–7 days. Hypervolemia (increased blood volume) may occur, leading to hypertension. Urea and creatinine levels continue to increase. Blood electrolyte imbalance is observed during this phase: hyperkalemia (high potassium), hyponatremia (low sodium), and hypocalcemia (low calcium) are common. Urine output decreases. Occasionally, metabolic acidosis will occur. Severe cases are accompanied by cardiac failure, pulmonary edema, and cerebral bleeding. Fifty percent of all HFRS fatalities will be due to such complications at this stage.

Recovering patients enter the diuretic phase; they experience drastic increases in urine output, with daily excretion volumes often ranging from 3 to 6 liters. This may be associated with dehydration and/or electrolyte imbalance, which, along with infection, may prolong lethargy and anorexia to 3–4 weeks. Finally, the patient enters the convalescent phase; complete recovery may take as long as 2–3 months from that point. Kidney function is gradually regained, as marked by improved glomerular filtration rate, renal blood flow, and urine-concentrating ability.

This disease course applies best to HTNV- and DOBV-induced HFRS; slight variations exist for infections caused by other major HFRS-causing viruses, namely, PUUV and SEOV. PUUV is the predominant form of hantavirus in central and northern Europe, Russia, and the Balkans; it causes the mildest known form of HFRS, formerly referred to as nephropathia epidemica. PUUV-induced disease proceeds through the same five stages and temporal course as the HTNV- and DOBV-induced HFRS; however, the stages are not as clearly defined. Mortality rates of PUUV are significantly lower than of HTNV or DOBV (<1%). Severe hemorrhaging does not occur, nor does clinical shock. Many cases are managed without hospitalization. SEOV, recognized mainly in Southeastern Asia, has a mortality rate between those of HTNV and PUUV (1–2%). Its clinical presentation is similar to that of HTNV and DOBV; however, a significant proportion of patients also develop hepatitis.

THE THREE PHASES OF HPS

Following a 1–5-week incubation period (median, 14–17 days), a 3–5-day, nonspecific prodromal phase occurs. During this stage, the patient has abrupt fever onset, chills, malaise, and myalgias (diffuse muscle pain). Anorexia, headache, nausea, vomiting, and abdominal pain may occur during this period. Respiratory symptoms are often absent, and sore throat and coryza are not common. Patients might seek medical attention at this point; however, physical examination, laboratory data, and chest radiographs may appear normal. Physicians might prescribe analgesics during this stage. If a blood analysis is performed, thrombocytopenia might be found.

The patient rapidly proceeds into the cardiopulmonary phase of the disease. Cough, dyspnea, and gastrointestinal symptoms are common. Within hours of presentation, pulmonary edema and respiratory failure sometimes develop. Affected individuals often have tachypnea (a rapid, shallow respiratory rate), tachycardia (unusually high heart rate), and postural hypotension, which are readily detectable on physical examination. At this point, a number of circulatory abnormalities will be manifest. The platelet count is a laboratory test that is of use during this stage; thrombocytopenia will most likely be pronounced, and will generally result in further examination of the patient. The presence of thrombocytopenia may help distinguish HPS from differential diagnoses, such as rickettsial infections, plague, tularemia, and relapsing fever, which have similar prodromal characteristics. Leukocytosis and hemoconcentration may be detected through laboratory testing.

In addition to circulatory system abnormalities, pulmonary evaluation will yield abnormal results as well. Low $P(O_2)$ levels may be accompanied by low CO_2 levels. In severe cases, patients will have metabolic acidosis with elevated lactate levels and severe oxygen debt. Forty-eight hours after admission to the hospital, nearly all patients will show abnormal chest radiographs, indicating interstitial edema; furthermore, two-thirds of these individuals will have developed extensive airspace (portion of the lungs distal to the bronchi) disease. Pleural effusions will occur in patients to varying degrees. The convalescent phase is accompanied by diuresis, increased oxygenation, and decreased hemoconcentration. Complete recovery can be rapid.

DIAGNOSIS

While clinical findings may strongly implicate hantavirus, laboratory results are needed to confirm the presence of hantaviral infection. Serology is the prime diagnostic indicator. Patients with both HPS and HFRS have circulating IgM antibodies that can often be detected as early as within 24 hours of infection; virtually all patients will have circulating antibodies within the first week of infection. IgM will be present in the bloodstream for at least 3–6 months after infection, and IgG can be detected from years after infection to the duration of the infectee's life; factors determining this duration are unknown. Methods that successfully identify IgG and IgM antibodies directed toward hantaviral proteins are thus of diagnostic use. Indirect immunofluorescence with native hantavirus is a sensitive, specific assay that can detect these antibodies. Enzyme immunoassays, including ELISA, have also been developed with both native and recombinant antigens, particularly nucleocapsid proteins of different hantaviruses.

The IgM ELISA using recombinant SNV antigen is of particular use; broad cross-reactivity allows for detection of most HPS hantavirus. The general population of the United States does not have a seroprevalence of hantavirus antibodies; therefore, tests utilizing highly cross-reactive antibodies are used.[5] However, in Paraguay, for example, seroprevalence of hantavirus antibodies is as high as 40% in individuals with no history of HPS; the reason for this high seroprevalence is unknown, but it is likely to involve asymptomatic infections by nonpathogenic strains. In such cases, results of laboratory testing may not be useful.

Reverse transcriptase (RT) PCR is a technique in which cDNA is first synthesized from RNA by reverse transcription and then amplified using PCR. RT-PCR using type-specific primers has been particularly useful in the detection of hantaviral RNA and in the overall examination of hantavirus. Since many genotypic nucleotide base sequences are known, RT-PCR allows for confirmation of infecting-virus genotype, and makes possible the exponential amplification of a segment of nucleic acid that may be sequenced for further study. This method played a major role in the discovery of SNV as the causative agent of the 1993 Four Corners outbreak.

A strip immunoblot assay has been developed and shown to be effective in identifying all cases of SNV infection from early clinical samples with no false-positive results in controls.[2] Antigens are combined forms of recombinant-expressed SEOV N protein, and synthetic and recombinant forms of SNV glycoprotein and N protein.

MANAGEMENT

Because bacterial illnesses are much more common than hantavirus disease, patients thought to have HPS should be placed on general antibiotics such as ceftriaxone and an aminoglycoside until HPS diagnosis has been confirmed by serological testing. Important supportive measures include timely adjustment of electrolyte, pulmonary, and hemodynamic (involving blood components) abnormalities to correct levels. While fluid administration may be considered to relieve hypotension, such treatment should be administered cautiously to avoid aggravation of pulmonary edema. Positive inotropic agents, which act to increase cardiac muscle contractility, may increase delivery of blood to vital organs and are therefore strongly encouraged. *Vasopressor* drugs, *peritoneal dialysis*, and *hemodialysis* are also commonly administered.

WEAPONIZATION

With the discovery of HPS in the Americas, hantavirus has gained recognition as a worldwide threat. The 50–60% mortality rate of HPS, the low natural immunity to hantavirus in the general population, and the lack of vaccine availability in the public domain have all contributed to the increasing concern about the potential use of hantavirus as a biological weapon. Previous outbreaks, particularly the 1993 Four Corners outbreak, have confirmed that SNV is highly lethal in aerosolized form. Aerosolization begins when infected mice shed urine, feces, or saliva; hantavirus virions are stable in most indoor environments for up to several days. When the virions are moved (for example, by sweeping to clean rodent droppings or by moving boxes or furniture), they are swept into the air within tiny particles of the rodent excreta. These particles, when inhaled, are infectious. Because aerosolization occurs so readily, it may be a mode of dissemination of hantavirus as a bioweapon. Recent suggestions of the person-to-person transmissibility of the Andes virus, discussed earlier, indicate that it too may be manipulated as a weapon of mass destruction.

ENHANCEMENT OF TRANSMISSIBILITY

If Andes virus is, in fact, transmissible via person-to-person interactions, the genetic feature that allows such transmissibility can be studied and manipulated. Particularly because RT-PCR allows sequencing of viral RNA, it is conceivable that one possible mode of manipulation might involve comparison of Andes virus with other hantaviruses to determine the genetic feature of Andes virus that enables such transmissibility; enhancement of this factor might result in increased interpersonal communicability. Andes virus is a renal variant of HPS; it causes not only the highly lethal HPS but also renal dysfunction, and therefore has the potential to be devastating if disseminated on a large scale.

EFFECT OF INCREASE IN RODENT POPULATION

While hantavirus virions can be destroyed by simple treatment with detergent, periods of natural disasters or war generally result in the disruption of basic hygiene; it is possible that during such periods an infected rodent population might increase and/or spread, which will likely result in increased virus dissemination. Furthermore, as hantaviruses seem not to pose a disadvantage to their rodent host, the spread of an infected population of rodents is just as likely as that of a noninfected population. Virions are stable in indoor environments for several days, so they may be stable in outdoor environments at moderate temperatures as well; if disseminated, they might persist in the environment for several days. Since HPS has an incubation period ranging from 1 to 5 weeks, infected individuals may not demonstrate symptoms for a period during which others might enter the area and become infected with infectious virions.

Much remains to be learned regarding the virus' ability to survive outside its rodent host or the human body; such information will undoubtedly allow for more accurate assessment of possible modes of transmission. One prominent indication of the use of hantavirus as a bioweapon would clearly be the presence of HFRS in the Americas or the presence of HPS in a region where this disease has historically been absent.

POSSIBLE ROLE OF EBOLA GLYCOPROTEIN

The center portion of the Ebola virus glycoprotein has immunosuppressive activity; in particular, it has been demonstrated that this portion suppresses cytotoxic T-lymphocyte responses, monocyte chemotaxis, and cytokine gene expression. If introduced into and expressed by a hantavirus, the genetic sequence that codes for this glycoprotein segment might result in an increased mortality rate, especially when combined with the effects of the highly lethal HPS.

POTENTIAL DEFENSES

No agents have yet been discovered that provide effective postexposure treatment for hantavirus infections, although ribavirin has not been excluded as a potential treatment. HFRS and HPS vaccines are currently undergoing testing on both humans and animals in Europe, Asia, and the Americas. In the absence of prophylactic vaccines or antiviral treatments, potential defenses have been suggested. These defenses include, first and foremost, avoidance of exposure to infected rodents. Extracorporeal membrane oxygenation (ECMO) has been suggested as a rescue treatment for adult HPS patients. The role of nitric oxide during hantaviral disease progression has come into question as well. The recent identification of the β-3 integrin receptor has also introduced a new aspect of study.

PROPHYLAXIS

Because the mode of hantavirus transmission involves aerosolized or inhaled rodent saliva, feces, or urine, rodent excrement should be avoided. Individuals at an occupational risk include those who work outdoors, particularly in fields or forests; attention should be paid to nearby animal populations. In addition, monitoring environmental factors that might increase rodent populations, such as increased precipitation, may be useful.

Numerous hantavirus infections, including some deaths, have been attributed to exposure to infected laboratory rodents. Although thorough decontamination measures may be costly and inconvenient, such measures are essential. All infected animals should be exterminated, facilities must be entirely disinfected, and new animals must be from a proven, uncontaminated source. Laboratory work entailing hantavirus propagation in either cell culture or animals requires Biosafety Level III (BSL III) conditions.

Respiratory filters such as the N-100 respirator are thought to be effective in preventing inhalation of hantavirus virions <5 μm. While such respirators have not undergone extensive testing as to their effectiveness, they may be useful in suspected contaminated areas.

RIBAVIRIN

Ribavirin is a widely-used RNA virus mutagen. In-vitro and in-vivo studies indicate that hantavirus replication is inhibited by ribavirin. This drug is often administered in the treatment of hantavirus disease in China, and studies conducted there indicate that it may be effective in treating HFRS if administered 5 days after onset of the disease.[6]

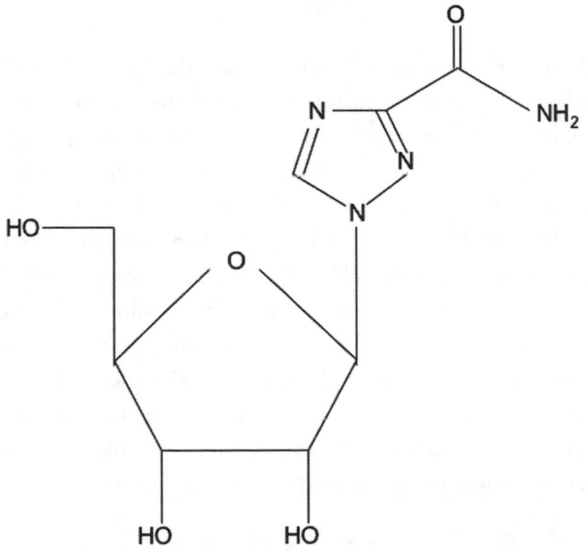

FIGURE 7.4 Chemical structure of ribavarin, a widely used RNA virus mutagen suggested for the treatment of HFRS and HPS. It has been shown to inhibit hantavirus replication, both in vitro and in vivo. Its effectiveness in the teatment of HFRS, if administered within 5 days of onset of disease, has been demonstrated. Its suitability for treating HPS remains undertermined; studies have been conflicting and/or inconclusive.

The potential of ribavirin as an antiviral agent in HPS treatment is as yet undetermined. Its use in China against HFRS instigated a trial of the agent for individuals thought to have HPS. This trial, conducted in 1993, enrolled 140 human subjects; hantavirus infection was confirmed in 30.[2] Mortality rate with ribavirin was 47%; comparison with 34 untreated HPS patients, who were not enrolled in the study, indicated no difference in survival between those who had received ribavirin and those who had not. The efficacy of ribavirin in treating HPS is still being examined; previous studies have been inconclusive.

EXTRACORPOREAL MEMBRANE OXYGENATION

A case report from the University of New Mexico School of Medicine describes results of ECMO as a rescue treatment in adults with cardiopulmonary failure from HPS.[1] Diagnosis of HPS was based on hemodynamic and oxygenation data, and on serological studies confirmed by both Western Blot and PCR. Three adult patients were placed on ECMO as a rescue therapy; in two of these patients EMCO provided sufficient cardiopulmonary support, and these patients survived. However, ECMO was shown to have its own complications; furthermore, these results have not since been reproduced.

NITRIC OXIDE

A joint study by the University of Alabama and the University of New Mexico hypothesized that, in addition to immune-mediated mechanisms, the pathogenic effects of SNV involve reactive oxygen/nitrogen species (RONS) as mediators. It was shown that viral infection in the human lung is related to an upregulation of RONS generation. The effect of RONS in altering epithelial and endothelial cell permeability has been demonstrated. The results of the study suggest that pharmacological blockage of an enzyme responsible for nitric oxide (NO) synthesis may be therapeutically beneficial in HPS cases. However, although no experimental control existed, inhalational NO therapy was beneficial in the case of a single, 16-year-old patient.[2] Nitric oxide is a vascular relaxant; local administration of NO to the pulmonary vascular epithelium reduced pulmonary artery pressure and improved edema. The role of NO in mediating various effects of hantavirus disease requires clarification by further study.

β-3 INTEGRIN

A possible treatment strategy may take advantage of the recent identification of the β-3 integrin receptor; such a treatment should not alter the general function of this integrin, as it is important to the immune response. However, targeting the interaction between the hantavirus protein (thought to be G1 or G2) and the integrin would likely prove valuable. The possibility of passive immunization, with antibodies that target G1 or G2, needs further research.

VACCINES

The design and testing of formalin-inactivated cell-culture and rodent-cerebral vaccines for HFRS (from SEOV and HTNV infection) are known. More than one of these vaccines are licensed for use in Asia; they are apparently well tolerated and induce antibody responses detectable by ELISA. Controlled efficacy studies have not yet been reported. One setback of these vaccines is their inability to provide significant protective efficiency against PUUV, which is responsible for over 90% of the total HFRS cases in Russia, the country with the second highest HFRS infection rate. The reason for this limitation is that the vaccines are derived from SEOV (one monovalent and one bivalent vaccine) and HTNV (six monovalent and one bivalent vaccine).

In the United States, a recombinant vaccinia virus was developed and tested in phase I and phase II clinical trials. The vaccine elicited a neutralizing-antibody response in vaccinia-naive individuals but did not elicit such a response in vaccinia preimmunes. In addition, an investigational vaccinia HTNV vaccine is currently offered to laboratory workers at the U.S. Army Medical Research Institute of Infectious Diseases; however, no hantavirus vaccine is widely administered in the United States.

Neutralizing Antibodies It has been demonstrated that animals may be passively protected from hantavirus infection through administration of neutralizing antibodies; thus, eliciting neutralizing antibodies may be a major aspect of future vaccine development. While the humoral response is evidently a key element of hantavirus protection, the cell-mediated response may also be a factor influencing recovery or protection.[8] In support of this was the finding that cross-reactive cell-mediated immune responses to multiple hantaviruses were observed in mice; however, this area must be further examined, and may lead to useful strategies in combating hantavirus infection.

Aerosolized DNA Vaccine The Research Centre for Toxicology and Hygienic Regulation of Biopreparations, in Serpukhov, Moscow reg., Russia is currently attempting to design an aerosolized vaccine preparation against hantaviral infection. They maintain that DNA vaccine design, although not yet completed, would allow for multivalent preparations that would protect individuals from a range of hantavirus strains. Aerosolized delivery, they suggest, would eliminate the possibility of bacterial and viral contamination. They also suggest that induction of mucous immunity, into the very region of hantavirus entrance (the respiratory tract), would result in higher protective efficiency. These researchers note that an effective vaccine preparation must be able to reach the lungs.

REFERENCES

1. Crowley, M. R., et al. 1998. Successful treatment of adults with severe hantavirus pulmonary syndrome with extracorporeal membrane oxygenation. *Crit Care Med* 26(2): 409–414.
2. Fabbri, M. and M. J. Maslow. 2001. Hantavirus pulmonary syndrome in the United States. *Curr Infect Dis Rep* 3: 258–265.
3. Hart, C. A. and M. Bennett. 1999. Hantavirus infections: epidemiology and pathogenesis. Microbes Infect 1: 1229–1237.
4. Hjelle, B. and G. E. Glass. 2000. Outbreak of hantavirus infection in the Four Corners region of the United States in the wake of the 1997–1998 El Nino–southern oscillation. *J Infect Dis* 181: 1569–1573.
6. McCaughey, C. and C. A. Hart. 2000. Hantaviruses. *J Med Microbiol* 49: 587–599.
7. Padula, P. J., et al. 1998. Hantavirus pulmonary syndrome outbreak in Argentina: Molecular evidence for person-to-person transmission of Andes virus. *Virology* 241: 323–330.
8. Schmaljohn, C. S. 1999. Hantaviruses (*Bunyaviridae*). In A. Granoff and R. Webster, eds., *Encyclopedia of Virology*, Vol. 1, pp. 621–630. New York: Academic Press.

CHAPTER 8

■ ■

ANTHRAX

(BACILLUS ANTHRACIS)

Anuj Mehta

Bacillus anthracis (anthrax) is an organism that is of prime importance when considering bioterrorism issues. It is considered one of the most dangerous and most likely agents that would be used in a bioterrorist attack. When deprived of nutrients, the bacteria revert to a dormant spore form that is able to withstand a great deal of environmental stress. As such, the spores are the optimal agent for biological weapons purposes.

B. anthracis, though, is a naturally occurring bacterium found in the soil in most parts of the world. It has historically been linked to several animal and human epidemics, although it was initially much more common in animals. It occurs in one of three forms: cutaneous, inhalational, or gastrointestinal (GI). Anthrax's virulence is associated with three toxins—lethal factor, edema factor, and protective antigen—and with a protein capsule that protects the bacteria from attack by the host's immune system.

HISTORY

PLAGUES

Anthrax is one of the earliest recorded animal diseases; Moses' description of a disease that afflicted the animal population in Exodus 9:9 is thought to refer to anthrax. Witkowski and Praish went further in analyzing the descriptions of the 10 plagues that befell Egypt in the Old Testament.[51] Based on the text and what is currently known about anthrax infections, the paper concludes that the fifth, sixth, and tenth plagues that struck Egypt could have been caused by B. anthracis. The fifth plague affected only herbivores, with symptoms resembling those that have been observed in livestock. The sixth plague was said to have caused boils and other ulcers, resembling the signs of cutaneous anthrax. Finally, the tenth plague occurred after a supposed sandstorm in which spores from dead animal carcasses would have become airborne and led to widespread casualties in humans and animals (inhalational and gastrointestinal anthrax). Although this theory is interesting, much of it is based on supposition, interpretation, and extrapolation from relatively short descriptions that may or may not be connected to actual events. No conclusive evidence can be found to prove such a theory, but the significance is that anthrax has been part of human history for thousands of years.

The story continues throughout Europe and the rest of the world. Widespread livestock epidemics that are now thought to have been caused by *B. anthracis* were documented throughout Europe beginning in 1100 A.D. The "Black Bane" struck Europe in the 1600s and killed well over 60,000 cattle in a very short period of time with symptoms typical of anthrax.[9]

Anthrax's significant role in human history placed it at the forefront of research that has served as the basis of modern bacteriology and immunology. It was the first disease to be conclusively linked to a microorganism. Subsequently, in 1876, Robert Koch isolated the organism in pure culture for the first time. Following these initial discoveries, Louis Pasteur created a bacterial vaccine in 1881 for use in animals that was proven to be effective.

WOOL-SORTER'S DISEASE

Anthrax in humans was a major problem for textile and tannery workers. Historically, it had been known as "wool sorter's disease," Bradford disease, and "rag picker's disease." These names referred to the fact that many laborers who handled skins and wool became infected when spores left on the carcasses became airborne. Those who had to beat the wool and other materials in order to remove dirt and dust frequently fell victim to the disease. Laborers knew of the risks of handling contaminated hides and often drew lots to see who would have to sort batches of mohair and alpaca.

In 1879 John H. Bell, a physician from Bradford, England, conclusively proved that wool sorter's disease was actually caused by *B. anthracis*. Following this discovery, he pushed for preventive measures in the textile industry, and manufacturers finally began to accept his recommendations in the early 1880s. The Bradford Rules were enacted into law in 1897 and led to the creation of the Anthrax Investigation Board. The Anthrax Prevention Act, passed in England in 1919, established a wool-disinfecting complex in Liverpool. Such acts in England and other countries greatly reduced the number of anthrax cases linked to occupational hazards.

During the twentieth century, owing to governmental actions such as those in England and improved disinfectant methods, natural occurrences of anthrax in all its forms had become fairly rare in the developed world. Only 224 cases were reported in the United States between 1944 and 1994,[35] and currently about 2000 cases of cutaneous anthrax are reported each year worldwide. The last naturally occurring case of inhalational anthrax in the United States occurred in 1976. Based on the rarity of the disease in the United States, even a single diagnosed case of inhalational anthrax is cause for suspicion.

USE OF ANTHRAX AS A BIOLOGICAL WEAPON

World Wars I and II The history of *B. anthracis* as a biological weapon began around World War I. It is believed that German agents in the United States infected horses, cattle, and other animals with anthrax on their way to Europe. Evidence indicates that German troops also infected livestock in Europe with anthrax in hopes that the Allied soldiers would consume the tainted meat and develop GI anthrax.

Research into anthrax weaponization was widespread during the decades that followed. Interest was especially high during World War II, especially within the Japanese government. A branch of their military known as Unit 731 developed a research unit at Pingfen comprising 150 buildings, five satellite camps, and a research staff estimated to number 3000. Prisoners of war were used as test subjects, and according to some reports thousands of deaths were caused by infection with anthrax and other biological agents and, after the war, execution.[7] Furthermore, Japan allegedly employed anthrax in its campaign against Manchuria, releasing spores into the atmosphere over the area during the war. No concrete information exists as to how many individuals actually died at Pingfen or the extent to which infection occurred in Manchuria following the use of anthrax, but some reports indicate that casualties due to the biological agents and related illnesses were extensive.

Many governments launched biological weapons initiatives following World War I. Britain developed an intensive anthrax research division and in 1942 conducted extensive testing at Gruinard Island, off the coast of Scotland. Many of the tests consisted of detonating bombs hung on scaffolding structures and examining the extent of contamination of the surrounding area. The testing was so extensive and the island became so contaminated that it was quarantined for 48 years. For most bacteria, the fear of further infection would abate after a short time because the bacteria would most likely not be able to grow outside of a host. Anthrax, though, poses a much greater threat because the spores can remain dormant in the soil for decades, if not centuries. Decontamination efforts began in 1986 when the English government commissioned a firm to undertake the task of cleansing the island, which took several years.[5] To this day, many are afraid to set foot on the island because they are convinced that no decontamination effort could have completely eliminated the bacteria.

On orders from President Franklin D. Roosevelt, the United States launched its own biological weapons research in 1943. It was not in response to the Japanese threat but rather to a perceived threat from the Germans. The research was carried out at Camp Detrick (now Fort Detrick). During World War II, a pilot plant established at Camp Detrick to produce biological weapons manufactured 5000 bombs filled with anthrax spores.[7] These weapons, however, were never used in the war effort.

The Cold War The U.S. biological weapons program expanded during the Korean War, and a new facility was developed in Pine Bluff, Arkansas. This compound provided many advantages over Camp Detrick—it was much more advanced and could provide proper isolation and storage, and it had a more elaborate containment system for large-scale production and research. Production of weapons-grade biological agents (as opposed to weapons with crude unpurified spore mixtures) began in 1954. At the same time, though, preventive and treatment measures, such as vaccines and antibiotics, received a great deal of attention. In 1953, the U.S. military established a biological defense research center known as the U.S. Army Medical Research Institute of Infectious Diseases (USAMRIID), where many of the current countermeasures are developed and studied. Research continued at this facility until 1969, when President Richard Nixon, via executive order, ended the U.S. venture into biological weapons. From 1971 to 1972 the U.S. government systematically destroyed its stockpile of biological agents, officially keeping only that needed for research purposes.

The Soviet Union also developed an extensive biological weapons research program called *Biopreparat*. Their program was almost on a par with the U.S. one until 1969, when the latter was officially dismantled. Kenneth Alibek, the former deputy director of the Soviet program, defected to the United States in 1992. He claimed that the Soviet initiative was 5 years behind the U.S. effort in 1969.[2] Following the termination of the U.S. program, the Soviet Union pressed even harder to develop stronger and more deadly biological agents. They quickly surpassed U.S. technology, creating one of the largest and most virulent biological stockpiles in the world.

In 1972, the United States, the USSR, and the United Kingdom signed the Convention on the Prohibition of the Development, Production and Stockpiling of Bacteriological (Biological) and Toxin Weapons and on Their Destruction, commonly called the Biological Weapons Convention. Over 140 countries have since joined the pact, pledging not to use or develop biological agents for military purposes. Despite the historic nature of the treaty, research in foreign nations, and perhaps the United States itself, continued beyond 1972, and was most notably confirmed by the 1979 incident in Sverdlovsk, Russia.

Sverdlovsk, Russia: 1979 Knowledge of the 1979 incident in Sverdlovsk, Russia, began to surface in the Western press early in 1980. These reports merely mentioned that an anthrax epidemic had broken out the previous year in Sverdlovsk, a town of 1.2 million citizens 1400 km east of Moscow. Subsequent articles in Soviet medical and legal journals discussed an anthrax incident in which livestock and individuals in Sverdlovsk were infected, many of whom died. Several of the Soviet journals cited GI anthrax as the culprit, arising from the consumption of contaminated meat. In 1986 Matthew Meselson, the principal author of the definitive 1992 epidemiological study of the incident, renewed efforts to bring a panel of independent investigators to Sverdlovsk. In response, two Soviet physicians traveled to the United States in 1988 and presented the findings from their investigation of the incident in 1979. They claimed that an outbreak of anthrax occurred in late March 1979 and resulted in 96 cases of anthrax with onsets from April 4 to May 18. There were 79 instances of GI and 17 cutaneous anthrax cases, with 64 deaths among those with the GI form and no fatalities among the cutaneous population.[32] Many of those who had interest in the incident now believed the official claim to be more plausible, but were still concerned by the lack of pathological, anatomical, or epidemiological evidence to support the case for GI anthrax. Meselson finally won approval to bring an independent team to Sverdlovsk in 1992.

The team of scientists met with many problems during their initial visit in June 1992. Many of the hospital and public health records from the anthrax outbreak had been confiscated and destroyed by the KGB in 1979. However, the group did obtain a list of patients who died during that period from illnesses thought to have been anthrax. Meselson's group conducted household interviews with friends and family of the deceased in order to establish their whereabouts before they became ill. Based on the administrative list and other sources, the group investigated 77 individuals, 66 of whom died from the anthrax. The locations of those who had been affected were mapped, and a region southeast of Compound 19, long thought to be a biological weapons facility for Biopreparat, was highlighted as the main region of infection. Only nine of the individuals being investigated lived and worked outside this region. However, the interviews revealed that five of these nine were on a military training exercise at Compound 32, an army base in the affected area, during the first week of April 1979. The group then created a map of daytime locations of the affected individuals for the week of April 2, 1979. They were able to map 57 of the 66 individuals for whom they had information to a narrow region 4 km long, extending from the military Compound 19 to the southern city limit. Of the remaining nine individuals, three lived within the contaminated area and three held very mobile jobs, such as truck drivers or telephone workers, that would have been likely to take them into the affected region during the high-risk period.[32]

The next step in the investigation was to analyze the meteorological data from that time period and assess whether it would support a conclusion that aerosolized release of anthrax spores from Compound 19 caused the epidemic. On April 2, 1979, the wind was moving in a southeasterly direction, which would have taken any aerosolized release from the compound in the exact direction of the infected individuals. On other days during that week, the wind was moving in different directions that would have led to a different profile for infections. Meselson's group concluded that an aerosolized release of anthrax spores occurred on April 2, 1979, from Compound 19, and the wind carried the pathogens in a southeasterly direction.[32] His discoveries were preceded by a single statement without clarification from President Boris Yeltsin in May 1992: "the KGB admitted that our military developments were the cause" of the epidemic.[32] The study and the admission confirmed that the Soviet Union was, in fact, in violation of the 1972 Biological Weapons Convention.

The First Gulf War The next period of major concern, one that continues to the present day, arose during the Gulf War with Iraq in 1991. There was considerable—and eventually shown to be justifiable—fear that Iraq had extensive biological weapons programs aimed at attaining military supremacy. In 1995, information was made available to United Nations (UN) inspectors that confirmed the earlier fears. In 1990 Iraq had 100 R400 bombs filled with botulinum toxin, 50 with anthrax, and 16 with aflatoxin. In all, they produced 8500 L of anthrax, 6500 L of which was weaponized into rockets and bombs.[53] Luckily for the international forces and surrounding nations, though, Iraq had problems with the targeting systems and feared that prevailing winds might direct the agents back onto Iraqi soil or infect their own soldiers. There is no indication that any of the agents was used during the war. However, this information upset much of the defense community because it demonstrated conclusively that Iraq had the ability to create and weaponize biological agents, a major concern for UN weapons inspectors.

Japan: 1995 A major terrorist incident occurred in 1995, when the Aum Shinrikyo cult released sarin nerve gas into the Tokyo subway system, resulting in numerous deaths. However, what was not known until the arrest of Seiichi Endo, Aum Shinrikyo's head of germ development, is that from 1990 to 1993 the cult released aerosolized anthrax and botulinum toxin on several occasions at the Diet (the legislature), the Imperial Palace, the American military base at Yokosuka, and other places throughout Tokyo.[4,6] Fortunately, no one became infected. Forensic evidence showed that the strain of anthrax used most closely resembled Sterne 34F2, the strain used in animal vaccination, which poses little health danger to humans. Luckily for the general public in Japan, Aum Shinrikyo made the mistake of choosing a strain of anthrax with fairly low virulence.

The United States: 2001 The most recent incident concerning terrorist use of anthrax occurred in 2001, when an as yet unknown individual or group sent mail containing refined anthrax spores in the form of a highly concentrated dry powder to a variety of media institutions and governmental offices. Of the 22 confirmed cases of anthrax, 11 were inhalational and five resulted in casualties. At least five letters sent to Florida, New York City, and Washington, DC, were identified as containing the anthrax powder, and it is thought that all the contaminated letters were mailed from Trenton, New Jersey. The fatalities were spread over four states: Florida (1), Washington, DC/Virginia (2), New York (1), and Connecticut (1).

The initial letter containing anthrax spores, postmarked September 18, 2001, was addressed to television newsman Tom Brokaw and the *New York Post*. The later recipient in Florida was an employee of American Media, Inc. (AMI), and one of the tainted letters was addressed to that building. On October 5, 2001, he became the first fatality attributed to the anthrax attacks. The two fatalities in the Washington, DC–Virginia area were postal workers who presumably became infected when tainted mail passed through the Brentwood mail

FIGURE 8.1. (Left) An envelope addressed to Senator Tom Daschle's Washington, DC, office. (**Right**) An envelope addressed to Tom Brokaw of NBC News. Both letters were part of the 2001 anthrax mail attacks that led to 22 confirmed cases of anthrax infections, with five fatalities. Each envelope contained a fine white powder that easily became aerosolized when the envelopes were opened. The letter addressed to Senator Daschle had 2 grams of the powder at a concentration of around 1 trillion spores per gram. The letters originated from Trenton, New Jersey. No one has yet been arrested for the attacks.

Source: Ref. 47, courtesy of G. Strieker.

facility in Washington, DC. How the remaining two individuals—a hospital worker in New York City and retired woman in Connecticut—became infected is still unclear. The death of the Connecticut woman was the final fatality on November 21. The two had had no connection to any of the tainted letters that were eventually discovered, nor did mail facilities around them test positive for anthrax spores. These unrelated cases as well as the other 20 victims who developed some form of anthrax infection demonstrated the ease with which even a small amount of anthrax spores can be spread across half a nation and cause widespread fear.

The investigation into the source of the anthrax spores has revealed that the Ames strain of *B. anthracis* was used in the attacks. Interestingly, this strain was not developed on foreign soil, but rather by scientists associated with the USAMRIID. Several versions of the Ames strain exist, with multiple labs throughout the nation possessing samples and conducting research on it. This information plus the familiarity that the perpetrator was thought to have had with Trenton has led investigators to focus on domestic suspects rather than foreign terrorists.

No one has claimed responsibility for the attacks, nor have there been any arrests. Because of the domestic nature of the attack, government officials are concentrating on individuals who have access to labs that have anthrax cultures. They have narrowed the field to a list of 30 potential suspects.[31] Steven J. Hatfill, a medical doctor and virologist, has received more attention than others but has not been named an actual suspect. Hatfill is a former Army scientist who had access to Fort Detrick and used to teach bioterrorism response tactics to police and paramedics. Hatfill's apartment at Fort Detrick has been searched several times, but nothing that links him to the attacks has been made public. In fact, the investigators have offered little explanation about why they are interested in him. The attention that he has received due to investigation has resulted in Dr. Hatfill's being fired from his position at Louisiana State University. He denies any connection to the attacks, and claims that the intense focus on him is unwarranted and has thrown his life into shambles. In 2003, he filed a civil suit against the federal government for the continuous invasion of his privacy and the damage to his reputation, career, and life.

MOLECULAR BIOLOGY

Bacillus anthracis is an aerobic, gram-positive, spore-forming, nonmotile bacterium. It is found ubiquitously in the soil and generally infects herbivores. The vegetative bacterial cell ranges from 1 to 8 μm long and 1 to 1.5 μm wide, while the inert spore form is approximately 1 μm. Figure 8.2 offers a microscopic view of the bacteria.

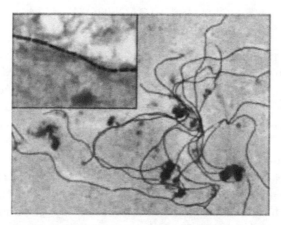

FIGURE 8.2. *Bacillus anthracis* is a gram-positive, spore-forming, nonmotile bacterium that, in its vegetative state, can range from 1 to 8 μm long and 1 to 1.5 μm wide. The spore form seldom is larger than 1 μm. (Inset) A ×20 magnified picture of the gram-positive staining of the *B. anthracis* bacteria. Note the signature rod shape of the bacilli family. Further magnification shows the jointed bamboo rod structure typical of *B. anthracis*.

Source: Ref. 19.

The shift between the active bacteria and spores depends on the nutrient content of the surrounding environment. Anthrax spores readily germinate (develop into an active bacterial form) when presented with an environment rich in amino acids, nucleosides, and glucose, but will revert to the spore form when the nutrient supply has been exhausted. Based on the needs of the active bacteria, *B. anthracis* has a very poor survival rate outside of a human or animal host. The spore, however, is extremely hardy; it can survive heat, cold, rain, drought, radiation, and other severe conditions. The amazing resilience of the spore form is one of the primary reasons that anthrax is such an attractive choice for use as a biological weapon.

As stated, once in the bacilli form, the bacteria begin to produce their two chief virulence factors: a gamma-linked poly-D-glutamic acid capsule and a tripart protein exotoxin composed of protective antigen (PA), edema factor (EF), and lethal factor (LF). The capsule protects the bacteria from phagocytosis, and thus is essential to the initial onset of infection. PA and EF are collectively called edema toxin; PA and LF are collectively called lethal toxin. There is evidence that each toxin contributes to anthrax's ability to avoid phagocytosis, but their principal action is to cause edema and other symptoms, such as the hemorrhaging associated with the disease. They are the principal cause of the pathology of *B. anthracis*.

THE ROLE OF EXTRACHROMOSOMAL
PLASMIDS pXO1 AND pXO2

Virulent strains of *B. anthracis* contain two large plasmids—pXO1 and pXO2—in addition to the bacterial chromosome. The genes for the toxins have been localized to the pXO1 plasmid, while the capsule genes have been traced to pXO2. The presence of both plasmids is required for full virulence, as is evident from analysis of strains that have lost either or both plasmids. Strains that have lost pXO1 do not produce the toxins and, as such, are essentially avirulent. *B. anthracis* strains that have lost pXO2 are not able to produce the capsule and exhibit at least a 10^5-fold decrease in virulence, most likely due to the bacteria's inability to avoid phagocytosis and lysis by the host's immune system.[30] Although both plasmids are present in all virulent strains of anthrax, the actual copy number of plasmids varies greatly between different strains. Some strains have many copies of pXO1 and few of pXO2, others the reverse. It is thought that higher copy numbers of pXO1 lead to increased virulence and lethality.[10]

Early studies by Louis Pasteur and Max Sterne demonstrated the relative ease with which variants of *B. anthracis* with reduced virulence could be isolated, which indicated that the genes encoding the toxins and capsule were extrachromosomal. The plasmids can be selectively cured—pXO1 by repeated passage at 42°C and pXO2 by growth on novobiocin.[25] Identification of the location of the genes was achieved by comparing various strains that had been selectively cured of pXO1 or pXO2. It was later established that pXO1 is 170–185 kb long and pXO2 is 90–95 kb long.

CONTROL OF GENE TRANSCRIPTION

Toxin gene transcription is controlled by a positive regulator, the *atxA* gene. *atxA* is located between the PA gene (*pag*) and the EF gene (*cya*), but transcribed in the opposite direction. The product of the *atxA* gene causes at least a 10-fold increase in transcription from an otherwise silent start site, P1, at bp −58 relative to the start codon. Constitutive low-level transcription initiates at another site, P2, at bp −26 relative to the start codon. Both P1 and P2, in addition to the −10 and −35 regions, are located in a potential 58-bp stem-loop structure. Studies with disruption of the *atxA* gene and complementation with a plasmid containing the *atxA* gene have shown that this positive regulator is required for transcription of all three toxins, and that anthrax strains lacking the gene have less virulence in mice.[11]

Based on the toxin sequences PA is 82.7 kd, LF is 90.2 kd, and EF is 88.8 kd, all with similar charges. PA facilitates the binding of LF and EF to the cell surface whereupon they can be internalized via endocytosis. PA is then able to serve as a membrane channel for the vesicle to transport LF or EF to the cytosol. Almost all eukaryotic cells have receptors on their cell surfaces for the PA. Most cells have anywhere from 5000 to 50,000 receptors for PA located on the plasma membranes. Only two cell lines—one derived from mice and a type of human lymphoma—are known to lack PA-binding receptors. Low nonspecific binding of PA and linear binding curves indicate that the cell-surface receptor is most likely a single cell-surface protein. In certain studies, researchers isolated mutant cells lacking functional PA receptor. The mutants' growth rates were significantly lower than those of cells with working PA receptors, implying that the PA receptor plays an important role in other essential cell processes.

CLEAVAGE OF PA TO PA20 AND PA63

Upon *B. anthracis* spore germination, the mature bacilli begin to produce the three toxins. PA is released into the host's body and binds to cell-surface receptors. Furin, a cell-surface protease, is one enzyme that can cleave PA between aa 164–167, which is extremely sensitive to proteases that recognize basic residues. The cleavage releases a 20-kd fragment (PA20) and leaves a 63-kd (PA63) attached to the receptor. The PA63-receptor complex then joins others to form a heptamer, which serves as the functional unit that binds LF and EF. The heptamer is then internalized by endocytosis, and on acidification of the vesicle, inserts itself into the vesicle's membrane. Once in the membrane, it serves as a channel to transfer LF or EF to the cytosol. Figure 8.3 illustrates the general pathway by which EF and LF are introduced into the cell.

THE FOUR DISTINCT DOMAINS OF PA

The structure of PA was established by X-ray diffraction, which has shown the toxin to be a long, flat protein rich in β-sheet structures. PA contains four distinguishable functional domains.[40] Domain 1 (aa 1–258) has two calcium ions tightly bound to it and a large flexible loop (aa 162–175) that acts as a cleavage site during proteolytic activation. Domain 2 (aa 259–487) consists of several long β-strands that form the foundation for the membrane-inserted channel. Domain 2 also contains a large flexible loop (aa 303–319) implicated in membrane insertion. The precise function of domain 3 (aa 488–595) has not yet been established, but it is thought to be involved in the formation of the heptamer. Domain 4 (aa 596–735) is associated with the function of the other three domains and is extremely important in receptor binding.

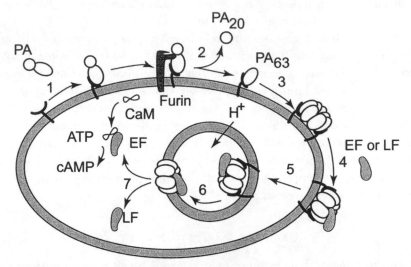

FIGURE 8.3. The general pathway by which anthrax attacks human cells is based on three toxins produced by the bacteria. Protective antigen (PA) binds to ubiquitous cell-surface receptors and is then cleaved by cell-surface proteases such as furin into two sections, PA20 and PA63. PA20 is released, but PA63 remains attached to the receptor. The PA63-receptor complex then forms a heptamer with other PA63 complexes. EF or LF then binds to the heptamer, which is internalized via endocytosis. Upon acidification of the vesicle, the heptamer inserts itself into the vesicle's membrane and acts as a channel through which EF or LF is transported to the cytosol of the cell, where they are able to carry out their biological function.

Source: Ref. 40.

Action by trypsin and chymotrypsin yields three separate fragments, and analysis of them has indicated that residues 168–312 comprise part of the binding site for LF and EF spanning domains 1 and 2.

Attention has also been given to the specific part of domain 4 that constitutes the receptor-binding domain for PA. Initially, it was thought that two large flexible loops, aa 679–693 and aa 704–723, were responsible for PA's affinity for the receptor. It was later discovered that the actual functional region was on the opposite side of domain 4. In particular, substitution of N657 or N682 completely prevents PA from binding to the cell surface. These two residues are adjacent to each other on the protein and appear to be the critical region for receptor binding. This conclusion was supported by the discovery that a PA fragment containing residues aa 663–735 competes with wild-type PA for binding to cell-surface receptors.[49]

EF: A CALMODULIN-DEPENDENT ADENYLYL CYCLASE

Protective antigen is essential for introduction of LF and EF into the cell. The N-terminal 250 amino acids of EF and LF have extensive homology, which is probably connected to their ability to bind to the PA63 heptamer. The binding of EF and/or LF to the PA63-receptor heptamer initiates endocytosis. In the cytosol EF acts as a calmodulin-dependent adenylyl cyclase, catalyzing the breakdown of adenosine triphosphate (ATP) to cyclic adenosine monophosphate (cAMP), one of the cell's principal secondary messengers involved in a myriad of signaling systems.

High levels of cAMP are known to be able to activate protein kinase A, which can phosphorylate (add a phosphate group) several proteins involved in cell signaling and other processes. In macrophages, high levels of cAMP inhibit NFκB, a nuclear transcription factor that normally stimulates production of several inflammatory factors essential to the immune response. Associated with its inhibition of NFκB, EF also inhibits neutrophil function, another part of the immune response responsible for attacking foreign bacteria. EF induces edema because cAMP also controls water homeostasis (the balance of water inside cells and tissue). High levels of cAMP upset water homeostasis within the cell, thus leading to massive localized edema in infected tissue. In laboratory experiments, localized tissue edema results when PA and EF are directly introduced into a culture.[30]

THE STRUCTURE OF EF AND THE EF-CALMODULIN COMPLEX

Recent studies have resolved EF by x-ray crystallography. Using the x-ray structure, the process of calmodulin binding to EF has been investigated.[13] EF's catalytic portion spans three distinct globular domains, and the active site lies at the border of C_A (aa 294–349 and aa 490–622) and C_B (aa 350–489), two of the domains. The third domain, aa 660–800, is helical and connected to C_A by a linker region, aa 623–659. In addition to the domains already mentioned, three other regions of EF, termed switches A, B, and C, have been identified. Switch A (residues 502–551) consists of amino acids that bind both calmodulin and the nucleotide substrate (ATP). Switch C (aa 630–659) resides within the linker region that connects C_A to the helical domain. In switch C, two β-strands and a connecting loop (aa 642–652) come together to form a helix upon binding of calmodulin. Switch B (aa 578–591) is located between switch C and the nucleotide substrate-binding region and, when not bound to calmodulin, maintains a highly disorganized conformation.[30]

Calmodulin, a protein that accounts for 1% of the many different cell types' total protein content, is the principal mediator for Ca^{2+} signaling, and EF's activity is dependent on binding to this protein. It conveys intracellular Ca^{2+} through two large globular domains connected via a central α-helix. Each globular domain is composed of two helix-loop-helix motifs. One globular domain has a high affinity for calcium and binds two Ca^{2+} molecules almost all the time, while the other domain has a low affinity for Ca^{2+} and binds it only when cellular calcium levels rise. Binding of Ca^{2+} to the globular domains forces a conformational change in calmodulin that helps expose certain hydrophobic residues.

EF binds to calmodulin in a very atypical manner. First, when EF binds to calmodulin, the N-terminal low-calcium-affinity region on calmodulin does not bind Ca^{2+} even when cellular levels are high. EF binds to calmodulin when two of the protein's arms attach themselves to calmodulin's two lobes, which causes a conformational change in calmodulin and results in the exposure of its hydrophobic residues (those that lack an affinity for water). More than 63 hydrophobic and basic EF residues come into contact with an estimated 53 hydrophilic (having an affinity for water) calmodulin residues, creating a very large contact surface. The contact region is divided into two portions: EFs aa 501–540 (part of C_A), and aa 616–798 (part of switch C and the helical domain). The large binding surface provides enough energy to stabilize various structural alterations that contribute to EF's activity, such as the stabilization of switch B and the reorientation of switch C, and makes the process essentially irreversible.

EF-CALMODULIN BINDING

Rather than a single step, though, EF-calmodulin binding is thought to occur in several stages. Initially, calmodulin-binding residues of EF are exposed in the calmodulin free-state. One such residue is Lys525 in switch A, which, when mutated, significantly increases the concentration of calmodulin necessary for wild-type activity.[29] Switch C on EF undergoes a rearrangement such that Asp647 is then able to form a salt bridge with calmodulin's Arg90. An Asp647Ala mutation does not affect calmodulin's binding but significantly lowers EF's activity.[29,30] This result implies that the switch C residue is more important for enzymatic activation rather than calmodulin binding. Since switch A mutations affected calmodulin binding, it has been proposed that calmodulin binds switch A residues before those from switch C.

This process provides enough energy to cause significant conformational changes (changes in the shape and structure) in the EF-calmodulin complex. In the EF-alone structure, switch B has a highly disorganized structure. However, upon binding to calmodulin, switch B reorients itself such that extensive contacts form between it and the N-terminal portion of switch C, stabilizing switch B. Switch C also undergoes extensive rearrangements. Two β-strands and their connecting loop are transformed into a new helical structure in the EF-calmodulin complex. The large binding surface also provides enough energy such that the helical domain moves and rotates so C_A, the linker, and the helical domain itself act as a clamp that encircles calmodulin.

CONVERSION OF ATP TO cAMP

Once bound to EF, calmodulin can no longer interact with its normal intracellular targets or perform its normal cellular functions, one of which is playing a part in the conversion of cAMP back to ATP. The EF-calmodulin complex also forces a rearrangement of EF that allows it to bind to ATP and act as an adenylyl cyclase. Therefore, EF not only catalyzes the formation of cAMP but also inhibits the reverse reaction.

Residues from six sections of the EF-calmodulin complex come together to form numerous hydrogen bonds and ionic interactions that create a pocket where ATP binds to EF. One of the most important steps in the binding process occurs at Lys 346, which contacts oxygens from all three phosphate groups on the nucleotide. Arg 329 binds to the β-phosphate, while Lys 372 and Ser 354 bind to the γ-phosphate. C_A and C_B form a cleft where a single metal ion interacts with the oxygens from the α- and β-phosphate groups. The metal ion in the cleft is controlled by two conserved residues, Asp 491 and Asp 493. Mutation of either residue alters the coordination of the metal ion and inhibits enzyme activity. Although definitive evidence has not yet been obtained, it is believed that the physiological ion is probably Mg^{2+}, but Yb^{3+} and Ni^{2+} have been observed in the cleft in crystallization efforts.[13,29,30] The metal ligand has not been observed in the EF-calmodulin complex, so it is thought that binding of the nucleotide and the metal ion may be interrelated. The crystal structure of the EF-calmodulin complex also shows His 351 in a position to interact with the 3′ hydroxyl of the ribose ring (a five-carbon sugar ring). Furthermore, it is situated perfectly for an in-line nucleophilic attack of the α-phosphate group. The nucleophilic attack is a key step in the breakdown of ATP and leads to the formation of cAMP.

The reaction scheme outlined shows that the metal ion and Lys 346 work together to activate the α-phosphate and stabilize the charge that develops on the oxygen, bridging the α- and β-phosphates. What differentiates EF from other adenylyl cyclases is that it requires only one metal ion whereas most other mammalian adenylyl cyclases require two ions.

CLEAVAGE BY LF

Although EF contributes to anthrax's virulence, it is not the primary agent responsible for the pathogen's high mortality. Strains of *B. anthracis* that have been mutated such that they lack the LF gene are able to cause localized edema, but they seldom result in death. These studies have shown LF to be the primary toxic agent.

Unlike EF, LF is known to affect only one cell type, macrophages. LF is internalized into the cell through the same process as EF. A fragment of LF, residues 1–254, actively competes for PA binding with wild-type LF, indicating that the N-terminal 250 residues constitute the PA binding domain. LF is a zinc-dependent metalloprotease that has been shown to cleave members of the mitogen-activated protein kinase kinase (MAPKK) family, namely, MAPPK1, MAPPK2, and MAPPK3 (or MEK1, MEK2, and MKK3, respectively).[27] Additional studies have also shown that LF can cleave other members of the MAPKK family, such as MKK4, MKK6, and MKK7, under certain laboratory conditions.[50]

The MAPKK family of kinases are essential to the cell signaling process that deals with mitotic and stress signals from the cell's environment. MEK1, MEK2, and MKK3 play crucial roles in activation of macrophages and production of cytokines. LF's primary cellular effect is upsetting secretion of tumor necrosis factor–α (TNF-α) and interleukin-1 (IL-1). Both TNF-α and IL-1 are proinflammatory signals important in initiating the immune response against foreign invaders. However, there have been conflicting reports as to whether LF inhibits the release of these compounds or induces the cell to release more of them.[30,39] In actuality, though, it may be a combination of both. It is possible that in the initial stages of infection, LF acts to inhibit the cell's secretion of TNF-α and IL-1 in order to impede the immune response and allow the bacteria to proliferate. At later stages of the infection, LF may actually promote the inflammatory response by inducing the release of TNF-α and IL-1, which could be connected to the massive system-wide edema and hemorrhaging observed in patients. The manner in which cleavage of MAPKK proteins elicits the effect on TNF-α and IL-1 is still unclear. The final result, though, is that LF eventually causes cell lysis of macrophages.

As stated, the earliest studies into LF revealed that it only has an acute effect on macrophages In rat and mice models, within 90–120 minutes of exposure to PA and LF, the toxins would cause lysis of the host's macrophages. The earliest cellular effect is observed at 45 minutes, with an increase in K^+ and Rb^+ ion fluxes across the membrane. At 60 minutes decreases in ATP concentrations and the release of a superoxide are detectable, followed by changes in cell morphology at 75 minutes and lysis at around 90 minutes.[30]

BIOLOGICAL ACTIVITY OF LF

LF's catalytic activity began to be understood when the HEXXH (aa 686–690) region typical of zinc metalloproteases was identified.[27] This realization was connected to the fact that zinc-dependent aminopeptidases have the ability to inhibit LF activity in macrophages. During the early stages of LF research, it was screened against a wide array of cell lines. It had the ability to inhibit growth in many cells but caused the most severe lysis in mouse macrophages. Years later, a database including the results of the cell-line screen was queried to find chemical agents that had activity similar to that of a compound known as PD98059, known to be a chemical specific inhibitor of the MAPKK pathway at MEK1. LF was to found to produce the results most similar to those of PD98059.[1]

X-ray crystallography has revealed that LF is also composed of four domains.[38] Domain 1 (aa 1–250) is responsible for LF's ability to bind to PA, which facilitates its entry into the cell. Domains 2–4 form a long deep groove that holds the N-terminal 16 residues of MEK2 before proteolytic cleavage. Domain 2 resembles the catalytic domain of the *B. cereus* toxin VIP2 and ADP-ribosylating toxin. However, domain 2 lacks several residues that are conserved throughout the family of ADP-ribosylating proteins, and is therefore not thought to possess such properties. Domain 3 acts as a protector in that it severely restricts access to the active site to any substrate other than MEK1 and MEK2. Other than its physical structure, domain 3's sequence also contributes to specificity because several residues in domain 3 interact directly with residues in MEK2. Domain 4 is the primary catalytic center as MEK2 binds in between it and domain 3. Furthermore, domain 4 binds the Zn^{2+} ion necessary for activation of its proteolytic activity. The only similarity between domain 4 and other zinc metalloproteases is the HEXXH motif common to all such proteins.

LF is a 90-kd protein with a structure typical of zinc metalloproteases toward the C-terminal at aa 686–690. Mutations of H686, E687, and H690 to alanine partially or totally impair LF's proteolytic activity.[30] As stated, the N-terminal 250 residues constitute the PA binding domain.[3,46] Distal to this region is a 19-residue sequence repeated five times with 60% homology between the repeats that does not appear to serve any catalytic function. It is thought, though, that the repeats may serve to physically separate the two terminal ends.

Following other tests and based on its homology to other zinc-dependent aminopeptidases, LF's biological activity was determined to be cleavage of MEK1, MEK2, and MKK3 (recent research, though, has indicated that it may have an effect on other proteins in the MAPKK pathway).[14,39,50] LF cleaves MEK1 and MEK2 both in vivo and in vitro within the proline-rich region near the N-terminal tails that is followed by the kinase domain. All MAPKKs are structurally related in that they are characterized by conserved kinase domains and divergent N-terminal proline-rich extensions that contain signal peptides that determine cellular localization. This class of proteins is extensively involved in cellular responses to mitogenic and stress signals, and they mediate a variety of essential protein–protein interactions. The exact manner, though, in which LF's cleavage of MEK1, MEK2, and MKK3 leads to rapid cell lysis is still unclear, but it is thought to be related to the eventual release of TNF-α and IL-1. As of yet, MEK1 and MEK2 have not been directly implicated in any pathways that hold strong control over short-term cellular homeostasis, the general property of the cell that is upset when LF initiates cellular lysis. This fact may allude to a further class of proteins that are cleaved by LF, although this has not been investigated or confirmed.

CLINICAL DIAGNOSIS AND RESPONSE

Anthrax is a disease associated mostly with herbivores, but historically it began to infect humans when they ate tainted meat or when they handled the hides, hair, wool, or other parts of infected animal carcasses. Depending on the method of infection, anthrax can manifest itself in one of three forms: cutaneous, inhalational (also known as pulmonary), and gastrointestinal.

CUTANEOUS ANTHRAX

Cutaneous anthrax occurs when the exposure occurs at the skin surface, with mucous membranes and previous cuts or wounds that have not fully healed being especially susceptible. A human host provides a nutrient-rich environment in which spores can readily germinate. As soon as the spores germinate within the skin, the bacteria begin to produce the anthrax toxins, leading to slight local edema. A small, slightly pruritic red papule (a small, solid, circumscribed elevation characterized by an intense itching sensation) appears at the initial site of infection, usually on the first day of infection. This papule then advances into a round ulcer by the second day. By the third day the ulcer further develops with tense vesicles or even bullae (hard elevations of the skin often containing fluids) filled with serosanguineous (containing both serum and blood) fluid forming on the plaque that has developed around the edema site. The next significant shift occurs when the central papule necrotizes (dies), ulcerates (breaks open, releasing the fluids within), and forms a usually painless black eschar (dry scab that forms on skin normally exposed to corrosive agents). This stage, though, is also associated with extensive local edema. The eschar dries and eventually falls off within 1 to 2 weeks, leaving a shallow ulcer that heals by the third week.

Cutaneous anthrax is also associated with lymphangitis (inflammation of lymphatic vessels) and painful lymphadenopathy (swelling of lymph nodes). Often the lymph nodes themselves become painful and tender upon infection. Patients usually experience headaches and, occasionally, fevers up to 102°F. In very rare cases, the inflammatory signs progress further, with multiple necrotic areas, massive edema spreading beyond the initial site of infection, and multiple bulbous formations. These symptoms can lead to systemic infection, attacking the pulmonary pathways and internal organs. Internal hemorrhage, shock, and death can occur within days of system-wide infection, although progression to this stage is very rare. Mortality for untreated cutaneous anthrax is about 20%, but with appropriate antibiotic treatment death has been seen only in extremely rare cases. None of the cutaneous cases from the 2001 mail attacks resulted in fatalities.

GASTROINTESTINAL ANTHRAX

Unlike the cutaneous form, GI anthrax is thought to occur more from the deposition of vegetative bacilli from uncooked meat in the upper or lower portion of the GI tract than from just spore germination. Regardless, however, exposure in this manner results in the development of oral or esophageal ulcers at the initial site of bacterial deposition. Infection leads to the development of regional necrotic ulcers, lymphadenopathy, edema, and sepsis (infection of the blood). Disease in the lower GI tract appears primarily as intestinal lesions occurring mainly in the terminal ileum or cecum (regions of the small intestine), which cause nausea, vomiting, and malaise initially and then bloody diarrhea, acute abdominal pain, or sepsis. Advanced infection resembles the sepsis seen in inhalational and cutaneous anthrax. The actual case numbers for GI anthrax are extremely low, so forming mortality statistics is not possible. Many, though, believe that, like inhalational anthrax, mortality for GI anthrax will decrease when it is treated with a rigorous regimen of antibiotics, but this has not been conclusively shown.

INHALATIONAL ANTHRAX

Inhalational anthrax is the most lethal form of the disease; it leads to mortality rates of 80% if untreated. The disease is contracted when spores are inhaled and deposited in the alveolar spaces of the lungs. Macrophages ingest the spores and induce lysis. Surviving spores are then transported to mediastinal lymph nodes (region behind the sternum and between the two pleural sacs containing the lungs), where the spores are able to germinate into active bacilli. The time lapse between deposition in the lymph nodes and germination is variable, ranging from 1 to 43 days. Patients in Sverdlovsk developed symptoms anywhere from 2 to 43 days following exposure.[32] Viable spores, however, have been found in monkeys 100 days after exposure. Extrapolations from animal data indicate that the LD_{50} for humans—the dose sufficient to kill 50% of people exposed to it—ranges from 2500 to 55,000 spores.[19]

The popular term anthrax pneumonia is misleading because true broncho-pneumonia does not occur in inhalational anthrax. Postmortem studies of victims from Sverdlovsk show that all patients had hemorrhagic thoracic lymphadenitis, hemorrhagic mediastinitis (inflammation of the mediastinum, which includes the heart, major vessels, trachea, thymus, and other connective tissue), and pleural effusions (escape of liquids from the serous membranes covering the lungs and the lining of the pleural lining), with half the patients showing hemorrhagic meningitis. Additionally, most people afflicted with inhalational anthrax develop arteritis (inflammation of the arteries) and arterial rupturing. *B. anthracis* was recovered in concentrations of 100 million colony-forming units per milliliter in the blood and spinal fluid.[20]

The pathogenesis of inhalational anthrax and development of all of these symptoms can be broken down into two stages. In the first stage, which normally lasts a few days, there are no clinically significant signs. Patients often exhibit only symptoms similar to those of flu and colds., making early diagnosis extremely difficult unless there is prior knowledge of an anthrax outbreak. The anthrax attacks in 2001 provided keener insight into the symptoms associated with the disease. During this period, 11 individuals came down with inhalational anthrax, and five died. Of the first 10 patients identified with inhalational anthrax, all noted malaise and fever with prominent cough, nausea, and vomiting. Drenching sweats, dyspnea (shortness of breath or labored breathing), chest pain, and headaches were also observed in the majority of patients. The second stage, however, develops rapidly, often seizing the patient within hours. Onset of acute dyspnea and subsequent cyanosis (bluish discoloration, especially of the skin and mucous membranes, caused by decreases in oxygenated hemoglobin) occur during this stage. Furthermore, physical examination reveals signs of pleural effusion in the chest, and chest x-rays during this period will also reveal the widened mediastinum and pulmonary edema. Most of the hemorrhagic symptoms mentioned become severe during the second stage, which usually lasts less than 24 hours and leads to death. Because of the extremely small sample of people infected with inhalational anthrax with which scientists have to work, conclusive numbers regarding the mean period between exposure and onset of symptoms have not been established.[20]

However, in the 2001 attacks, the median time between presumed exposure and onset of symptoms for the six patients for whom conclusive information is available was 4 days. All four patients who sought treatment after the full onset of symptoms died prior to the initiation of antibiotic treatments. Although these numbers come from a very small group, they provide insight into the speed with which anthrax can work in its inhalational form. Moreover, if left untreated, inhalational anthrax has a mortality rate of greater than 80%. Luckily, however, human-to-human transmission of the disease, unlike smallpox, is extremely rare, and would occur only through direct transfer of fluids containing the bacteria from one individual to another.

Following the attacks in 2001, an attempt was made to statistically analyze data regarding symptoms in patients with anthrax and symptoms from influenza and ambulatory community-acquired pneumonia, and thus develop a more accurate picture of inhalational anthrax that would enable it to be distinguished from influenza and pneumonia early in the disease progression.[18] Hupert et al. compared 28 cases of inhalational anthrax, both modern and past occurrences, with more than 2700 cases of influenza and 149 cases of ambulatory community-acquired pneumonia. The study revealed that abnormal lung examination, dyspnea, and nausea or vomiting are statistically greater indicators of a presumptive diagnosis of anthrax whereas sore throat and rhinorrhea are statistically greater predictors of a diagnosis of influenza. As indicators, cough, chest pain, abnormal temperature, and headache did not demonstrate a statistical difference between anthrax and influenza.[18] Even with these limited data, careful physical exam of patients early in the case of the disease may still be able to yield the proper diagnosis.

When inhalational anthrax was compared with ambulatory community-acquired pneumonia, though, researchers were unable to establish a clear way to distinguish the two. Dyspnea, nausea or vomiting, chest discomfort, cough, abnormal temperature, and headache were all found in both diseases, preventing their comparison from producing statistical significance one way or the other.[18] Therefore, there is still a great need to develop a fast way to diagnose anthrax, one that is sensitive and specific enough to identify anthrax and only anthrax early in the disease course while antibiotics are still effective in reducing mortality.

CONFIRMATION OF ANTHRAX INFECTION

Almost all the attention in diagnostic procedures has focused on inhalational anthrax, as it is the most likely manifestation of anthrax infection following an aerosolized release of spores. As has been made evident, anthrax is an extremely rare disease, and, until the incidents in 2001, few doctors were aware of its clinical signatures. In the United States, a Laboratory Response Network (LRN) has been established through collaboration between the Association of Public Health Laboratories and the Centers for Disease Control and Prevention (CDC). Eighty-one clinical labs in the LRN are now equipped to diagnose bioweapon pathogens. Simple cultures and staining of *B. anthracis* reveal that it is a gram-positive, nonhemolytic (does not cause rupture of red blood cells), encapsulated, spore-forming bacillus, and such tests can be performed by almost any hospital lab. More advanced diagnostic procedures such as *immunohistochemical staining* and *polymerase chain reaction (PCR)* tests (for the rpoB gene), however, must be performed by one of the qualified labs in the LRN that are equipped with the proper instruments and facilities.

Detection of anthrax in the environment is extremely difficult. The first sign that an attack has occurred will most likely be the arrival of the first patient to the emergency room. Diagnosis will occur on four levels. The epidemiology will not refer to a specific patient, but will recognize as suspicious the sudden appearance of several patients with severe acute febrile disease with a full set of symptoms that quickly results in death. The other possibility, though, is recognition of acute febrile illness in several individuals of a specific population deemed at risk, such as postal workers from Trenton during the 2001 anthrax attacks. The diagnostic tests become more patient-focused. In most cases, a chest x-ray that reveals a widened mediastinum, infiltrates, pleural effusion, and other signs will be indicative of anthrax infection. A chest compound tomographic (CT) scan will usually reveal hyperdense hilar and mediastinal nodes along with extensive mediastinal edema. A microbiological approach will reveal gram-positive, nonhemolytic bacteria that can be readily identified as being of the *Bacillus* genus. The most rapid molecular biology tests, as stated, are available only at LRN labs. The pathology of a patient with inhalational anthrax will normally reveal hemorrhagic mediastinitis, hemorrhagic thoracic lymphadenitis, and hemorrhagic meningitis.

Collectively these are the signs that indicate that an individual has been infected with inhalational anthrax. However, because of the rapid onset of the disease, waiting for confirmation from each test before beginning an antibiotic regimen would be extremely risky. The speed at which anthrax works necessitates that treatment start at the first sign of infection. Because anthrax's virulence is linked to the toxins it produces, it is possible to completely eradicate the bacteria from the blood, but if the toxins have reached sufficient levels death will follow regardless of whether new toxins are being produced.

IMPORTANCE OF EARLY TREATMENT

Recommendations for treatment of *B. anthracis* infection prior to October 2001 were based on the very rare human case, with much more emphasis on animal models and cell cultures. Even after the recent attacks, the number of cases of people with anthrax who have been treated is still extremely low. Thus, it should be kept in mind that any recommendation for treatment is based on a very small amount of data.

Given the rapid progression with which inhalational anthrax can move from cold and flu-like symptoms to massive edema, hemorrhaging, and death, early diagnosis and initiation of antibiotic treatment are essential to patient survival. A delay of even a single day can severely reduce a patient's chances for survival. If the antibiotic regimen begins too late in the process, the toxins will reach critical concentrations such that even if all the bacteria were killed the toxins would still cause death.

EVOLUTION OF TREATMENT

At the beginning of the twentieth century, the most common therapy for cutaneous anthrax was surgery before the infection became systemic. The pustule (inflamed cuticle containing pus), along with a great deal of the surrounding healthy tissue, was surgically excised and the wound treated with a bichloride of mercury. Carbolic acid in both high and low concentrations was used as a secondary treatment.

The next development was antibiosis treatment (treatment with bacteria) recommended by Pasteur, Jourbert, and eventually Fortineau. They were aware that certain bacteria, such as *B. pyocyaneus*, could inhibit the growth of *B. anthracis*. Sometimes these bacterial injections were accompanied by sodium sulfate or hyposulfate and quinine. There were, however, clear threats of secondary infections from a treatment course using live bacteria.

Local treatments such as excision and surgery, though, fell into disfavor because of fears that the action itself could induce septicemia (disease associated with the persistent presence of pathogenic bacteria or their toxins in the blood). Thus, modern therapy has evolved to the point that the focus is entirely on antibiotics. Treatment for *B. anthracis* now occurs on a system-wide level rather than a local level, even for cutaneous infections.

Natural strains of anthrax are currently resistant to many classes of antibiotics. However, in the 1970s and early 1980s, the primary and most recommended compound was penicillin. Penicillin G would have been administered in doses of 1–6 million units intravenously for at least 2 weeks. In the event of penicillin allergies, chlortetracycline, streptomycin, or chloramphenicol were the antibiotics of choice. For systemic or inhalational anthrax, the therapy was slightly different; 2 million units of crystalline penicillin were administered every 6 hours.[12,20,23,35]

CURRENT ANTIBIOTICS AND RESISTANCE

Current therapy has moved beyond penicillin because of the development of resistant strains and because stronger and more effective antibiotics have been developed. These recommendations are based on guidelines from the CDC and the Working Group on Civilian Biodefense, an organization consisting of 23 experts from academic medical centers, research organizations, and governmental military, public health, and emergency management institutions and agencies that in 1999 published guidelines in the *Journal of the American Medical Association*[19] for dealing with the use of anthrax as a biological weapon. The recommendations were updated in 2002.[20] In conjunction with doxycycline, ciprofloxacin, a fluoroquinone, is the first line of antibiotic attack. For cutaneous infections, 500 mg ciprofloxacin or 100 mg doxycycline twice daily via oral administration is the standard recommendation. Second-line therapy is amoxicillin at 500 mg orally three times a day. Recommendations for children employ the same antibiotics but at much lower doses based on the child's age and weight. Use in children has mostly been extrapolated from the use of the antibiotics for other infections in children.

Treatment for inhalational anthrax is similar to that for cutaneous infection but has certain distinct differences. For both, ciprofloxacin, doxycycline, and penicillin are the only drugs approved by the U.S. Food and Drug Administration (FDA); however, other drugs are routinely employed to combat illness. For inhalational anthrax, the Working Group on Civilian Biodefense recommends 400 mg ciprofloxacin and/or 100 mg doxycycline via continuous IV every 12 hours. Inhalational cases, though, warrant the use of one or two additional antimicrobials such as ampicillin, rifampin, vancomycin, and chloroamphenicol.[36] See Table 8.1 for complete treatment recommendations.

Current standard therapies are normally continued for 60 days after initial treatment. In inhalational cases in which IV administration is initially utilized, the patient should be switched over to oral antibiotics when clinically appropriate. Experts feel that the long-term antibiotic course is warranted based on evidence that anthrax spores can remain dormant in the lungs for more than 40 days after initial exposure. The longest incubation period for a fatal case in Sverdlovsk was 43 days. Cutaneous victims are also advised to continue therapy for 60 days for fear of additional exposure of their lungs.

POSTEXPOSURE PROPHYLAXIS

To date, no specific recommendations of postexposure prophylaxis for those merely at risk of developing infection have been established. In the case of the 2001 attacks, in addition to the 22 confirmed cases of infection, many more were exposed to the anthrax spores, including postal workers and those in the Hart Senate Office Building. The current belief is that these individuals should be given the same course as if they had been infected, but through oral administration. If an individual may have been exposed to anthrax spores, it is thought to be safer to begin treatment rather than risk delay for even a day. In the aftermath of

TABLE 8.1 TREATMENT RECOMMENDATIONS
FOR INHALATIONAL ANTHRAX

CATEGORY	INITIAL IV THERAPY	DURATION
Adults	Ciprofloxacin 400 mg every 12 hr *or* Doxycyline 100 mg every 12 hr *and* 1 or 2 additional antibiotics	When clinically appropriate switch to oral therapy: Ciprofloxacin 500 mg 2× daily *or* Doxycycline 100 mg 2× daily Continue oral or IV therapy for 60 days
Children	Ciprofloxacin 10-15 mg/kg every 12 hr *or* Doxycycline: >8 y and >45 kg: 100 mg every 12 hr >8 y and ≤45 kg: 2.2 mg/kg every 12 hr <8 y: 2.2 mg/kg every 12 hr *and* 1 or 2 additional antibiotics	Switch to oral when clinically appropriate Ciprofloxacin 10–15 mg/kg every 12 hr *or* Doxycycline: >8 y and >45 kg: 100 mg 2× daily >8 y and ≤45 kg: 2.2 mg/kg 2× daily <8 y: 2.2 mg/kg 2× daily Continue oral or IV therapy for 60 days
Pregnant women	Same for nonpregnant adults	Same for nonpregnant adults

Source: Adapted from Ref. 36.

Note: Several classes of antibiotics are recommended in the event of anthrax infection. The primary line of defense is ciprofloxacin and/or doxycycline. In addition to these two antibiotics, a secondary line of treatment with drugs such as penicillin may be used. The ideal situation would be for antibiotics to be given via IV during the initial phases and then continued for 60 days via oral therapy once symptoms abate. It is important to note that normally ciprofloxacin and doxycycline are not recommended for children or pregnant women, as little clinical testing has been done on these groups. However, in life-threatening situations, these therapies have proven to be the most effective, and the risk of adverse reactions to the antibiotics are outweighed by the risk of the infection.

the October 2001 attacks, 30,000 people were prescribed antibiotics, with 10,000 individuals taking ciprofloxacin or a related antibiotic for 60 days or more. These numbers do not reflect only the people exposed—many anxious people not at risk asked their physicians to prescribe antibiotics—but the number of individuals who legitimately received antibiotic treatment was nonetheless staggering.

Although diagnostic procedures for anthrax infection have improved in the last 20 years, mortality for inhalational anthrax is still around 80%. In the event of widespread infection, the initial few cases of inhalational anthrax may be misdiagnosed or not diagnosed in time. The individual(s), not knowing that they have been exposed to the pathogen, most likely would not seek treatment until they have passed beyond the critical stage at which antibiotics may be helpful. Only after medical personnel had knowledge of a potential exposure to anthrax would they rapidly begin treatment and postexposure prophylaxis on patients with the symptoms of anthrax. The first few individuals to fall ill, though, would not have much hope for survival because detection would not occur until late-stage infection.

TREATMENT MODIFICATIONS IN A MAJOR ATTACK

In the event of a massive attack, the previously described antibiotic regimen might not be practical. A significant widespread attack would place a significant strain on the public health infrastructure. The first concern is the availability of sufficient quantities of antibiotics. The recommended regimen requires that patients remain on the medication for 60 days. However, enough antibiotics to treat a large infected population may not be readily available and shorter periods of treatment may be the only answer. Furthermore, although IV therapy is recommended for those actually infected with the bacteria, in the event of a large-scale attack supplying IVs to everyone infected may not be feasible. Substitutes for IV therapy, such as oral treatment, may be necessary even though they would not be the ideal choice. The Working Group on Civilian Biodefense has recommended that, in the event of a mass-casualty setting, 500 mg ciprofloxacin be administered orally every 12 hours in addition to 100 mg doxycycline every 12 hours and 500 mg amoxicillin every 8 hours if the anthrax strain proves susceptible.[20]

VACCINES

Early Vaccines In addition to antibiotics, vaccines against anthrax have existed since the late nineteenth century. The first anthrax vaccine was developed by Louis Pasteur in 1881. His was a veterinary preparation that was adopted for general use in cattle and sheep. It was widely used in Europe and South America until it was modified in 1937. The modifications led to the development of the Sterne live-spore vaccine, an attenuated nonpathogenic strain of *B. anthracis* administered in the dormant-spore form. Derivatives of this vaccine are still widely used in livestock and have had remarkable success in safeguarding animal populations and controlling the spread of the bacteria. However, because of the morbidity associated with live spores, this vaccine line has not been used in humans in the United States or throughout most of Europe. The former Soviet Union and associated countries still use a live attenuated version of the virus for human vaccination purposes.

Newer Vaccines Attention has long been focused on developing another vaccine that utilized something other than an attenuated spore. Much of the research advances came after the 1950s and 1960s, the period during which PA,

EF, and LF were established as the primary cause of anthrax's virulence. Based on PA's role in anthrax pathogenicity and its relatively innocuous nature in and of itself, it became the target for developing a new vaccine.

The vaccine currently used in the United States, AVA (anthrax vaccine adsorbed), was licensed in 1970 and is produced and marketed by Bioport Corp. (Lansing, Michigan) as BioThrax. The vaccine consists of a cell-free filtrate that concentrates the amount of PA, the antigenic component of the vaccine to which the immune response reacts. It is derived from the V770 *B. anthracis* strain. AVA combines PA with aluminum as the adjuvant (substance that when administered with a pharmacological agent speeds its absorption or increases/ quickens its activity). Aluminum is used as an adjuvant in several vaccines, such as tetanus toxoid. The United Kingdom uses a similar vaccine derived from the Sterne strain with PA and LF as the antigenic compounds.

Historically, vaccination was limited to individuals at high risk for anthrax infection, namely, laboratory workers who routinely dealt with anthrax cultures. In response to the growing threat of biological agents being used in war or for terrorism purposes, the U.S. Department of Defense mandated in 1998 that military forces be vaccinated against anthrax, an endeavor that has not yet been successful. The program, known as the Anthrax Vaccine Immunization Program (AVIP), has met with resistance due to fears on the part of the individuals designated to receive the vaccine. Some military personnel have left the service; others have been threatened with court-martial for their refusal to participate in the program.

In addition to its preventive properties, the AVA has also been recommended for postexposure prophylaxis, in which the vaccine works with the antibiotic therapy to fight anthrax infection. There exists considerable fear that, following exposure to anthrax spores, some of the spores may remain dormant in the victim's lungs. In such situations the vaccine, if administered at weeks 0, 2, and 4, is able to bolster the immune system's response in the event of future germination and infection.

One of the main problems with the U.S. and UK vaccines is the complex regimen required to confer full immunity. The indications for AVA (BioThrax) mandate six doses of 0.5-mL subcutaneous injections at weeks 0, 2, and 4 and at months 6, 12, and 18, followed by annual boosters. The UK vaccine is similar in that it entails 0.5-mL injections at weeks 0, 3, 6, and 32, with annual boosters.

Another major problem with AVA is the threat of localized adverse events following administration. Based on data from USAMRIID, significant local reaction (defined as inflammatory reaction greater than 5 cm in diameter around the site of injection) occurred in about 2.4–3.9% of cases. Other studies cite even higher rates of adverse reactions, which are sometimes severe. Although these numbers are not unreasonable, such effects, in combination with the oppressively long vaccination regimen, have kept very active the search for new anthrax vaccines.

WEAPONIZATION

The essential question is why *B. anthracis* is such an attractive pathogen for a biological weapon compared with other bacteria and viruses. The bacteria itself is highly pathogenic but not very stable; it cannot survive outside of a host body for very long. Once exposed to ambient air, colony counts begin to fall at a rapid rate. Furthermore, when inoculated into water, the number of colonies begins to decline to virtually undetectable levels within 24 hours.

HARDINESS OF THE SPORES

Although the vegetative bacteria are unstable, the spore is extremely resistant to environmental stresses. As stated, *B. anthracis* bacilli will grow rapidly in a nutrient-rich environment, but once the nutrients are exhausted, rather than dying, the bacteria will form dormant spores, shedding most of the cytosolic portion of the cell and even decreasing the size of the nucleus. The spore form is essentially a method of preserving the DNA until conditions return to an optimal state for bacterial growth.

The spore is resistant to heat, cold, many chemical disinfectants, long dry spells, and low levels of ultraviolet light. In ambient conditions, the anthrax spore can survive for decades, if not longer. Evidence of the hardiness of the spore form can be seen in the efforts needed for decontamination. For example, although the exact cost is not known, 280 tons of formaldehyde and over 2000 tons of sea water were required to decontaminate Gruinard Island off the coast of Scotland; the process took 8 years.[5] Still, to this day, some individuals refuse to set foot on the island because they fear that it remains contaminated.

The most recent decontamination efforts were directed at areas exposed to anthrax spores during the 2001 attack. Extensive measures have been taken to remove any trace of anthrax spores from buildings used by government and media organizations, as well as postal facilities. The Hart Senate building was closed October 17, 2 days after spores were discovered, and did not reopen until January 18, 3 months later. The many months of sterilization efforts cost an estimated $23 million, and several attempts were required to attain full decontamination.[20] The task took so long because the Environmental Protection Agency employed chlorine dioxide, to which the spores were fairly resistant. Although chlorine dioxide is extremely hazardous to humans, anthrax spores seem to be able to survive a certain level of exposure. Decontamination of the Brentwood, California, mail-sorting facility in Washington, DC, took even longer. This history demonstrates that anthrax spores can survive through most naturally occurring conditions and still remain viable.

Additionally, anthrax is extremely attractive as a bioweapon because the aerosolized form has no odor, is essentially colorless (depending on the method of aerosolization), and is virtually undetectable. The first sign that an attack has occurred will probably be the first diagnosis of a patient in a hospital. It would be extremely difficult to detect an aerosolized release of anthrax spores. Only by interrogating a captured cult member did Japanese authorities learn that the Aum Shinrikyo cult had attempted an anthrax attack in Tokyo.

POTENCY

Beyond the hardiness of the spore form, anthrax, as discussed above in "Clinical Diagnosis and Response," is an extremely potent and deadly bacteria with mortality rates as high as 80%. In 1993 the U.S. Congressional Office of Technology Assessment undertook the task of estimating the casualties from a hypothetical bioterrorist attack utilizing aerosolized spores of *B. anthracis*. They concluded that in the event of an aerosolized release of 100 kg of anthrax spores upwind of Washington, DC, an estimated 130,000 to 3 million casualties would result.[37] This staggering statistic, based on numerous assumptions about such factors as weather patterns and spore concentration, shows that an organized attack utilizing *B. anthracis* could cause more casualties than the detonation of a nuclear device. Fortunately, such an attack has never occurred.

AVAILABILITY

Anthrax is readily available throughout the world. Its presence in soil led to the widespread incidents of anthrax in livestock before the animal vaccine was developed. Although difficult, it is possible to isolate the bacteria or the spore from the soil and begin to culture it; this is how samples were first obtained. *B. anthracis* will grow relatively easily on most laboratory media, another attribute that makes the bacteria attractive as a biological agent. Anthrax spores can also be aerosolized for mass dissemination.

The main issue in anthrax's use as a biological agent, though, is clearly its high mortality. If left untreated, inhalational anthrax has a mortality exceeding 80%, whereas that of an agent such as smallpox is slightly more than 30%. Furthermore, the recent attacks have demonstrated that, even when the victims are treated, the mortality can still be as high as 50%, depending on the time between initial exposure and initiation of antibiotic treatment.

Although anthrax has several characteristics that make it an ideal biological weapon, there are aspects of the bacteria that make it more manageable from a biodefense perspective. First, anthrax is not known to spread from person to person unless there is a direct transmission of bodily fluids. In terms of dealing with an anthrax attack, this feature is one of the few positive aspects of the bacteria. One of the major fears regarding organisms such as smallpox is that if it were released in a subway or on an airplane it could infect everyone in the area, and they in turn would go on to infect everyone with whom they had contact. Colonel Erik A. Henchal, one of the U.S. Army's top biodefense officers, stated in *The New York Times* that releasing smallpox among hundreds of thousands of travelers at Frankfurt International Airport could "create a worldwide epidemic of smallpox pretty quickly," because smallpox can easily be transmitted from person to person.[43] The same threat does not exist with anthrax because it cannot be spread from one individual to another.

Second, in terms of biodefense, anthrax is slightly more controllable than other contagions because there is very little risk from secondary aerosolization. One of the primary methods by which a terrorist group could carry out an attack would be to release an aerosolized form of anthrax spores from either a plane or the top of a tall building. This would be the primary aerosolization. When any aerosolized material settles to the ground, there is the threat of secondary aerosolization—the mixture is resuspended, carried on the wind or by other environmental factors. If that were to happen, in addition to the group infected from the primary release of the agent, there would also be a second group of individuals who became infected from the secondary aerosolization, increasing the weapon's lethality. However, aerosolized forms of anthrax spores have been shown to have a very small chance of being reintroduced into the atmosphere once they settle to the ground.[20,52] Therefore, military and health professionals need to be concerned only with individuals infected in the initial release. Although this could be a large number, it is certainly smaller than would be the case if anthrax became reaerosolized.

STATUS OF THE THREAT

The threat of anthrax also stems from aspects of the bacteria that are not related to its biological activity. Weaponized forms of anthrax spores are known to have been developed by several countries during the Cold War. Currently, it is thought that at least 13 and possibly more than 20 countries posses stockpiles of anthrax, including but not limited to Iran, Iraq, North Korea, Japan, and Russia.[34]

Iraq In recent times, Iraq's capabilities with regard to biological weapons and other weapons of mass destruction have been of great concern. Proof of Iraq's foray into weapons of mass destruction arose even before the First Gulf War in 1991. During the Iran–Iraq war (1980–1988), Iraq used chemical weapons to gain a decisive advantage, an action that arguably saved the Iraqis from being overwhelmed by the larger Iranian army.[15] Furthermore, Iraq unleashed chemical weapons on its own citizens—Kurdish rebels in the north—in retaliation for their insurgence. Some reports claim that nerve gas initially killed thousands of people but that the lasting effects of the toxins resulted in more than 100,000 deaths.[15,24] Attempts have even been made to bring genocide charges against Saddam Hussein's regime for specific attacks using chemical agents.[42]

Based on discoveries made during the first Gulf War, Iraq had developed an extensive bioweapons research program. As already stated, Iraq had a great deal of weaponized anthrax loaded onto bombs and missiles ready for launch. However, after the first Gulf War, UN edicts and sanctions required Iraq to destroy all its stockpiles of weapons of mass destruction. Evidence shows that Iraq did in fact destroy many stockpiles immediately following the war, and later UN weapons inspections supported that claim.[16] Iraq's denials that they possessed any weapons of mass destruction, however, were weakened by their refusal to allow unconditional access by UN inspectors, a violation of the agreement made after the first war. On November 8, 2002, the UN issued a new declaration calling for the resumption of open inspections in Iraq, threatening the use of diplomatic and military might if necessary.[48] However, based on intelligence reports and U.S. assessment of Iraq's refusal to give open access to UN inspectors, the administration of George W. Bush believed that the Hussein regime posed a serious threat to U.S. national security and that Iraq still maintained large stockpiles of chemical and biological weapons. As a result, a U.S. invasion of Iraq began on March 20, 2003, with missile strikes against Baghdad.[16]

To date, U.S.-led arms inspectors have failed to find any evidence of stockpiles of weapons of mass destruction (anthrax or otherwise) or research and development for such programs. Assessment of reports provided by U.S. inspectors has shown that Iraq's former anthrax capabilities were completely defunct. The remnants of the facilities that existed prior to the first Gulf War were rendered ineffective and, to a large extent, destroyed by that war and 12 years of internal conflict, technology and arms embargos, and economic and political sanctions. Some reports even conclude that Iraq's anthrax capabilities just before the U.S. invasion in 2003 are inferior to what they possessed before 1990.[16]

With the failure to find weapons of mass destruction in Iraq, the Bush administration has been forced to justify its initial claims about Iraq's defiance of UN sanctions. Prior to the invasion, Secretary of State Colin Powell and Central Intelligence Agency Director George Tenet, in a session of the UN Security

Council, claimed that intelligence data added up to "facts" and "not assertions" about Iraq's inventory of chemical and biological weapons.[24] However, after a year of occupation, David A. Kay, the Bush administration's chief weapons inspector, reported to the Senate Armed Services Committee that nothing had been found. According to him, the failure was not motivated by political ends but rather stemmed from misinterpretation of intelligence information. Kay testified, "The limited data we had led one to reasonably conclude this [that Iraq possessed such weapons]. I now see that there's another explanation for it."[26] These issues became the basis of heated political debates about the intentions of the Bush administration prior to the war.

One difficulty in assessing Iraq's capabilities solely on the existence of large stockpiles of anthrax, however, is that biological weapons laboratories need not be huge complexes nor must active stockpiles be kept. Small amounts of anthrax can be maintained, and large quantities produced quickly using small mobile facilities. Such facilities would be almost impossible to detect because, having not yet been used, they would not have any traces of anthrax. Capability in and of itself ought to be reason for concern. Some of these questions about Iraq's capabilities may never be answered. Nonetheless, it is important to acknowledge Iraq's formidable biological weapons capacity prior to the first Gulf War and be vigilant for any traces of programs that could potentially fall into the hands of rogue nations or terrorists.

THE FORMER SOVIET UNION

As already discussed, there are vaccines and antibiotics that can be used in the event of anthrax exposure. In 2001, such treatment saved the lives of 17 people and probably many more who were exposed but never showed symptoms because of early treatment. One of the major problems in dealing with bioterrorism is related to treatment and stems from a fundamental dichotomy between the U.S. biological weapons program (terminated in 1969) and that of the Soviet Union. The U.S. program mandated that some form of treatment— whether it be vaccination, antibiotics, or antitoxins—exist for every pathogen being researched. As such, almost everything developed before the termination of the program could be treated in some way. Often the treatments were not complete, or they would be only partially effective, but such a requirement provided a certain level of safety.

According to Kenneth Alibek, the Soviet Union had a completely different philosophy. Their belief that the best agents would be those for which no vaccine or treatment existed led Biopreparat to attempt to engineer strains of *B. anthracis* resistant to most antibiotics and vaccines that existed at the time.[2] Strains of anthrax that are resistant to penicillin and doxycycline have been identified.

Luckily, these strains have remained in the laboratory setting, and knowledge of their existence comes from information-sharing between U.S. scientists and former Biopreparat scientists. Strains resistant to ciprofloxacin have not yet been identified, but it is not far-fetched to think that such a strain was engineered. The real fear is that a nation that has developed antibiotic-resistant strains of anthrax, mostly thought to be only the former Soviet Union, would sell the bacteria to rogue nations or individual terrorist groups. Because of the current economic situation in Russia, there are many unprotected military installations and unpaid scientists throughout the region.

Global Threat Regarding the use of biological agents by governments, a form of "protection" is already in place. One question remaining from the Gulf War is why Iraq chose not to use its extensive biological arsenal against U.S. troops. One theory, among many, is that they feared an even greater reprisal, perhaps nuclear. When planning retaliation against a specific country, a nation like the United States could focus on defined military installations, cities, and other targets. Terrorist organizations, however, are not confined to national boundaries, nor do they have defined locations at which a country could direct retaliation. Terrorist organizations are by definition stateless entities; they have members dispersed throughout the world, and it can be difficult if not impossible to completely eliminate their threat. How can one strike back against an enemy that does not have an address? The U.S. "War on Terrorism," specifically the attempts to destroy all remnants of al-Qaeda and capture Osama bin Laden, highlights the difficulty in waging a war against a terrorist organization that does not adhere to national boundaries.

As a result, the threat of biological weapons is much more potent from a terrorist organization. Terrorists have used conventional weapons to intimidate countries and individuals throughout history, but biological weapons offer an even more appealing option. They can achieve a much higher casualty rate over a much larger area; their efficacy can last weeks rather than the duration of a single attack; they are often harder to detect and easier to hide; they can be disseminated without the victims' even knowing that they have been exposed until it is too late; and they are an even more powerful method of intimidation. Furthermore, even a small attack, like the mail attacks of 2001, can cause considerable widespread panic, and that, in and of itself, may be the goal of the attacker.

ACQUISITION AND DEVELOPMENT
OF BIOLOGICAL WEAPONS

Terrorists could obtain biological weapons in a variety of ways. First, they could simply purchase the bacteria/spores from nations that already possess them. The current status of the Russian biological weapons division is of considerable concern in this regard. No one can deny that the Russian government is in dire economic straits. As a result, numerous experts from Biopreparat have been fired and have resorted to becoming street vendors to make ends meet. Those who are still employed either are paid very poorly or have gone many months without being paid. The possibility therefore exists for a well-funded government or terrorist organization to approach a disgruntled scientist and purchase his or her expertise, equipment, or the agent itself. To date, none of the anthrax strains that have been used for terrorist purposes appears to have been engineered or obtained from the former Soviet Union, but the existence of such strains is a major source of concern.

Furthermore, nations such as Iraq and Iran—countries that President George W. Bush has labeled parts of the "Axis of Evil"—are known to have developed anthrax for military use and could conceivably sell it to terrorist organizations that they have been reported to support. Another possibility arises from the fact that hundreds of laboratories around the world currently have strains of *B. anthracis* for research purposes. Although these labs usually have heavy security and removal of the pathogens would be fairly difficult, the events of 2001 show that it is not impossible. Authorities now ascribe the mail attacks to a domestic terrorist who had expertise in molecular biology, access to the bacteria, and the technical know-how for weaponization. Former Soviet Union scientists and stockpiles, nations such as Iraq and Iran, and disgruntled workers from laboratories that maintain anthrax cultures are avenues that a terrorist organization could take in order to obtain viable anthrax cultures.

Well-funded terrorist organizations could also develop biological agents themselves, as is believed to be the case with the Aum Shinrikyo attacks in Japan. As stated, *B. anthracis* can be found in the soil throughout the world. It is believed that the Aum Shinrikyo cult obtained a sample of the bacteria from either the soil or a modified version of the animal vaccine.[6] With spores so readily available in nature, it is not beyond the realm of possibility that a group could obtain natural strains of anthrax and simply grow them in a laboratory. In fact, Gruinard Island was so contaminated until 1987 that many feared that a terrorist group could use soil from the island as a source of anthrax.

PROCESSING FROM BACTERIA TO REFINED SPORES

Luckily, weaponization of anthrax is more difficult than merely growing bacteria on a culture plate. Beyond the difficulty of first obtaining the bacteria, the subsequent steps of weaponization require more work, money, and expertise than most terrorist organizations have available. The CNN website (www.cnn.com), in its extensive coverage of the 2001 anthrax attacks, described a probable

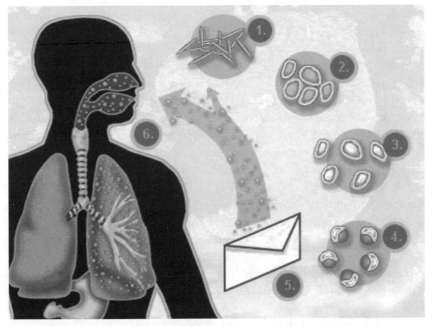

1. *B. anthracis* is grown on common laboratory media as vegetative bacilli.
2. Nutrients are removed and the bacteria revert to the spore form.
3. The spores are separated from the media, purified, and concentrated.
4. The spores are mixed with extremely fine dust particles to maintain their separation and to increase the time they will remain suspended in the air.
5. The resultant powder is then placed into an envelope and mailed.
6. The recipient opens the letter and the powder is released into the air, exposing anyone nearby to anthrax spores.

FIGURE 8.4 Anthrax, before it can be used for terrorist purposes, must be developed into a weaponized form. Two methods exist that result in either a wet or a dry formulation. The dry fine powder is the more desirable, more effective, and more easily disseminated form. The process described below has been proposed as the process used to develop the anthrax spores used in the 2001 mail attacks in the United States. It is a very rough sketch but provides insight into the steps necessary to create a dry fine powder with concentrated spores.   *Source*: FBI Press Releases, 2001.

process of weaponization (obtained from the University of Arizona).[47] Because the bacterial form is not tolerant of ambient environmental conditions, any successful biological agent utilizing *B. anthracis* would employ the spore form. The most efficient dissemination method is to aerosolize the spore and release it as a mist or fine dust. The 2001 anthrax attacks used a remarkably efficient aerosolization method with fine dust particles to create the powder in the contaminated letters.

This process produced a concentration of about 1 trillion spores per gram in the powder sent to Senator Tom Daschle's office; the letter had a total of 2 grams of powder. Remember that the LD_{50} for inhalational anthrax ranges between 2500 and 55,000 spores. Simple division reveals that an extremely high casualty rate could have resulted from such a powder widely disseminated, for example, though the ventilation system of a building. The other threat noted in the 2001 mail attacks was that the mail-sorting machines themselves caused the release of the some of the powder and thus aided in its aerosolization, contaminating the machines and infecting postal workers. Five known contaminated letters were able to infect 22 people, cause five deaths, and yield positive tests for anthrax spores in eight states, extending as far west as Missouri. Although it does not seem that the intent of the terrorist, who has not yet been identified, was to cause mass casualty, the individual or group was able to cause widespread panic, a rush on ciprofloxacin, and a disruption of the mail system. Terrorist groups need not seek large numbers of casualties—although, based on the estimate from the U.S. Congressional Office of Technology Assessment, anthrax could achieve such numbers—rather, they could intend merely to spread panic throughout a nation. Such was the case in the 2001 attacks.

DELIVERY SYSTEMS

The anthrax attacks of 2001 illustrated one of two delivery systems thought to be effective in the use of anthrax as a biological weapon. The first uses a wet dispersal method, which involves suspending the spores in a wet solution that, upon dispersal, forms droplet-size particles. The main problem with this method is that most conventional "spraying devices," like Windex bottles, form droplets that are quite large (with diameters on the order of 100 μm) and would therefore not stay suspended in the air for very long. Crop-dusting equipment produces droplets that are slightly smaller, but, as previously noted, anthrax does not normally resuspend into the air once it has settled to the ground. The ideal method would be to reduce the droplet size to less than 10 μm, but that would require extensive equipment not readily available.[51]

For these reasons, the ideal weaponized form of anthrax spores is dry powder. The first step in forming a dry powder is to allow the bacteria to grow on basic media. When the liquid culture has matured, the bacteria must be forced to convert to the spore form. This step involves processing the liquid culture in a fermentation tank. Because no new nutrient sources are provided in the tank, the live bacteria will revert to spores. This process results in a slurry-like liquid that contains large amounts of spores. The slurry must then be dried and processed in a milling machine that will create a fine, dry powder. Tiny spore particles—optimally between 1 and 5 µm—must then be separated, purified, and concentrated. Following the isolation of the spores, the powder must be combined with fine dust particles to maintain spore separation and to increase the time that they will remain suspended in the air.[47]

This process, while seemingly simple, has many complicating pitfalls and thus necessitates a high level of expertise in processing bacterial spores. Following the fermentation procedure, the slurry often has the consistency of peanut butter, which makes the drying process very difficult. The resulting product is often a solid brick that is very unwieldy to work with. When it is milled down to fine particles on the order of 1 µm, the spores will obtain a surface charge, which causes the spores to form small clumps that are useless for weapons purposes. To deal with the surface charge, the spores are usually coated with some sort of silica gel or other product that can neutralize the charge (understandably, the exact materials used are kept secret). As a result, the spores, when combined with the dust, maintain their separation and are less likely to stick to surfaces. Weaponized anthrax does not have good reaerosolization properties, but non-weaponized anthrax is even less well suited because the static charge makes it stick to surfaces, making resuspension more difficult.

THE COST OF DEVELOPMENT

The U.S. Department of Defense has reported that three employees with technical skill in molecular biology but no expertise in anthrax or biological weapons developed and weaponized bacteria with characteristics similar to those of anthrax (except for the pathogenicity in humans) in less than a month for slightly over $1 million.[33] Applying the same process to *B. anthracis* would require much more money and time because the bacteria would be difficult to obtain, tighter security precautions would have to be taken, and more careful steps to ensure worker safety would be necessary. The cost of developing anthrax into a potent agent for bioterrorism is high, but not out of the reach of several terrorist organizations known to be well funded. The threat exists, and precautionary measures should be taken. As a result of the attack on the World Trade Center and the anthrax attacks in 2001, several airports now swab every piece of luggage going on a plane and run simple tests for traces of pathogens and chemicals. As a primary step to prevent the transport of the agent, it shows promise.

EFFECT ON PUBLIC HEALTH INFRASTRUCTURE

In addition to the dangerous characteristics of anthrax itself, there is also the major question of whether the current public health infrastructure could handle a large-scale attack. One of the key problems is that the treatment for anthrax exposure, described previously, is not feasible in the event of widespread infection. The primary method of treatment mandates IV antibiotics, but if thousands of people were infected, which is quite possible in a large-scale attack, not everyone could be treated intravenously. The CDC recommends in such cases that oral antibiotics be given. However, it is not entirely clear how oral antibiotics will affect the survivability of patients who are already infected. Furthermore, there is the problem of the extremely limited supply of the vaccine. As noted earlier, six doses of BioThrax are required to provide full immunity. In the event of an attack in a major city, there are simply not enough doses of the vaccine to treat or immunize everyone.

A widespread anthrax attack would pose a serious public health and economic threat to the country. The current public health system has never been forced to deal with such an attack, and it is questionable whether, in its current state, it could be effective. In addition, a terrorist attack would have a crippling impact on the economy because there would be widespread panic, people would not be able to go to work, there would be a rush demand for antibiotics, and many other disruptions. The full economic impact cannot even be predicted because numerous unforeseen problems would develop. Anthrax is such a potent biological weapon not only because of its innate lethality but also because it poses an immense threat to the economy and the health care system itself.

To avoid attack or massive casualties, eternal vigilance of intelligence agencies and the medical community is necessary. Only the intelligence world and the government have the capacity to track terrorist organizations and predict the probability of an attack. Their ability to act on clear-cut evidence is essential in preventing an attack. However, education in the medical community is also essential. Given the rarity of anthrax, many physicians are not readily able to recognize the symptoms of anthrax infection. Delay in the first diagnosis will also delay the forensic effort of finding the source of exposure as well as the prophylaxis treatment of people who have been exposed. Anthrax is currently classified as a Category A pathogen for bioweapon purposes by the CDC. With such a recognizable threat, it is essential for the medical community to be able to identify anthrax immediately and begin the process of treating those who have been exposed. Only then can heavy casualties be avoided after an attack.

POTENTIAL DEFENSES

With the increased threat of a biological attack, new defenses are currently being developed. As stated, several classes of antibiotics are recommended as the standard treatment for anthrax infection, starting with ciprofloxacin, a member of the fluoroquinone family. Additionally, doxycycline is also fairly effective, and penicillin is still recommended against naturally occurring strains of anthrax.

Beyond antibiotics, various vaccine formulations that have been developed are currently in use. As noted above, the vaccine used in the United States is BioThrax, a cell-free filtrate that uses high concentrations of PA to elicit an immune response. Other countries use different formulations, with some employing PA with attenuated versions of EF and/or LF. Also, vaccines using attenuated bacteria still exist. Among the main problems with current vaccines, though, are that it often takes more than 6 months to confer full immunity and the regimen calls for six doses over the course of 18 months and annual booster shots thereafter.

New treatments and vaccines are always being pursued. One current avenue of research has moved away from traditional antibiotics and investigated the usefulness of lysins in the fight against anthrax. Lysins have the ability to lyse bacterial cell walls by binding to certain carbohydrates on the wall that are essential for viability, leading to rapid cell death. Schuch et al. investigated the γ-phage associated with *B. anthracis* as a source of lysins because of their inherent binding specificity and bacteriolytic properties. Called PlyG (for phage lysin γ), it has been shown to be an effective means of selectively killing colonies of *B. anthracis* in vitro.[44] Although still in its initial stages, research along this line is promising.

DOMINANT NEGATIVE MUTANTS OF PA

Another avenue that has been investigated utilizes a dominant negative mutant of PA to disrupt the heptamerization process. A dominant negative protein is not only inactive itself, but when mixed with other active forms of that protein it inactivates all of them. The dominant negative does not actually alter wild-type PA produced by the anthrax bacteria; rather, when the PA63 heptamer forms, it disrupts the conformation of the final product, rendering the heptamer inoperative. Theoretically, if the dominant negative mutant were given to an individual infected with anthrax, the mutant protein would combine with the wild-type PA and inhibit the translocation of EF and LF into the cell. Sellman et al. investigated four mutants that exhibited the dominant negative phenotype, a double-mutant K397D and D425K, a mutant with the 2^β_2–2^β_3 loop deleted, and two single-mutant proteins, F427A and D425K.[45] The point mutations were able to prevent translocation of EF and LF by disrupting the heptamer structure. The other mutant with an entire region deleted was unable to be activated by furin, unable to oligomerize, and unable to bind the ligand. However, inactivation was achieved because it competed with wild-type PA for cell-surface receptors, virtually blocking wild-type PA binding to the cell.

Although still a relatively new concept, dominant negative variants of PA provide a novel method of combating anthrax.

NEW VACCINE FORMULATIONS

In addition to postexposure treatment, extensive research has been conducted into new vaccines. Cieslak et al. from the USAMRIID have identified three major approaches that could yield new more effective vaccines.[8,28] The first involves recombinant vaccines that clone the PA gene into organisms of low pathogenicity such as *B. subtilis*. The second approach would employ mutant-strain vaccines whose virulence is dependent on aromatic compounds (derived from the benzene molecule) not normally found in human tissue. Since human tissue would lack the compounds necessary for the bacteria's virulence, the immune system could develop resistance against antigens unique to anthrax without having to fight off a deadly infection at the same time.[21] The third major approach identified by the USAMRIID utilizes the same method as the AVA: purified preparations of PA combined with various adjuvants.

Along these avenues, several promising new vaccine candidates have been developed. One employs purified PA with monophosphoryl lipid A (MPL) as the adjuvant. One of the main benefits of this combination is that lyophilization apparently has no effect on the efficacy of the vaccine; this potentially offers an advantage over the current AVA, which requires a cold-storage chain. Furthermore, in initial testing in guinea pigs, the MPL combination was more effective than current AVA in conferring immunity.[17,22]

Another avenue deals specifically with the expression of the cereolysine AB gene in *B. anthracis*. The cereolysine AB gene in *B. cereus* confers some hemolytic properties on the bacteria, properties that anthrax lacks. The gene was cloned into three strains of anthrax: the virulent H-7 strain, which contains both plasmids (pXO1 and pXO2); the STI-1 strain, which only has pXO1 and has therefore been used in some vaccines to confer immunity on the virulent strains; and the 221 strain, which lacks any plasmids and pathogenicity. Hamsters vaccinated with the STI-1 strain were resistant to infection by the H-7 strain, but when the recombinant H-7 strain (with the cereolysine AB gene) was administered several of the hamsters developed the infection and later died. Essentially, the cereolysine AB gene gave the H-7 strain the ability to overcome the immunity conferred on the hamster by the recombinant STI-1 vaccine. It has not been fully determined whether such a strain could overcome the AVA vaccine, but the possibility warranted further research. It was discovered that, although the recombinant H-7 strain infected hamsters vaccinated with STI-1, it could not infect those immunized with a recombinant STI-1 strain containing the cereolysine AB gene. The interesting feature here is that, although a method of vaccine evasion was developed, when the same method was applied to the vaccine itself it was able to grant immunity against both the wild-type and the recombinant strains. The cereolysine AB gene, when cloned into a vaccine utilizing an attenuated spore form, confers immunity against anthrax strains with certain hemolytic properties.[41]

Other vaccine candidates exist, such as strains of anthrax that depend on aromatic compounds not found in human tissue, but much more research needs to be done in these areas. The MPL preparation has shown promise and is still being investigated. At the same time, the USAMRIID is conducting extensive research into more effective vaccines. However, recent research has shown that the current AVA may be more effective than previously thought. In one study, just two doses of AVA conferred 100% protection to a vaccinated group of rhesus monkeys, whereas the placebo group experienced 100% mortality.[17,22] Such evidence implies that there may be a better formulation of the current AVA, an avenue that should not be overlooked.

POLICY

Defense against bioweapons, while having a significant basis in science, is also linked to policy. One of the primary lines of defense for the U.S. military is the AVIP, whose purpose, as discussed above, is to preemptively immunize all members of the armed forces against anthrax before it is used in a wartime situation. Out of fear of adverse reactions to the vaccine, however, more than 400 members of the military have either left the service or refused to be vaccinated, thereby facing court-martial. Although specific reports of serious side effects to the vaccine are rare, there is a perceived threat.

REFERENCES

1. Aless, D. R., et al. 1995. PD098059 is a specific inhibitor of the activation of mitogen-activated protein kinase kinase in vitro and in vivo. *J Biol Chem* 270: 27489–27495.
2. Alibek, K. 1998. Terrorist and Intelligence Operations: Potential Impact on U.S. Economy. Joint Economic Committee, U.S. Congress, May 20.
3. Arir, N. and S. H. Leppla. 1993. Residues 1–254 of anthrax toxin lethal factor are sufficient to cause cellular uptake of fused polypeptides. *J Biol Chem* 268: 3334–3341.
4. Atlas, R. M. 2002. Bio-terrorism: From threat to reality. *Annu Rev Microbiol* 56: 167–185.
5. BBC News. 2001. *Britain's 'Anthrax Island.'*
6. Broad, W. 1998. Sowing death: how Japan germ terror alerted world. *New York Times*, May 26.

7. Christopher, G. W., et al. 1997. Biological warfare: A historical perspective. *JAMA* 278(5): 412–417.

8. Cieslak, T. J., et al. 2000. Immunization against potential biological warfare agents. *Clin Infect Dis* 30: 843–850.

9. CNN. 2001. *Timeline: Anthrax Through the Ages*, Oct. 6. Online: http://www.cnn.com/2001/health/conditions/10/16/anthrax.timeline.

10. Coker, P. R., et al. 2003. *Bacillus anthracis* virulence in guinea pigs vaccinated with anthrax vaccine adsorbed is linked to plasmid quantities and clonality. *J Clin Microbiol* 41(3): 1212–1218.

11. Dai, Z., et al. 1995. The atxA gene product activates transcription of the anthrax toxin genes and is essential for virulence. *Mol Microbiol* 16: 1171–1181.

12. Dixon, T. C., et al. 1999. Medical progress: Anthrax. *N Engl J Med* 341(11): 815–826.

13. Drum, C. L., et al. 2002. Structural basis for the activation of anthrax adenylyl cyclase exotoxin by calmodulin. *Nature.* 415: 396–402.

14. Duesbery, N. S. and G. F. Vande Woude. 1999. Anthrax lethal factor causes proteolytic inactivation of mitogen activated protein kinase kinase. *J Appl Microbiol* 87: 289–293.

15. Ekeus, R. 2003. Why Saddam's arsenal has not been found. *Sunday Times* (London), Aug. 3.

16. Gellman, B. 2004. Arsenal was only on paper. *Washington Post*, Jan. 7.

17. Hambleton, P. and P. C. Turnbull. 1990. Anthrax vaccine development: a continuing story. *Adv Biotechnol Processes* 13: 105–122.

18. Hupert, N., G. Bearmen, et al. 2003. Accuracy of screening for inhalational anthrax after a bioterrorist attack. *Ann Intern Med* 139(5): 337–345.

19. Inglesby, T. V., et al. 1999. Anthrax as a biological weapon: Medical and public health management. *JAMA* 281(18): 1735–1745.

20. Inglesby, T. V., et al. 2002. Anthrax as a biological weapon, 2002: Updated recommendations for management. *JAMA* 287(17): 2236–2252.

21. Ivines, B. E., et al. 1990. Immunization against anthrax with aromatic compound dependent (Aro-) mutants of *B. anthracis* and with recombinant strains of *B. subtilis* that produce anthrax PA. *Infect Immunol* 58(2): 303–308.

22. Ivines, B. E., et al. 1995. Experimental anthrax vaccines: Efficacy of adjuvants combined with protective antigen against an aerosol *Bacillus anthracis* spore challenge in guinea pigs. *Vaccine* 13(18) 1779–1784.

23. Jamie, W. E. 2002. Anthrax: Diagnosis, treatment, prevention. *Prim Care Update* 9(4): 117–121.

24. Jehl, D. and D. Sanger. 2004. The struggle for Iraq: Intelligence. *New York Times*, Feb. 1.

25. Kaspar, R. L. and D. L. Robertson. 1987. Purification and physical analysis of *Bacillus anthracis* plasmids pXO1 and pXO2. *Biochem Biophys Res Commun* 149: 362–368.

26. Kay, D. 2004. Transcript of testimony before the Senate Armed Services Committee, Jan. 28.

27. Klimpel, K. R., N. Arora, and S. H. Leppla. 1994. Anthrax toxin lethal factor contains a zinc metalloprotease consensus sequence which is required for lethal toxin activity. *Mol Microbiol* 13: 1093–1100.

28. Kortepeter, M., et al. 2001. *Medical Management of Biological Casualties Handbook*, 4th ed. Fort Detrick, Maryland U.S. Army Medical Reseach Institute of Infectious Diseases.

29. Leppla, S. H. 1984. *Bacillus anthracis* calmodulin-dependent adenylate cyclase: Chemical and enzymatic properties and interactions with eukaryotic cells. *Adv Cyclic Nucleotide Protein Phosphorylation Res* 17:189–198.

30. Leppla, S. H. 1999. The bifactorial *Bacillus anthracis* lethal and oedema toxins. In J. E. Alouf and J. H. Freer, eds., *The Comprehensive Sourcebook of Bacterial Protein Toxins*, 2nd ed., pp. 243–263. London: Academic Press.

31. Locy, T. and L. Parker. 2002. Anthrax case remains frustrating mystery. *USA Today*, Oct. 1.

32. Meselson, M., et al. 1994. The Sverdlovsk anthrax outbreak of 1979. *Science* 266(5188): 1202–1208.

33. Miller, J. 2001. Next to old rec hall, a "germ-making plant." *New York Times*, Sept. 4.

34. Monterey Institute of International Studies. 2002. *Chemical and Biological Weapons Resource Page*. Online: http://cns.miis.edu/research/cgw/possess.htm.

35. *Morbidity and Mortality Weekly Report*. 1994. Summary of Notifiable Diseases, 1945–1994. *MMWR Morb Mortal Wkly Rep* 43: 70–78.

36. *Morbidity and Mortality Weekly Report*. 2001. Update: Investigation of bio-terrorism-related anthrax and interim guidelines for exposure, management, and antimicrobial therapy. *MMWR Morb Mortality Wkly Rep* 50: 909–919.

37. Office of Technology Assessment. 1993. *Proliferation of Weapons of Mass Destruction*, Washington, DC: U.S. Congress.

38. Pannifer, A.D., et al. 2001. Crystal structure of the anthrax lethal factor. *Nature* 414: 229–233.

39. Pellizzari, R., et al. 1999. Anthrax lethal factor cleaves MKK3 in macrophages and inhibits the LPS/IFNgamma induced release of NO and TNFalpha. *FEBS Lett* 462: 199–204.

40. Petosa, C., et al. 1997. Crystal structure of anthrax toxin protective antigen. *Nature* 385: 833–838.

41. Pomerantsev, A. P., et al. 1997. Expression of cerolysine AB gene in *B. anthracis* vaccine strain ensures protection against experimental hemolytic anthrax infection. *Vaccine* 15(17/18): 1846–1850.

42. Sachs, S. 2003. The struggle for Iraq: Toward a trial. *New York Times*, Dec. 17.

43. Schmitt, E. 2003. Military says it can't make enough vaccines for troops. *New York Times*, Jan. 8.

44. Schuch, R., D. Nelson, and V. A. Fischetti. 2002. A bacteriolytic agent that detects and kills *Bacillus anthracis*. *Nature* 418: 884–889.

45. Sellman, B. R., M. Mourez, and R. J. Collier. 2001. Dominant-negative mutants of toxin subunit: An approach to therapy of anthrax. *Science* 292(5517): 695–697.

46. Singh, Y., et al. 1999. Oligomerization of anthrax toxin protective antigen and binding of lethal factor during endocytic uptake into mammalian cells. *Infect Immunol* 67: 1853–1859.

47. Strieker, G. 2002. *Making Anthrax into a Weapon*. University of Arizona and Associated Press. Online: http://www.cnn.com.

48. United Nations. 2002. Text of U.N. Resolution on Iraq, Nov. 8. Online: http://www.cnn.com/2002/US/11/08/resolution.text/index.html.

49. Varughese, M., et al. 1999. Identification of a receptor-binding region within domain 4 of the protective antigen component of anthrax toxin. *Infect Immunol* 67: 1860–1865.

50. Vitale, G., et al. 2000. Susceptibility of mitogen activated protein kinase kinase family members to proteolysis by anthrax lethal factor. *Biochem J* 352: 739–745.
51. Witkowski, J. A. and L. C. Praish. 2002. The story of anthrax from antiquity to the present: A biological weapon of nature and humans. *Clin Dermatol* 20: 336–342.
52. Zajkowska, J. and T. Hermanowska-Szpakowicz. 2002. Anthrax as biological warfare weapon. *Medycyna Pracy* 53(2): 167–172.
53. Zilinskas, R.A. 1997. Iraq's biological weapons. *JAMA* 278: 418–424.

CHAPTER 9

■ ■

SEVERE ACUTE
RESPIRATORY SYNDROME (SARS)

Joseph Patrick Ward and Maria E. Garrido

As of early 2003, people thought that a new virus spreading throughout the world faster than the Internet could happen only in bad dreams or modern science-fiction movies. Then we learned about SARS-CoV, the virus that causes severe acute respiratory syndrome (SARS). Less than 7 months after it was first reported, SARS had taken the lives of approximately 800 people and infected a total of more than 8400 people worldwide.

SARS-CoV belongs to the family of coronaviruses. The coronaviruses (order *Nidovirales*, family *Coronaviridae*, genus *Coronavirus*) are members of a family of large, enveloped, positive-sense, single-stranded RNA viruses that replicate in the cytoplasm of host cells. Coronaviruses are highly species-specific, and the family had only included two human-infecting viruses prior to the discovery of SARS-CoV. Coronaviruses are divided into three groups, and exactly where SARS-CoV fits into the three groups has been debated. It has even been suggested that SARS-CoV represents a new, fourth group of coronaviruses.

No vaccine or effective treatment currently exists for SARS.[11] The natural host has not yet been identified despite widespread speculation that the host is the civet, a cat-like mammal from which a virus similar to SARS-CoV has been isolated. Because of the circumstances surrounding the SARS outbreak of 2003 and the very small number of cases that have been reported since, it has been proposed that the 2003 outbreak was strictly a one-time event. Although much research was conducted during and immediately after the 2003 outbreak, little research has been done on the disease and its causative agent in the latter half of 2004. The disease-control community is in disagreement about whether time and money should be committed to studying SARS in the face of limited resources and many other infectious agents that appear to be more of a threat.

No one can definitively declare that another large SARS outbreak will not occur in the future. There were four cases in January 2004 in the same Pearl River Delta region of China where the 2003 outbreak began. In April 2004, a small outbreak in eastern China was traced to contamination in a research laboratory in Beijing. These most recent cases were quickly and successfully contained. Still, SARS poses a possible threat to global health. The disease was highly contagious during the 2003 outbreak, and there is currently no effective treatment. Additionally, SARS presents with very nonspecific symptoms, making it difficult to diagnose. With early case detection being a necessity for containment of the disease, another large-scale outbreak of SARS or a bioterrorist attack using SARS-CoV would be a frightening event. However, such an outbreak or attack seems unlikely. SARS-CoV remains classified as a Category C list emerging pathogen by the Centers for Disease Control and Prevention (CDC). The Category C list represents newly identified pathogens and those that are least likely to be used by bioterrorists.

HISTORY

It should be noted that the term *severe acute respiratory syndrome* did not exist until March 10, 2003. Before then, the disease was usually described as "atypical pneumonia" and is thus referred to as such in the chronology below. Another key point is that the chronology of the disease has been retrospectively determined.

Epidemiologists have traced the outbreak of SARS to 11 index cases of "atypical pneumonia" that occurred in the Pearl River Delta region of China between November 2002 and January 31, 2003. Although there was no history of contact among those 11 patients, all the later cases of SARS can be traced back to them. Much of the history of the 2003 outbreak consists of "super-spreader" events, in which many people are infected by the disease in a short period of time (G.-P. Zhao, personal communication, April 9, 2004). This is how 11 index cases led to the infection of thousands.

ORIGIN AND SPREAD

November 2002 to January 31, 2003. The first epidemiological outbreak of the twenty-first century was retrospectively determined to have begun in November 2002 in Foshan, a city of Guangdong Province, China. The Guangdong Center for Disease Control and Prevention has records of 11 index cases of "atypical pneumonia" during this 3-month period with no contact history between the cases. The cases were in different cities or different parts of cities in the Pearl River Delta region of China and were spread over the 3-month period. Seven of the patients had documented contact with wild animals.

January 12, 2003. The complicated cases of atypical pneumonia from different cities began to be transferred to the major hospitals in Guangzhou, a large city in the Guangdong Province, in order to receive better medical care.

January 31, 2003. The first super-spreader event occurred in hospitals in Guangzhou. The outbreak featured more than 130 primary and secondary infections, of which 106 were determined to be hospital-acquired.[5]

February 10, 2003. The Beijing office of the World Health Organization (WHO) received an e-mail message from Guangdong Province about a "contagious disease" that had already infected more than 100 people in the course of a week. A day later, the Chinese Ministry of Health reported to the WHO about an outbreak of "atypical pneumonia" in Guangdong, with 300 cases and five deaths. Residing in Guangdong Province was a 64-year old nephrologist from Zhongshan University. After treating patients in the university hospital, the doctor contracted the "atypical pneumonia" and subsequently traveled to attend a wedding in Hong Kong.

February 21, 2003. The doctor stayed on the ninth floor of the Metropole Hotel in Hong Kong. Even though he had developed symptoms of a respiratory illness, he initially felt well enough to see the city. The next day, he sought medical attention at a hospital in Hong Kong and was admitted to the intensive-care unit with respiratory failure. He warned the staff that he had been treating patients with a "very virulent disease" and that he was afraid he might have contracted it. It was later determined that the doctor infected at least 12 other guests and visitors to the ninth floor of the hotel, including a woman traveling to Toronto, a Chinese-American businessman traveling to Vietnam, and a stewardess traveling to Singapore.

February 28, 2003. Dr. Carlo Urbani of the WHO began to take care of the Chinese-American businessman in Vietnam. He was concerned that this "atypical pneumonia" might be avian influenza and notified the WHO Regional Office for the Western Pacific. On March 11, he departed for Bangkok to attend a meeting but fell ill upon arrival and was immediately hospitalized. The ability of this "atypical pneumonia" to spread from a small town in China to several countries in less than one day is indicative of a new era: flights that can take you from one continent to another in less than one day and people that carry the disease with them to other countries without even knowing about it.

March 5, 2003. The Toronto woman died of the disease and spread it to five members of her family. The Guangdong doctor had died the day before at a hospital in Hong Kong. The Chinese-American businessman was in critical condition and had spread the disease to seven health care workers at a Vietnamese hospital.

March 10, 2003. The new syndrome was designated *severe acute respiratory syndrome*, or SARS.

March 15, 2003. The WHO issued a travel advisory to countries where SARS had been reported.

March 17, 2003. The WHO called on 11 leading laboratories in nine countries to join a network for multicenter research into the etiology of SARS and to simultaneously develop a diagnostic test. The network takes advantage of modern communication technologies (e.g., e-mail, a secure website) so that the outcomes of investigations of clinical samples from SARS cases can be shared in real time. On the secure WHO website, network members share electron-microscope pictures of viruses, sequences of genetic material for virus identification and characterization, virus isolates, various samples from patients, and postmortem tissues.[6] Samples from one patient were analyzed in parallel by several laboratories and the results shared in real time.

March 18, 2003. Canada, Germany, Singapore, China, Vietnam, the United Kingdom, and Thailand reported more than 219 suspected cases and four deaths.

March 20, 2003. The total number of suspected cases reached 306, with 10 deaths in 11 countries, including one case in the United States.

March 22, 2003. The exact identity of the virus remained unknown. Anthrax, pulmonary plague, leprospirosis, and hemorrhagic fever were ruled out. *Chlamydia pneumoniae* was suspected because of the findings of *Paramyxoviridae* bacteria in tissue samples from SARS patients.

March 24, 2003. Scientists at the CDC and in Hong Kong announce that a new virus had been isolated from patients with SARS. The virus responsible was isolated from the lung tissue of a patient who eventually died of SARS.

March 26, 2003. The first global "grand round" on the clinical features and treatment of SARS was held by the WHO. The electronic meeting united 80 clinicians from 13 countries. A summary of their discussions and conclusions was made available on the SARS page of the WHO website.

March 27, 2003. The WHO announced that the SARS causative agent was suspected to belong to the family *Coronaviridae*.[10]

March 30, 2003. In Hong Kong, a steep rise in the number of SARS cases was detected in Amoy Gardens, a large housing complex consisting of 10 35-story blocks, home to around 15,000 people. The Hong Kong Department of Health issued an isolation order to prevent the further spread of SARS. The order required residents of Block E of Amoy Gardens to remain in their apartments until midnight on April 9. Residents of the building were subsequently moved to rural isolation camps.

April 4, 2003. Primers for the SARS genome became publicly available on the Internet.

April 14, 2003. Canadian researchers completed the genome sequence of the new virus.

April 16, 2003. WHO officials announced that the new virus was a member of the family *Coronaviridae*. The virus was named SARS-CoV.

April 17, 2003. Australia and Mongolia reported their first cases. SARS had now reached 25 countries, with 3389 probable cases and 165 deaths.

April 19, 2003. A scientific paper demonstrating that a coronavirus was the cause of SARS was published in *The Lancet*.[24]

April 23, 2003. The total number of suspected cases surpassed 4000 worldwide.

April 28, 2003. For 20 consecutive days (twice the number of days of the incubation period), there were no reported cases in Singapore, which became the first country to have successfully contained SARS. By now the total number of suspected cases had surpassed 5000 people, with 321 deaths.

May 1, 2003. The United Kingdom and the United States were removed from the list of countries with recent local transmission. Poland reported its first SARS case. The number of worldwide cases reached 27 countries, with 5965 probable cases and 391 deaths.

May 6, 2003. Twenty-nine countries, including Colombia and India, had reported SARS cases. The WHO reported that the case-fatality ratio of SARS depends on the patients' health and age and ranges from less than 1% in patients 24 years or younger to more than 50% in patients 64 years old, with an average case-fatality ratio between 14% and 15%.

May 8, 2003. The number of probable cases reached 7183 worldwide, with 514 deaths in 29 countries on six continents.

May 21, 2003. The number of probable cases passed 8000 worldwide. SARS-CoV was isolated from wild civets in northwestern China, suggesting that the civet or a similar animal may be the host for the virus.

May 31, 2003. Singapore was removed from the list of local chains of transmission.

June 3, 2003. The WHO reported that the number of new cases had decreased to somewhere between two and three cases per week and that local chains of transmission were found mainly in Toronto and several parts of China.

June 17, 2003. The first global conference on SARS took place in Kuala Lumpur, Malaysia.

June 23, 2003. Hong Kong was removed from the list of areas with local chains of transmission, and Beijing was removed a day later.

July 2, 2003. Toronto was removed from the list of areas with recent local transmission.

July 5, 2003. Finally, Taiwan was removed from the list of suspected areas of transmission, leading the WHO to declare that SARS outbreaks had been contained worldwide.

THE AFTERMATH OF CONTAINMENT

September 8, 2003. The WHO reported a SARS case in Singapore due to an accidental laboratory contamination.

December 5, 2003. The WHO reported a SARS case caused by accidental laboratory contamination in Taiwan.

December 16, 2003 to January 31, 2004. Four independent laboratory-confirmed cases of SARS occurred in Guangzhou, China. The index cases were a 20-year-old waitress, a 32-year-old male resident of Guangzhou, a 35-year-old businessman, and a 40-year-old hospital director in Guangzhou. Over 100 known contacts for each index case were found and put under surveillance by the WHO, but it was determined that no human-to-human transmission had occurred. All four patients made full recoveries.

April 22–29, 2004. The Chinese Ministry of Health reported a total of nine cases of SARS in China; seven of the patients were from Beijing and two were from the Anhui Province in east-central China. One of the patients died. Two of the nine patients were graduate students who worked at China's National Institute of Virology Laboratory (NIVL) in Beijing, which is known to conduct research on SARS-CoV. Of the seven other SARS cases, two were directly linked to close personal contact with one of the graduate students who worked at NIVL— the student's mother (who died) and a nurse who provided care to the student. The remaining five cases were linked to close contact with the nurse.

A full listing of reported cases and dates may be obtained at the WHO SARS website at http://www.who.int/csr/sars/archive/en.

THE REMAINING QUESTION: THE HOST

Because seven of the 11 index cases had had contact with wild animals, many researchers have investigated wild animals as possible natural hosts for the virus. The Pearl River Delta region, where the 2003 outbreak originated, has enjoyed rapid economic development since the late 1970s, leading to the adoption of culinary habits requiring exotic animals such as ducks, rabbits, wild raccoon dogs, scorpions, and civets. The civet has been strongly suspected to be the natural host of SARS-CoV after WHO officials isolated a very similar virus from civets caged in live-animal markets and restaurants in Guangzhou in January 2004. A separate research team was also able to isolate a virus similar to SARS-CoV from wild civets in northwestern China. When this research was made public, the Chinese government banned the sale of civets at animal markets, and in early 2004 there was mass extermination of civets in China.

However, researchers and WHO officials insist that there is no conclusive evidence that the civet is the natural host for the disease. There have been prob-

lems with reproducing the experiments that isolated SARS-CoV from the civets. Also, other coronaviruses have been isolated from the civets. In June 2003,
one research group with members from the University of Hong Kong and the
CDC isolated a SARS-CoV-like virus from civets in live-animal markets in
Guangzhou. The viral RNA isolated from the market animals was 29 nucleotides longer than the RNA isolated from human SARS patients. This has led
to speculation that SARS-CoV became more adept at infecting humans after
losing a piece of its genome. Furthermore, a competing research group from
the Chinese Agricultural University in Beijing claimed that they had isolated a
different coronavirus from civets bought at live-animal markets in Guangzhou
during the same time period as in the previous group's study. The sequence of
that coronavirus was said to be only 77% similar to the SARS-CoV isolated
from humans.[23]

The exact role of the civet in the SARS outbreak is thus unclear. Some researchers strongly suspect that it is the natural reservoir for the disease; others
argue that the civet is simply a clue that will lead to the true natural host. The
general belief among researchers is that the live-animal markets provide the essential arena in which animals and humans can swap infectious agents. Also,
because of the novel RNA sequence in SARS-CoV, it has been suggested that
the virus evolved for a long time before crossing the species barrier to infect
humans. Thus, some researchers believe that the civet is not a crucial part of
the natural life cycle of SARS-CoV, but that the civets acquired the virus from
another animal in the markets or holding facilities. Studies support the hypothesis that SARS-CoV has an animal origin. IgG antibody for the virus was detected in a statistically higher percentage among animal traders (9.1% in animal
traders vs. 1.2% for non–animal traders).[33] Other studies showed that ferrets,
civets, and domestic cats are susceptible to SARS-CoV infection,[13] and that
the virus can be transmitted to uninfected animals housed near them.[21] Still
another theory has proposed that it was the infected humans who transmitted
SARS-CoV to the civets and other animals.[19] The key point is that currently
there is no known natural host despite much research. At present, the involve-
ment of the civet is the best clue that researchers have in their search for the
natural reservoir of SARS-CoV.

EFFECT OF THE OUTBREAK
ON GLOBAL RESEARCH EFFORTS

Although the SARS outbreak did result in over 8000 cases worldwide, the biodefense system can be credited for successfully containing the epidemic. The
outbreak functioned as a real test for a new global network of disease-fighting
organizations that had been established in 2000.

In 2000, coordination efforts led to a meeting in Geneva with representatives from 67 institutions—including laboratories, universities, and charities—to create the Global Outbreak Alert and Response Network (GOARN),[6] the purpose of which was to direct the activities in future disease outbreaks. Prior to the establishment of the GOARN, typically a multitude of agencies would join the fray when there was a disease outbreak; there would be little or no coordination among the agencies because no one was in charge. As a result, the scene of a disease outbreak was chaotic and often characterized by rivalry among different groups. As Ray Arthur, associate director of the CDC, stated, "It was clear that there had to be a better way of doing business."[6]

The 2003 SARS outbreak gave the WHO and the GOARN credibility as the directors of global outbreak response. When the GOARN was established in 2000, many saw it as a bunch of well-educated administrators incapable of a hands-on approach. The SARS outbreak allowed the GOARN and the WHO to prove their effectiveness in a dire situation. They implemented classic isolation and quarantine procedures that were successful in containing the disease. The GOARN also utilized modern communication technologies so that laboratories worldwide could simultaneously investigate clinical samples in real time. This resulted in the isolation of the causative agent and the development of a series of diagnostic tests for the disease. Thanks to the SARS experience, the GOARN also built up an intensive working relationship with such countries as China, Vietnam, and Thailand with whom they had never collaborated in the past. Additionally, the SARS episode marked the beginning of a much stronger presence of the WHO in the media. When the SARS outbreak erupted, teleconferences were held almost daily, and the WHO's formerly dull website was continually updated with fresh information. The SARS outbreak proved that the GOARN and the WHO could coordinate the containment of a global disease outbreak.

MOLECULAR BIOLOGY

As mentioned above, SARS-CoV is a novel member of the family *Coronaviridae* and the causative agent of SARS.[18] Coronaviruses are large, enveloped, positive-sense, single-stranded RNA viruses.[2,32] The size of the coronavirus genome ranges from 27 to 32 kb. Once in host cells, the virus replicates in the cytoplasm.[9] The morphology of coronavirus, which is best depicted by electron microscopy, displays "pleomorphic spherical or elliptical virions";[25] the surface is covered with projections, narrow bases, and club-shaped ends that give the virus its distinctive crown-like appearance. Virions measure between 100 and 140 nm in diameter. Surface projections extend a further 20 nm from the surface.

Coronaviruses are divided into three groups based on serological cross-reactivity and genomic sequence homology.[16] Group I includes viruses such as canine coronavirus, feline infectious peritonitis virus, porcine transmissible gastroenteritis virus, porcine respiratory virus, and human coronavirus 229E. Group II includes bovine coronavirus, murine hepatitis virus, rat sialodacryoadenitis virus, and human coronavirus OC43. Group III includes avian viruses such as avian infectious bronchitis virus and turkey coronavirus.

CORONAVIRUSES IN HUMANS

Coronaviruses are highly species-specific. In animals, coronaviruses can lead to highly virulent respiratory, enteric, and neurological diseases, as well as hepatitis, with short incubation periods. In immunocompetent hosts, infection elicits neutralizing antibodies and cell-mediated responses that kill infected cells. Many coronaviruses such as mouse hepatitis virus and avian infectious bronchitis virus cause fatal systemic diseases. These viruses can replicate in the liver, lung, kidney, stomach, spleen, brain, spinal cord, retina, and other tissues.

SARS-CoV is the first coronavirus shown to regularly cause severe disease in humans.[3] Before the appearance of SARS-CoV, human coronaviruses (HCoVs) were associated only with mild diseases. HCoV-229E (group I) and HCoV-OC43 (group II) are associated with the common cold and cause mild respiratory illnesses.[14] They occasionally cause serious infections of the lower respiratory tract in children and necrotizing enterocolitis in newborns. Human coronaviruses, including SARS-CoV, are able to survive on environmental surfaces for up to 3 hours. HCoVs can be transmitted from person to person by droplets, hand contamination, fomites, and small-particle aerosols. The transmissibility of a human coronavirus coupled with the fact that it causes severe disease makes SARS-CoV a potent infectious agent.

THE SARS-COV GENOME

In April 2003, the Michael Smith Genome Sciences Centre in Vancouver and the National Microbiology Laboratory in Winnipeg were the first to complete the genome sequencing of SARS-CoV. Two days later, the CDC also completed the genome sequencing. The SARS-CoV genome contains five major open reading frames (ORFs) that encode five proteins: the replicase polyprotein (rep), the spike (S), the envelope (E), the membrane glycoprotein (M), and the nucleocapsid protein (N).[27]

The largest of the five ORFs, occupying 21.2 kb of the genome, is the *rep* gene. It encodes an RNA-dependent RNA polymerase that, along with a helicase, is responsible for replication of the genome and for the formation of transcripts needed in the synthesis of the viral proteins. The S protein, a 1225-amino-acid glycoprotein, has a surface projection that is inserted into the endoplasmic reticulum (ER)–Golgi intermediate compartment. The main function of the S protein is to bind species-specific host-cell receptors and to trigger a fusion event between the viral envelope and the cellular membrane of the host cell. The S protein is the principal viral antigen that elicits neutralizing antibody on behalf of the host. Mutations in the S protein have previously been associated with an alteration in the virulence and pathogenesis of other coronaviruses. The E protein consists of 76 amino acids that give the viral envelope its structure. Researchers have speculated that the E protein may not be required for viral infectivity. However, the E protein has been found to be vital for virus replication in other coronaviruses, such as the porcine transmissible gastroenteritis virus. The M gene encodes a membrane glycoprotein that consists of 221 amino acids. The M protein is the major component of the virion envelope. It is also the primary determinant of virion morphogenesis, selecting the S protein for incorporation into virions during viral assembly. Further evidence suggests that the M protein also selects the genome for incorporation into the virion. The N protein consists of 422 amino acids and is similar to other nucleocapsid proteins found in other coronaviruses. The N protein gathers with positive strands of RNA that have been fully replicated. This complex of RNA and proteins is later joined by the M proteins, which are inserted into the ER membrane. Viral particles are formed while the N-RNA complex buds into the ER of the lumen. The virus then migrates along the Golgi complex and exits the host cell by exocytosis.[19,26]

The organization of the SARS-CoV genome shows characteristic features of coronaviruses. In general, the 5′ half of the genome appears to be related to group II coronaviruses while the 3′ half seems more closely related to group III coronaviruses. This suggests that a recombination event may have occurred. The genome has a 29,751-nucleotide sequence with 13–15 ORFs. The discrepancy in the number of ORFs is due to two untranslated regions (UTRs) that may contain additional ORFs.

Examination of the SARS-CoV genome as a whole shows that nucleotides 1–72 contain a RNA leader sequence preceding a UTR spanning 192 nucleotides. Two overlapping ORFs (1a and 1b) downstream from the UTR encompass about two-thirds of the genome (nucleotides 265–21,485). The remaining one-third of the genome contains the S, E, M, and N structural proteins. The 3′ region also encodes additional accessory genes. The 3′ end of the genome (after the structural proteins) contains a second UTR of 340 nucleotides followed by the poly A tract. The 3′ UTR contains a 32-nucleotide stem-loop motif (a group II characteristic) also found in avian infectious bronchitis virus, which is in group III. Transcription-regulatory sequences (TRSs) are found at the 3′ end of the leader RNA and usually precede each translated ORF.

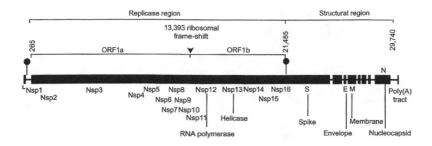

NSP=Nonstructural proteins

FIGURE 9.1. The SARS-CoV genome is composed of two main regions: the replicase region and the structural region. The figure shows the two large open reading frames (ORFs) of the replicase region and the respective positions of the leader sequence and the 3′ poly A tract. Filled circles: the nine transcription regulatory sequences specific to SARS-CoV. *Source*: Modified from Ref. 27.

The accessory genes found in the 3′ region are interspersed between the well-characterized structural genes. The accessory genes are a series of ORFs of unknown function. These ORFs are probably involved in viral RNA replication. It has also been suggested that deletions of some of these accessory genes result in a reduction in the virulence of the virus. Conceivably, these ORFs could be responsible for the high virulence of SARS-CoV.

THE REPLICATION CYCLE

Like other coronaviruses, SARS-CoV replicates in the cytoplasm of host cells. The replication cycle is outlined in detail below:

1. Virions bind to the plasma membrane of the host cell by interaction with specific glycoproteins.
2. Penetration occurs by S protein–mediated fusion of the viral envelope with the plasma membrane of the host cell.
3. The RNA genome is then released into the cytoplasm of the host cell, where replication takes place.
4. The translation machinery of the host translates the overlapping ORF1a and ORF1b by a ribosomal frame-shifting mechanism to produce a single polyprotein.

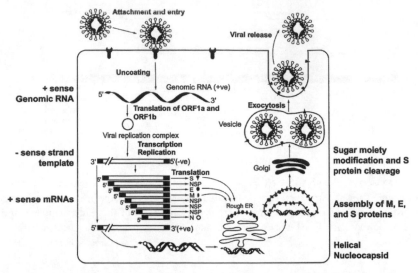

FIGURE 9.2. The SARS-CoV replication cycle: the general flow of genetic information is from positive-sense genomic RNA to negative-sense strand templates to positive-sense mRNAs to positive-sense genomic RNA. The negative-sense strand templates are produced during a discontinuous transcription process regulated by the nine TRSs in the viral genome. These discontinuously synthesized minus strands serve as templates for the positive-sense mRNAs. The mRNAs will then be translated to form the protein products necessary for viral assembly.   *Source: Ref. 27.*

5. Cleavage by virally encoded proteinases yields the components that are necessary to assemble the viral replication complex, which synthesizes full-length negative-strand RNA.

6. A discontinuous transcription strategy during negative-strand synthesis produces a nested set of subgenomic negative-sense RNAs.

7. In this process, the TRS is postulated to fuse the 3′ ends of the nascent subgenomic minus strands to the antisense leader sequence.

8. These minus strands then act as templates for the synthesis of positive-sense mRNAs.

9. The N protein and the genomic RNA assemble in the cytoplasm to form the helical nucleocapsid.

10. This core structure acquires its envelope by budding through intracellular membranes between the ER and the Golgi apparatus.

11. The M, E, and S proteins are transported though the ER to the budding compartment, where the nucleocapsid interacts with the M protein to trigger assembly.
12. During the transport of the virus through the Golgi apparatus, sugar moieties are modified and (usually) the S protein is cleaved into S1 and S2 domains.
13. Virions are incorporated into vesicles for transport.
14. Any S protein that is not incorporated into the virion is transported to the cell surface.
15. Finally, the virus is released from the host cell by fusion of virion-containing vesicles with the plasma membrane.

To summarize the SARS-CoV replication cycle, the general flow of genetic information is as follows:

Positive-sense genomic RNA → Negative-sense-strand template → Positive-sense mRNAs → Positive-sense genomic RNA

PHYLOGENY

The important question following the completion of the SARS-CoV genome is whether this new virus represents a completely new group of coronavirus, a variant of one of the three known groups, or a combination of these groups.

Phylogenetic analysis based on the first 300 (of 405) nucleotides of the highly conserved polymerase gene indicates that SARS-CoV is distinct from the three known groups and should be in its own group, group IV.[19,24] Amplification of the polymerase gene using reverse transcriptase polymerase chain reaction (RT-PCR) also suggests that SARS-CoV should be in its own group.[17]

Phylogenetic analysis of only the ORF1b, the most conserved region in the SARS-CoV genome, indicates that SARS-CoV represents an early split-off from group II. Using a consensus sequence, phylogenetic analysis of less conserved regions, such as the S, M, and N gene sequences, showed a statistically significant relationship with group II, suggesting a shared common ancestor. This study also showed that 19 of 20 cysteine residues in the S1 domain of the SARS-CoV S protein are spatially conserved when compared with group II.

A majority of scientists believe that SARS CoV is a new coronavirus that, although closely related to group II, should be the member of a new group, IV. Others place SARS-CoV into group II. This gives the two different phylogenetic trees shown in figure 9.3.

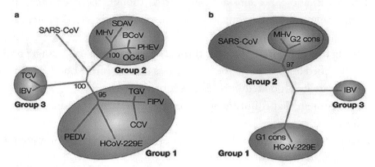

FIGURE 9.3. Phylogenetic trees showing the competing theories of how to group SARS-CoV as a coronavirus. (**a**) An unrooted tree based on the well-conserved polymerase protein sequence is shown. This approach shows SARS-CoV as the member of a new fourth group of coronaviruses. The tree was constructed using the protein sequences of the RNA-dependent RNA polymerase of other coronaviruses. (**b**) This tree was obtained using the sequence of the S1 domain of the spike (S) protein. Consensus sequences were generated from the three known groups of coronaviruses and SARS-CoV. This tree shows SARS-CoV as a member of group II coronaviruses, such as mouse hepatitis virus (MHV).

Source: Ref. 27.

CLINICAL ISOLATES

The clinical isolates of SARS-CoV have interesting differences. The GZ01 strain of SARS-CoV originates from Guangdong Province, where the 2003 outbreak began. Although all other SARS-CoV genomes lack a 29-nucleotide sequence in the 3′ domain of ORF8a, this 29-nucleotide sequence is present in GZ01. SARS-CoV-like strains isolated from mammals in China have been found to contain the same 29-nucleotide sequence.[5] This observation supports the hypothesis that the 29-nucleotide deletion observed in most human isolates could have increased the fitness of the virus in human hosts and allowed its spread to the human population.

Another strain of SARS-CoV isolated from an early case in Guangdong Province contains an 82-nucleotide deletion. An identical 82-nucleotide deletion was found in coronaviruses isolated from farmed civets and other animals in China.[5] This supports the theory that early human infection originated from wild animals.

Although the overall organization of SARS-CoV is similar to that of the known groups of coronaviruses, amino acid conservation is low. This has implications for phylogenetic analysis. RNA viruses usually have a high rate of mutation because they lack a proofreading mechanism. Since there are only a few clinical isolates of SARS-CoV, it is too early to draw any definitive conclusions about the mutation rate of the virus. However, two recurrent amino acid substitutions in the S1 domain of the S protein have been reported.[27] A glycine-to-asparagine substitution was found in the Hong Kong index case group, and there was an isoleucine-to-threonine substitution in the mainland isolates. It has been speculated that these mutations represent adaptations to the new host or its immune response. Analysis of 61 SARS-CoV genomic sequences from each phase of the epidemic indicates that the earliest genotypes were similar to the animal SARS-CoV-like virus. The S protein shows the strongest initial responses to positive selection pressures, followed by subsequent purifying selection and stabilization.

In light of the molecular biology of SARS-CoV, an important question remains: why was there no SARS outbreak during the winter and spring of 2004? There has been speculation that the ability of SARS-CoV to cross the species barrier could have been the result of a rare series of events that is unlikely to be repeated. In general, coronaviruses are not considered likely to undergo rapid evolution and infection of a new host species. SARS-CoV disproved that line of thinking. The effects of the 2003 outbreak have the world hoping that this crossing of the species barrier will not happen again.

CLINICAL PRESENTATION AND FINDINGS

During the 2003 outbreak, SARS infected a wide range of people, but there were some notable trends. The largest group of people infected were health care workers. They made up 21% of all cases, as the super-spreader events occurred most frequently in hospitals and health care centers. The disease tended to infect more females than males—the male-to-female ratio was 0.87. Although rare, asymptomatic cases of SARS were also reported. SARS-CoV RNA was detected in four asymptomatic quarantined people in Hong Kong, but the virus could not be isolated.

PROGNOSIS

The major prognostic factors in SARS are as follows: age, coexisting illness (especially diabetes and heart disease), increased lactate dehydrogenase

(LDH) level, high neutrophil count at the time of admission, and low lymphocyte counts. The overall death rate of SARS patients is 14–15%. The rate is less than 1% in people 24 years old or younger, 6% in people 25–44 years old, 15% in people 45–64 years old, and 50% in people 65 years or older. In general, the illness is milder in children and more severe in the elderly.

CLINICAL COURSE

The incubation period for SARS usually lasts 2–10 days but has been reported to last as long as 20 days. The severity of the disease ranges from mild flu-like symptoms to respiratory failure and death. The symptoms at initial exam have varied, but they usually include a high fever, chills, cough, myalgia, and malaise. After the onset of the illness, a lower-respiratory-tract condition with dry coughs or dyspnea ensues. Intubation and mechanical ventilation are required in approximately 10–20% of cases.[30] In many patients a chest radiograph will show focal infiltrates that progress into patchy, generalized interstitial infiltrates. The total lymphocyte count decreases during the first phase of the disease whereas the white-cell counts are usually normal or below normal. Some patients also show elevated levels of creatine phosphokinase and hepatic transaminase. In general, the disease has three phases: viral replication (phase 1), immune hyperactivity (phase 2), and pulmonary destruction (phase 3).

In phase 1, the patient presents with flu-like symptoms. The symptoms generally improve after a few days, and then the fever ends. At that point it is unclear whether the patient has actually recovered or will relapse with more severe symptoms. Phase 1 is also characterized by an increasing viral load due to viral replication and cytolysis.

Phase 2 begins approximately 8 days after the onset of symptoms. At around day 9, there is a recurrence of fever with onset of diarrhea and oxygen desaturation. At this point, the pneumonia progresses to a site other than that of the original presentation. IgG seroconversion correlates with a fall in the viral load between days 10 and 15. There is also a decrease in viral shedding and an increase in immune-mediated lung damage between days 10 and 21. At the end of phase 2, the patient's condition severely worsens.

Phase 3 involves pulmonary destruction. Only 20% of patients who initially present with SARS progress to this phase.[30] Patients who have reached this phase are now diagnosed with chronic acute respiratory distress syndrome (ARDS), meaning that they require ventilatory support. These patients suffer from end-organ damage and severe lymphopenia.

CASE DEFINITION AND DIAGNOSIS

Because there is no defining clinical characteristic of SARS, diagnosis is based on both clinical symptoms and contact history. Although the virus is found in fecal and nasal swabs as well as in blood and serum samples, it is difficult to isolate safely from an infected patient. The WHO definition of a suspect case of SARS is as follows:

Fever is 38°C (100.4°F) or above *and*

- Lower-respiratory-tract symptoms such as coughing or difficult breathing are present *and* one or more of the following:
- The patient has been in contact with a person believed to have SARS
- The patient has traveled to an area where transmission of SARS has been reported
- The patient resides in an area where transmission of SARS has been reported

or, retrospectively:

- The patient died from an unexplained acute respiratory illness after November 1, 2002, without an autopsy *and* one or more of the following exposures prior to the onset of symptoms:
- The patient had close contact with a person believed to have had SARS or
- The patient traveled to an area with recent local transmission of SARS or
- The patient resided in an area with recent local transmission of SARS

The WHO defines a probable case of SARS as follows:

- A suspect case with radiographic evidence of infiltrates consistent with pneumonia or respiratory distress syndrome (RDS) on chest x-ray
- A suspect case of SARS that is positive for SARS-CoV by one or more laboratory assays
- A suspect case with autopsy findings consistent with the pathology of RDS without an identifiable cause

The WHO has also declared that a case should be excluded from diagnosis as SARS if an alternative diagnosis can fully explain the patient's illness.[29]

In general, diagnosis of SARS is difficult. The disease has nonspecific symptoms, and a proper diagnosis relies on the patient's giving an accurate contact history. Additionally, there is no single diagnostic test that can definitively confirm the diagnosis of SARS. SARS-CoV RNA can be detected by RT-PCR and confirmed by a second PCR test. However, during the 2003 outbreak, PCR lacked enough sensitivity during the first 5 days of illness. Thus, negative PCR results do not mean that a physician can rule out infection by SARS. Another diagnostic test employed during the 2003 outbreak was enzyme-linked immunosorbent assay (ELISA), used to detect SARS-CoV antibodies. However, antibodies can be detected only a week after the onset of symptoms, making this a less than ideal way to confirm diagnosis of SARS. The most effective means of confirming the diagnosis of SARS in the laboratory involves the use of tissue culture.

A consequence of the difficulty of SARS to diagnose was that many physicians, nurses, and other health care workers contracted the disease while attempting to diagnose and treat infected patients. High-risk activities such as intubations caused many health care workers to be infected.[22] This resulted in super-spreader events within the hospitals and health care centers themselves. The problem was that physicians had to achieve early case detection to successfully contain the SARS outbreak. It eventually became clear that it was absolutely necessary to isolate all SARS patients (even suspected cases), and all health care workers had to exercise postexposure precautions (e.g., degowning) after treating any SARS patient.

The CDC established guidelines for the collection of specimens from potential victims of SARS. The guidelines highlight the three main sources of SARS-CoV for laboratory testing an stress the importance of collecting specimens from suspected SARS patients as quickly as possible after the onset of symptoms. However, one can also see how this need for early detection and quick collection resulted in health care workers' contracting the disease themselves. The CDC guidelines are listed below:

■ Respiratory-tract specimens: these should be collected as soon as possible because the possibility of recovering viruses starts to diminish substantially 72 hours after the onset of symptoms. Three types of specimens can be collected for viral isolation or PCR: nasopharyngeal wash or aspirates, nasopharyngeal swabs, or oropharyngeal swabs.

■ Blood collection: serum specimens should be collected and submitted to the laboratory as soon as possible. At least 5 mL of whole blood should be collected to obtain a serum sample of at least 200 μL. Whole blood can also be used for laboratory analysis.

■ Stool samples: stool samples can be collected, although the first two types are preferred.

TREATMENT

TREATMENTS USED IN THE 2003 OUTBREAK

There are no recommended treatments for SARS other than supportive care. However, during the 2003 outbreak, patients were treated empirically with antiviral drugs, antibodies, and steroids. The antiviral drugs ribavirin and lopinavir-ritonavir were the drugs most frequently administered.[25] As an RNA virus mutagen, ribavirin is a widely used antiviral drug, but its effect against SARS-CoV is unknown. Lopinavir-ritonavir is a protease inhibitor used to treat the human immunodeficiency virus (HIV). It blocks proteases, preventing the viral replicase polyprotein from replicating the viral RNA. A convalescent serum of antibodies used in hospitals in Hong Kong was believed to curb increase in the viral load.[28]

Patients throughout the world were also treated with corticosteroids. Treatment with steroids caused a decrease in fever, a resolution of radiographic infiltrates, and an improvement in oxygenation. Unfortunately, large doses of steroids can cause avascular necrosis, a debilitating bone disease. The steroids can also have an immunosuppressive effect.[15] Moreover, they facilitate coronaviral replication and invite bacterial sepsis and opportunistic infections.

TREATMENTS USED EXPERIMENTALLY IN ANIMAL MODELS

Interferons—cytokines important in intercellular immune response—were experimentally administered to macaques (monkeys) as a possible SARS treatment.[7] Interferon-α blocks replication of several viruses and activates the immune system. Administered to the macaques 3 days before infection with SARS-CoV, interferon-α caused the macaques to excrete less virus from their throats. Lung damage was also reduced by 80%. Interferon-β was separately administered to the macaques. It showed to be an even more potent inhibitor of SARS in vitro than interferon-α; interferon-β had a selectivity index 50–90 times higher. Glycyrrhizin, an extract of licorice roots, was also administered experimentally in macaques. The drug rids the monkeys' kidney cells of SARS-CoV by inhibiting absorption, penetration, and replication of the virus. This was confirmed in human cell lines as well, but the drug only works in very high doses.

TRANSMISSION

In October 2003, more than 30 SARS researchers attended a meeting of the WHO SARS Scientific Research Advisory Committee.[31] Some of the conclusions drawn from that meeting are:

■ Regarding the level of preparedness for another possible SARS outbreak, sporadic cases of SARS might not be detected early. Cases with mild or atypical symptoms might initially be missed because at least half the world's population has no access to hospital services such as chest x-rays. This clearly shows that the global alert system has its shortcomings.

■ SARS was able to spread quickly in the modern world, as demonstrated in the Metropole hotel incident, in which one person seeded an international outbreak. There is fear that this could happen again.

■ In poor countries with limited resources, the health infrastructure might not be able to cope with the demands of the disease.

■ The diagnosis of SARS is difficult to achieve and laboratory tests are limited. No "gold standard" was determined for laboratory diagnosis. Another problem with laboratory testing is the potential for cross-reaction of SARS-CoV with other human coronaviruses, such as human coronavirus 229E.

■ Regarding the evolution of SARS, a few coronaviruses have crossed from one species to another. Animal studies also showed that coronaviruses can spread mechanically. Specifically, it is possible that transmission among the residents of the Amoy Gardens apartments in China was effected via mice that had had contact with infected human fecal matter (G.-P. Zhao, personal communication, April 9, 2004).

■ In the laboratory, SARS-CoV should be cultured under CDC Biosafety Level 3.[4] Diagnostic activities that do not involve live virus should be carried out under Biosafety Level 2.

■ Cultures of SARS-CoV should be stored at a minimum of Biosafety Level 3. National governments are advised to license and register all laboratories studying or storing SARS-CoV.

WEAPONIZATION

Although considered a Category C list emerging pathogen by the CDC, SARS-CoV has the potential to be used in a bioterrorist attack. The following analysis looks at SARS-CoV as a potential bioweapon from multiple perspectives. First, there is an examination of how and why a bioterrorist might choose SARS-CoV as his means of attack. Then, in a response to the concerns, it is explained why the CDC and other governmental organizations do not view SARS-CoV as a likely choice of terrorists to use as a bioweapon. Although not every possible scenario for the use of SARS-CoV as a bioweapon is examined, this format allows for an in-depth examination of its potential.

THE BIOTERRORIST PERSPECTIVE

A bioterrorist might employ the one or more of these methods:

- Infect himself with SARS-CoV, then travel on multiple airplanes or otherwise put himself into contact with many people
- Contaminate food or water sources with SARS-CoV
- Disseminate SARS-CoV in an aerosolized form

His selection of SARS-CoV as his weapon might be based on the following characteristics:

- As demonstrated in the January 2003 outbreak in Guangzhou and the consequent outbreaks worldwide, SARS is highly contagious and very susceptible to super-spreader events in which one index patient can directly or indirectly infect over 100 people.
- Practicing proper quarantine and disease-containment procedures is difficult in the age of frequent global travel.
- It appears that, with the proper equipment, SARS-CoV can be readily obtained from civets or other animals in Chinese live-animal markets.
- There is no proven effective treatment against SARS.
- If the first cluster of SARS cases appeared in a place such as the United States where the disease had rarely been seen in the past, the cases would most likely not alert officials to the presence of a potent infectious disease because of the similarity of the symptoms to those of other diseases. This would give a possible pandemic a head start on health organizations.
- Although SARS has an average fatality rate of only 14–15%, the contagiousness of the disease would cause social and economic strain by wiping out the work force. Conceivably, the cost of canceled travel, heightened security, containment, and medical care would be in the billions of dollars. During the 2003 outbreak, the equivalent of approximately $100 billion was lost in southeast Asia alone as a result of canceled travel and declining investments in only 5 months.[1]
- If bioterrorists infected themselves with SARS in the hope of spreading the disease, they might not die, and they might not even be suspected of acting as terrorists. They could claim to have traveled to certain regions in China, and their illness would simply be explained as examples of the SARS cases that seem to mysteriously arise from that area of the world.
- SARS would be a greater threat to world leaders because they are generally older, and SARS has a greater fatality rate in people over age 65.

■ Less attention is being paid to SARS as a threat since it did not make a strong reappearance in 2004. For example, like many other officials, Stephanie Factor, an epidemiologist with the Bioterrorism Preparedness Response Program of the CDC, has said she is "not too worried" about a bioterrorist attack using SARS-CoV (S. H. Factor, personal interview, March 30, 2004).

■ Similarly, little work has been done to find an effective treatment since the burst of research in the wake of the 2003 outbreak. There are confusion and disagreement about the likelihood of another outbreak and whether time and money should be devoted to the study of SARS-CoV as an imminent threat to global health.

■ The origin of the virus is still unknown. Klaus Stohr, the WHO virologist who coordinated a network of SARS research laboratories until August 2003, has said that there has been "virtually no progress"[8] in the hunt for the origins of SARS-CoV.

THE GOVERNMENT PERSPECTIVE

■ Researchers and epidemiologists believe that the ability of SARS-CoV to cross the species barrier from animals to humans was the result of a rare series of events that is unlikely to be repeated. The high mutation rate of SARS-CoV is cited as evidence of this. It has been suggested that a 29-nucleotide sequence deletion observed in most human isolates was responsible for the virus' escaping human host defenses, and that this one-time event caused the virus to be spread within the human population.

■ The palm civet is neither the natural reservoir nor a reliable source of SARS-CoV. Thus, a potential bioterrorist would need to obtain the virus from a governmentally supervised research laboratory where the virus is studied under Biosafety Level 3.

■ It is unknown whether disseminating SARS-CoV in an aerosolized form would be effective. The viral load necessary for human infection is not known, and the size of the droplets by which the disease spreads is suspected to be quite large.

■ Although it is unknown whether the virus can be contracted through food or water, the connection with the Chinese live-animal markets and restaurants makes it seem likely. Yet it is unlikely that a terrorist would spend time and money on a plan to infect large amounts of food and water if the efficacy of this is unknown.

■ Considering the limited availability of the virus, the many unknowns, and the ample number of "better" choices, it is unlikely that a bioterrorist would choose SARS as his weapon.

POTENTIAL DEFENSES

Multiple vaccines already exist against animal coronavirus, and this gives hope for the development of a vaccine against SARS-CoV. The candidates for such a vaccine are inactivated whole virus, genetically engineered adenovirus containing genes from the SARS virus, and a mix of recombinant SARS protein. Advances have been made in vaccine development in animal models, but there is a dispute over the best animal model for the SARS vaccine: is it macaques or mice? Macaques develop SARS-like symptoms when injected with the virus, but because they are not inbred there is difficulty in reproducing data. Mice give consistent experimental results and the virus replicates, but the mice show no signs of clinical disease.

In early 2004, the National Institutes of Health tested a DNA vaccine in a mouse model.[34] The vaccine encoded the S glycoproteins of SARS-CoV. It induced T-cell and neutralizing-antibody responses and reduced viral replication in the lungs of the mice. A group at the University of Pittsburgh also developed a vaccine but tested it in a monkey model.[12] The Pittsburgh vaccine was an adenovirus with the SARS-CoV S1 fragment of the S protein as well as the M and N proteins. The vaccine induced T-cell and neutralizing-antibody responses in the monkeys.

There are legitimate concerns about vaccine development especially since it is unknown whether there will be another SARS outbreak. It has been debated whether it is too soon for human trials of SARS vaccines. Interestingly, Sinovac Biotech of Beijing has already tested a vaccine with an inactivated form of SARS-CoV in a human trial, which has led to some scientific dispute. There have already been multiple cases of SARS resulting from laboratory contamination, and it is believed that vaccine development might enhance the contagiousness and lethality of SARS-CoV. It has even been suggested that the vaccine may predispose humans to an accelerated form of the disease.[20] The development of live vaccine strains with defined mutations has definite risks because the virus probably has the ability to evolve rapidly. Theoretically, a live vaccine strain of SARS-CoV could mutate and cause an even more potent condition.

Will SARS return? This is the ultimate question. Perhaps the 2003 SARS outbreak was truly a one-time event in which a coronavirus crossed the species barrier and caused severe disease in humans. Then again, perhaps the disease will re-emerge, either naturally or as an act of bioterrorism. Although it seems unlikely that a bioterrorist would choose SARS-CoV as his weapon, it is difficult to know how the mind of a terrorist works. Either way, the world should be ready to see SARS again.

REFERENCES

1. Abbott, A. 2003. SARS: What have we learned? Are we prepared for the next viral threat? *Nature* 424: 121–126.

2. Anand, K., et al. 2003. Coronavirus main proteinase (3CL^pro) structure: Basis for design of anti-SARS drugs. *Science* 2003. 300: 1763–1767.

3. Bradburne, A. F. and D. A. J. Tyrrell. 1971. Coronaviruses of man. *Progr Med Virol* 13: 373–403.

4. Centers for Disease Control and Prevention. 2003. *Laboratory Testing and Specimens.* Online: http://www.cdc.gov/ncidod/sars/lab.htm.

5. Chinese SARS Molecular Epidemiology Consortium. 2004. Molecular evolution of the SARS coronavirus during the course of the SARS epidemic in China. *Science* 303: 1666–1669.

6. Enserink, M. 2004. A global fire brigade responds to disease outbreaks. *Science* 303: 1605–1606.

7. Enserink, M. 2004. Interferon shows promise in monkeys. *Science* 303: 1273–1274.

8. Enserink, M. and D. Normile. 2003. Search for SARS origins stalls. *Science* 302: 766–767.

9. Fields, B. N., et al. 2001. In D. M. Knipe and P. M. Howley, eds. *Fields Virology,* 4th ed. Philadelphia: Lippincott Williams and Wilkins.

10. Fouchier, R. A., et al. 2003. Koch's postulates fulfilled for SARS virus. *Nature* 423: 240.

11. Gandley, A. 2003. Is it flu or SARS? MDs gear up for a difficult winter. *Can Med Assoc J* 169(8): 821.

12. Gao, W., et al. 2003. Effects of a SARS-associated coronavirus vaccine in monkeys. *Lancet* 362: 1895–1896.

13. Guan, Y., et al. 2003. Isolation and characterization of viruses related to the SARS coronavirus from animals in southern China. *Science* Express. Online: http://www.scienceexpress.org/4september2003/page1/10.1126/science.1087139.

14. Holmes, K. V. and L. Enjuanes. 2003. The SARS coronavirus: A postgenomic era. *Science* 300: 1377.

15. Janeway, C. A., et al. 2001. *Immunobiology: The Immune System in Health and Disease,* 5th ed. New York: Garland.

16. Kamps, B. S. and C. Hoffman, eds. 2003. *SARS Reference: 10/2003,* 3rd ed. Flying Publisher. Online: http://www.sarsreference.com/index.htm. Accessed April 10, 2004.

17. Ksiazek, T. G., et al. 2003. A novel coronavirus associated with severe acute respiratory syndrome. *N Engl J Med* 348(20): 1953–1966.

18. Kuiken, T., et al. 2003. Newly discovered coronavirus as the primary cause of severe acute respiratory syndrome. *Lancet* 362: 263–270.

19. Marra, M. A., et al. 2003. The genome sequence of the SARS-associated coronavirus. *Science* 300: 1399–1403.

20. Marshall, E. and M. Enserink. 2004. Caution urged on SARS vaccines. *Science* 303: 944–946.

21. Martina, B. E. E., et al., SARS virus infection of cats and ferrets. Nature, 2003. 425: 915.

22. McDonald, L. C., et al. 2004. SARS in healthcare facilities, Toronto and Taiwan. *Emerging Infectious Diseases*, May. Online: http://www.cdc.gov/ncidod/EID/vol10no5/03–0791.html.

23. Normile, D. and M. Enserink. 2003. Tracking the roots of a killer. *Science* 301: 297–299.

24. Peiris, J. S. M., et al. 2003. Coronavirus as a possible cause of severe acute respiratory syndrome. *Lancet* 361: 1319–1325.

25. Rainer, T. H. 2004. Severe acute respiratory syndrome: Clinical features, diagnosis, and management. *Curr Opin Pulm Med* 10: 159–165.

26. Rota, P. A., et al. 2003. Characterization of a novel coronavirus associated with severe acute respiratory syndrome. *Science* 300: 1394–1399.

27. Stadler, K., et al. 2003. SARS—Beginning to understand a new virus. *Nature Rev Microbiol* 1: 209–218.

28. Wang, J. T., et al. 2004. Clinical manifestations, laboratory findings, and treatment outcomes of SARS patients. *Emerg Infect Dis* 10:818–824.

29. World Health Organization. *Case Definitions for Surveillance of Severe Acute Respiratory Syndrome*. Online: http://www.who.int/csr/sars/casedefinition/en.

30. World Health Organization. 2003. *SARS Case Fatality Ratio, Incubation Period*. Online: http://www.who.int/csr/sars/archive/2003_05_07a/en.

31. World Health Organization. 2003. *Situation Updates—SARS*. Online: http://www.who.int/csr/sars/archive/en.

32. Young, B., et al. 2003. Reverse genetics with a full-length infectious cDNA of severe acute respiratory syndrome coronavirus. *Proc Natl Acad Sci USA* 100(22): 12995–13000.

33. Yu, D., et al. 2003. Prevalence of IgG antibody to SARS-associated coronavirus in animal traders—Guangdong Province, China. *MMWR Morb Mortal Wkly Rep* 52(41): 986–987.

34. Zhao, P., et al. 2004. DNA vaccine of SARS-CoV S gene induces antibody response in mice. *Acta Biochim Biophys Sin* (Shanghai) 36: 31–37.

CHAPTER 10

■ ■

PLAGUE

(YERSINIA PESTIS)

Barbara Chubak

HISTORY

In October 1347, a Genoese fleet landed in the harbor of Messina, in northeast Sicily. Every member of its crew was dead or dying, afflicted with a mysterious disease from the East. Rumors of the pestilence had reached the major European seaports in previous months and the harbormasters quickly tried to quarantine the fleet, but to no avail. Rats from the ships carried the disease to Messina and its environs, where the pestilence spread: within 6 months, half the region's population either died or fled. This fleet, together with others carrying pestilence along trade routes to ports throughout Eurasia and North Africa, formed the vanguard of the Black Death.

THE BLACK DEATH

The Black Death is the greatest natural disaster in European history. It ravaged the Western world from 1347 to 1351, killing between 30% and 50% of Europe's population, and causing or accelerating major political, social, economic, and cultural changes. Few, if any, historians would deny the Black Death an important role in European history, but there is much debate over the nature and timing of this role. Some consider the Black Death to be the major turning point in the transition from medieval to modern Europe, while others regard it as but a part of the general economic and moral crises of the time. The most compelling histories of the plague straddle this divide, acknowledging the fundamental problems in European society before 1347 while concluding that the Black Death, especially its cyclic reoccurrence, was the primary impetus for change.

JUSTINIAN'S PLAGUE

Plague first struck Europe in 541, in an epidemic called Justinian's Plague after the Byzantine emperor who ruled at the time of the outbreak. It began in Egypt, and spread through most of the known world, including central and southern Asia, northern Africa, Arabia, and Europe. In Constantinople, the capital of the Byzantine Empire, the plague is thought to have killed 200,000 people, or 40% of the total population, within a 4-month period in 541 and 542. By the decline of Justinian's Plague in 544, between a fifth and a quarter of the European population had died. The political impact of the plague was enormous, halting Justinian's plans to reclaim the eastern part of the Roman Empire from its German conquerors, and contributing to the weakness that enabled the eventual defeat of Byzantium.[8]

Where plague was established in the time of Justinian, it recurred in 10- to 24-year cycles for the next 200 years; altogether, these periodic epidemics were the first plague pandemic. It has been estimated that this pandemic, which finally ended in the late eighth century and was largely confined to the Mediterranean region, was accompanied by an overall population loss of 50 to 60% from 541 to 700. There is little statistical evidence from the time, but it is likely that the epidemics disrupted trade routes, food production, and distribution, and generally added to the bleakness of the so-called Dark Age.

THE SECOND PANDEMIC

When the second pandemic began, with 5 years of Black Death, people were baffled, bewildered, and terrified. They struggled to make sense of the catastrophe, both to give meaning to so many deaths and to halt the contagion in its tracks. A combination of three kinds of plague—bubonic, pneumonic, and septicemic—the Black Death defied description. Witnesses could not agree on the symptoms of the pestilence, much less on a cause of it. What was certain was that it killed without regard to social status or personal virtue; some sought refuge in asceticism and some in hedonism, but none escaped the plague unscathed.

As in the first pandemic, where the plague appeared once, it recurred, in periodic outbreaks that continued through the seventeenth century. With the repeated outbreaks, patterns emerged, and theories of the disease abounded: it was caused by a poisonous miasma, polluted water, food poisoning, conformations of the planets, the will of God in retribution for sin. Its spread was blamed on those at the margins of society: vagrants, prostitutes, Jews, and heretics. It became axiomatic that plague was a disease of the poor, who were notoriously malnourished, overcrowded, and careless in the upkeep of both body and spirit, who could not afford to flee from the urban epidemics as became the habit of the wealthy.

News of the plague inevitably caused a collapse of employment and supply to cities. Desertion of the city by the wealthy led to a dependence on the unrespectable poor for the continuation of essential operations of urban life—specifically, in time of plague, fumigation, cleaning, and burial of the dead. With the urban population whittled away by flight and death, the social order overturned, and the economy suffering, it was clear that something had to be done.

THE ROLE OF GOVERNMENT

It became the duty of government to protect its subjects from disease, as well as from human enemies. Northern Italian city-states led the rest of Europe in adopting measures for dealing with the recurrence of plague. From the fifteenth century on, laws and institutions were permanently established to protect public health. Both shipping and overland commerce were controlled and occasionally banned; first the sick, and then the contacts of the sick, were isolated, either at home or in specially built "pest houses," to prevent contagion. The first boards of health were created, governing both emergency measures during epidemics and a general system of parochial poor relief, to mitigate two supposed causes of plague: poverty and divine wrath.

Our current understanding of plague—particularly the bubonic plague, which was its most common form—makes it clear that none of these measures could have been successful in dealing with that disease. But they would have been very effective in controlling other contagious diseases that were also common in early modern Europe, including dysentery, smallpox, and influenza.[2]

THE ROLE OF MEDICAL PERSONNEL

Doctors and physicians played little part in the development of these public health measures. Classical medicine proved inadequate in the face of the plague, as "many gallant gentlemen, fair ladies, and sprightly youths, who would have been judged hale and hearty by Galen, Hippocrates and Aesculapius ... having breakfasted in the morning with their kinsfolk ... supped that same evening with their ancestors in the next world."[1]

This failure to maintain the health of society brought medicine under intense criticism, which in turn stimulated new policies in medical education and practice. After the Black Death, physical science and surgery replaced philosophy at the center of medical school curricula, and surgeons joined university-trained physicians in the ranks of professionals. Hospitals adopted a commitment to curing the sick, rather than simply providing a place for them to die in isolation. Practical guides detailing diagnosis and treatment in the vernacular became extremely popular, allowing literate laypeople to take their health into their own hands. This new, often morbid, popular fascination with disease led people to consult with doctors more, so that physicians, surgeons, and apothecaries joined lawyers and wealthy merchants among the financial elite.[27]

All these changes brought the medical profession greater practical knowledge, skill, and esteem than it had previously had and paved the way for modern medicine.

THE EFFECT ON POPULATION

The consistent barrage of epidemics made demographic growth impossible, and the population of Europe declined steadily for at least a century after the Black Death. It was only in the mid-sixteenth century that Europe regained its thirteenth-century population levels, and by then years of diminished population had wrought drastic changes.

As urban populations were diminished by plague, they were replaced by country people eager to take advantage of economic opportunity. The consequent depopulation of rural areas, exacerbated by local disease, was fatal to the manorial system, which formed the structural basis of medieval society. Government, which temporarily collapsed in the wake of the Black Death, re-formed as centralized bureaucracies in order to rule most efficiently in the new, depopulated conditions. Cities and their denizens gained unprecedented social, economic, and political clout.

Mortality was particularly high among the clergy, many of whom fell ill and died in pursuance of their duties, which included comforting the sick and administering last rites. The church had provided Europe with its greatest patrons, scholars, and thinkers, so the clerical shortage had a massive cultural impact. Spiritual standards fell as the church hurried to fill vacant positions with priests who were ill prepared to take on their responsibilities. Education at every level was temporarily crippled by a lack of teachers, and the vernacular rapidly replaced Latin as the language of politics and scholarship.[27]

In sum, the second pandemic disrupted medieval social and political systems, and shook longstanding philosophical and religious convictions. The society that followed in its wake had new sources of wealth, new forms of authority, and new attitudes and ideas.

THE THIRD PANDEMIC

The third pandemic started in the Yunnan Province of China in 1855 and spread southward, reaching Hong Kong and Canton in 1894 and Bombay in 1898. By the turn of the century, the plague had spread via steamship to Africa, Australia, Europe, Hawaii, India, Japan, the Middle East, the Philippines, South America, and the United States, establishing many endemic foci and providing an early lesson in the potential dangers of globalization to human health. Statistics for India alone show that by 1903 plague was killing more than a million people each year, with more than 12 million deaths from 1898 to 1918.[26]

As horrifying as this death rate is, it represents a significant decrease in mortality from earlier epidemics. Nor was this the most destructive epidemic to afflict India in the late nineteenth and early twentieth centuries. Malaria and tuberculosis killed approximately twice as many people in about the same length of time, and in barely 4 months the influenza epidemic of 1918–1919 also killed twice as many. In addition, smallpox and cholera epidemics had death tolls numbering in the millions.[3]

No Indian epidemic evoked more fear than the plague. At first, the colonial state that governed India was reluctant to admit the existence of plague, insisting on calling it "fever plague" or "bubonic fever." When denial became impossible, a vigorous program of plague countermeasures was initiated, representing a dramatic and unprecedented intrusion of the colonial state into the private domain. The orders, which were often forcibly and violently executed by British soldiers, were to identify and isolate the sick and their contacts and then remove them to hospitals. Houses were disinfected and personal possessions burned. Bombay was "literally drenched in disinfectant solution," to such an extent that people carried umbrellas to avoid being sprayed.

The desperate zeal with which the colonial government attacked the plague stands in stark contrast to its passive, fatalistic attitude toward the other diseases that were epidemic in India. The government reaction can be attributed partly to the possible imperial consequences of plague: its threat to the commercial center of the British Empire and its international reputation. But antiplague measures remained in effect longer and were more severe than was strictly necessary to assuage concerns about the spread of plague abroad. The frenzy may also be explained as an artifact of the Black Death—bubonic plague was, after all, *the* plague, an appellation that should not be dismissed as meaningless.

In fact, reactions to the plague in India were in some ways uncannily similar to medieval reactions. Many colonial officials considered the plague a sort of divine retribution for failing to combat the filth, overcrowding, and disease that characterized Indian society. Plague was a punishment for the failure of the government to uphold its fundamental duty, as established with the Black Death, to protect its subjects from disease. As in the second pandemic, the burden of suspicion fell on those who could be defined as outsiders. The British blamed the plague on the Indians, insisting that the disease was a consequence of unsanitary living conditions. The Indians in turn blamed the British, attributing the plague to a colonial conspiracy that poisoned the native water supply.[3]

At the start of the third pandemic, contemporary scientific understanding of the plague had not surpassed the medieval. Authorities on the plague mistakenly favored the view that the disease was caused by a localized, contagious miasma around the beginning of the twentieth century.[3]

The first breakthrough was made in the Hong Kong epidemic, with the identification of the plague bacillus by Alexandre Yersin, after whom it was eventually named *Yersinia pestis* in 1970. Yersin was also the first to make the connection between rats and plague, noting that the black rat reservoir was affected prior to human epidemics. However, by the time the plague reached India, it was not yet clear how the disease was contracted by humans.

Scientists flocked to Bombay to study the disease, but their investigations were hampered by sociocultural bias. Noting that many patients were poor and unshod, they generally concluded that people were infected when abrasions on their feet came into contact with bacteria in rat excretions. In 1898, studies by Hankin and Simond refuted this theory, demonstrating that the plague bacillus survived only briefly outside the body and was seldom recoverable from supposedly infected objects.[3] In that same year, Ogata and Simond independently discovered the role of the flea as the vector of plague transmission.[26]

Despite these new findings, the colonial Indian Plague Commission remained convinced of the validity of the barefoot-Indian theory. Nor did the Commission acknowledge that bubonic plague was seldom transmitted from person to person, as was clear from the start of the epidemic, when rapid isolation of the sick had no discernible effect on the spread of the disease. The policies formulated on the basis of these mistaken assumptions were not only oppressive but also fatal. Suspected plague patients were harassed for no reason, and the destruction and decontamination of supposedly infected properties simply drove the rats and fleas that were the true carriers of the disease elsewhere, spreading the epidemic.[3]

TWENTIETH-CENTURY PUBLIC HEALTH MEASURES

Only later were public health measures correctly tailored to combat plague. In 1910, during the Manchurian outbreak, L. T. Wu recognized and characterized the pneumonic form of the plague, and established measures to prevent the spread of the disease by aerosol. Much of what is known today about the epidemiology and pathology of pneumonic plague is rooted in the work of Wu and other scientists during this epidemic. Also, the first effective vaccine to prevent plague was created and used during the Manchurian outbreak; the vaccine, designed by W. M. Haffkine, contained killed plague bacteria. Unfortunately, these measures were of limited help, and from 1910 to 1911 as many as 60,000 people developed pneumonic plague in Manchuria. Without antibiotics with which to treat the disease, all the cases were fatal.

The adoption of appropriate public health measures against plague and the availability of antibiotics make future pandemics unlikely. However, plague remains a risk to human health, with the World Health Organization (WHO) reporting an average of 1700 cases of plague worldwide each year for the past

50 years. In fact, since the 1990s, plague has been considered a reemerging disease, owing to an increase in the number of cases reported to the WHO, the outbreaks of several epidemics worldwide in 1994, and the slowly increasing number of foci in certain countries, including the United States.[7] In addition to the dangers of natural plague outbreaks, there is also the threat of plague from the use of a biological weapon.

THE HISTORY OF PLAGUE AS A WEAPON

The first recorded instance of the use of plague as a weapon was in 1346, when a Tatar army that was being decimated by the disease tossed infected corpses over the walls of the besieged city of Kaffa, in what is now Ukraine. Their hope that the plague would do their conquering for them was fulfilled—the ensuing epidemic resulted in the city's surrender.[25]

More recently, during World War II, a secret branch of the Japanese army known as Unit 731 infected Chinese water supplies with *Y. pestis* cultures. The bacteria could survive in the water for up to 16 days to infect anyone who drank it. More infamously, Unit 731 also dropped plague-infected fleas over populated areas of Manchurian China, causing several bubonic plague outbreaks. (See box 10.1.)

BOX 10.1. SURVIVORS SUE OVER
WORLD WAR II USE OF BIOWEAPON

In 1995, a group of 180 victims of the plague in Yiwu, China, an area in which Unit 731 had dropped infected fleas, sued Japan, charging that its forces had spread bubonic plague and other diseases in China during World War II, killing 300,000 people. Because there are no official tallies, this figure is questionable. However, survivors of the plague in Yiwu testify to its horrors, describing how the disease ravaged the village for months, killing 20 villagers a day at one point in 1942. The plaintiffs achieved a partial victory in late August 2002. Judge Kohi Iwata of the Tokyo District Court ruled that Unit 731 had "used bacteriological weapons under the order of the imperial Japanese Army's headquarters," but refused to award reparations, saying that it was against international law for the plaintiffs to demand money from Japan. The group is currently appealing this decision, under the dedicated leadership of Wang Xuan, many of whose relatives were victims of the plague in Yiwu.

The Japanese government still denies that its army ever used biological agents, and no other government has an interest in alienating Japan by acknowledging the plaintiff's cause. The struggle to break through the collective amnesia about state-sponsored biological warfare in the Pacific will continue.[19]

During the Cold War, both the United States and the Soviet Union developed techniques to aerosolize plague directly, eliminating reliance on the vulnerable and unreliable and maximizing the danger by ensuring the spread of pneumonic, rather than bubonic, plague. The United States did not make the weaponizing of plague as high a priority as the USSR did, and by 1970, when the United States ended its offensive military program, it did not have enough plague to make an effective weapon. However, the Soviet Union is reported to have had more than 10 institutes and thousands of scientists successfully working to weaponize the disease, and is said to have manufactured large quantities of *Y. pestis* suitable for use in weapons.

MOLECULAR AND CELL BIOLOGY

BACTERIOLOGICAL CHARACTERISTICS OF THE PATHOGEN

Y. pestis is a gram-negative, nonmotile, nonspore-forming coccobacillus that belongs to the family *Enterobacteriaceae*. The bacterium is 0.5–0.8 µm in diameter and 1–3 µm long. It is a facultative anaerobe, and its nutritional requirements make it an obligate parasite. It has a typical bacterial cell wall, and although it lacks a true capsule it forms a protein envelope, called capsular antigen or fraction 1 (F1), during growth above 33°C. *Y. pestis* grows optimally at 28°C and a pH of 7.2–7.6 but can tolerate temperatures from 4 to 40°C and pH 5–9.6.

The bacillus exhibits a distinctive bipolar, or "safety pin," staining with Giemsa's, Wright's, or Wayson's stains. Colonies, which grow best on blood or MacConkey agar, are distinctive in appearance, with a shiny surface similar in sheen to hammered copper and an irregular, slightly raised "fried egg" shape. Although these unique characteristics of *Y. pestis* can serve as useful diagnostic markers, their utility as such is minimized by the slow growth of the bacillus, which typically requires at least 48 hours of incubation before there is any visible growth.

THE THREE BIOTYPES

There are three biotypes of *Y. pestis*. Each is named for a different pandemic that it is thought to have caused. It is not possible to definitively confirm the suspected link between the first two pandemics and biotypes Antiqua and Medievalis with the available historical evidence. However, the fact that all strains isolated from areas that were unaffected by plague prior to the third pandemic are of biotype Orientalis, whereas isolates from older foci were of other biotypes, is strongly suggestive of a correlation between the Orientalis and the third pandemic.[10]

TABLE 10.1 *Y. PESTIS* BIOTYPES AND THEIR CHARACTERISTICS

BIOTYPE	FERMENTS GLYCEROL	NITRATE-NITRITE
Antiqua	Yes	Yes
Medievalis	Yes	No
Orientalis	No	Yes

The three biotypes are distinguishable by the conversion of nitrate to nitrite and the fermentation of glycerol (table 10.1). There is no difference in the virulence or pathology of these biotypes in animals and humans.

EVOLUTION OF *Y. PESTIS*

The genomes of two strains of *Y. pestis* have been sequenced: CO92 (biotype Orientalis) at the Sanger Center in Cambridge, England, and KIM (biotype Medievalis) at the University of Wisconsin. The genome of *Y. pestis* was found to contain 149 pseudogenes, an indication of reductive evolution, and to have 90% chromosomal DNA relatedness to *Y. pseudotuberculosis*, an enteropathogen in the genus *Yersinia*. From this evidence, it is thought that *Y. pestis* evolved from *Y. pseudotuberculosis* between 1500 and 4000 years ago, with genes required by the latter for adhesion, mobility, and colonization of the gut having decayed in the former.

The genome has other noteworthy characteristics, including an unusually large number of insertion sequences and an unusual bias toward guanine and cytosine nucleotides in certain areas, both of which are indicative of frequent intragenomic recombination. Using the G/C skew and other bioinformatics tools, 21 "adaptation islands" that show characteristics of pathogenicity were found. Scientists believe that many of these islands were acquired from other organisms through lateral transfer, approximately 1500 years ago.[5]

THE GENOME OF *Y. PESTIS*

The genome of *Y. pestis* consists of 4500 genes, distributed among one 4.65-Mb chromosome and three plasmids of different sizes. The largest, pFra, is 96.2 kb; pCD1 is 70.3 kb; and pPst is the smallest, measuring only 9.6 kb. Each part of the genome codes for known and putative virulence factors that enable the bacteria to survive and flourish in flea vectors and mammalian hosts.

The Chromosome The chromosome contains a 102-kb pathogenicity island, known as the pigmented locus (Pgm). Pgm mutants undergo a drastic loss of virulence, because of their inability to acquire iron. To cause infection, *Y. pestis*, like other pathogenic bacteria, must remove iron, an essential nutrient, from host iron- or heme-chelating proteins. Two important iron transport and storage systems are known to be encoded in the Pgm locus: yersiniabactin (*Ybt*) and the hemin storage (*Hms*) genes.

Yersiniabactin is a siderophore, a small, inorganic compound with an extremely high affinity for ferric iron, which is synthesized and secreted by the bacillus. It preferentially chelates iron bound to eukaryotic proteins and transports it to the bacterium. This transport system is thought to be essential for iron acquisition during the early, intracellular stage of plague infection, when the bacillus has first been taken up by the host macrophage.

The three *Hms* genes allow *Y. pestis* to store large quantities of hemin, an iron source in its outer membrane, forming greenish/brown or red colonies. However, experiments have shown that the iron stored by the *Hms* system is not used nutritionally, and that the *Hms* phenotype is not essential for the pathogenesis of bubonic plague in mammals.[21] *Hms* does play a very important role in the transmission of plague, changing *Y. pestis* from a harmless inhabitant of the flea vector's midgut to one that amasses in its foregut. The ensuing blockage causes the starving flea to go into a frenzy of blood-feeding in which it regurgitates the mass of bacteria and transmits the plague.[13] Transcription of the *Hms* genes is temperature-dependent, being higher at 26°C, the temperature of the flea, than at 37°C, mammalian body temperature. This suggests that, although *Hms* is essential in flea infection, its role in mammalian plague is secondary.

The Fra Plasmid The Fra plasmid contains genes for murine toxin (*Ymt*) and the F1 capsule (*Caf1*). The protein formerly known as murine toxin has recently been recharacterized as a phospholipase D (PLD). It had been known for some time that transcription of *Ymt* is threefold higher at 26°C (flea) than at 37°C (mammal), suggesting that the so-called murine toxin is important for the colonization of *Y. pestis* in the flea. In fact, PLD is essential for survival of *Y. pestis* in the flea, with PLD activity serving to protect the bacteria from a cytotoxic digestion product of blood plasma in the flea gut.[14]

The formation of the F1 capsule is also temperature-dependent, forming at mammalian but not flea temperatures. The capsule renders bacteria resistant to phagocytosis by monocytes in the mammalian host. When *Y. pestis* first makes the transition from flea to mammal, it is not yet encapsulated, and most of the vulnerable bacteria are phagocytosed and killed by polymorphonuclear leukocytes (PMNs). A few are taken up at their point of entry by macrophages, which provide a protected environment for the bacteria's synthesis of their protective F1 capsules, as well as other virulence factors. The encapsulated bacteria then lyse the macrophages that hosted them and are released into the extracellular space, where they are able to safely journey from the dermis to the lymph nodes and other viscera.

The Pst Plasmid The Pst plasmid encodes for plasminogen activator (Pla). Plasminogen is a cell-surface protease, which exhibits temperature-dependent coagulase and fibrinolytic activity. At temperatures below 30°C, as in the flea, it acts as a coagulase, initiating blockage of the flea by forming a fibrin matrix that anchors *Y. pestis* to the proventriculus spines in the flea foregut. At 37°C, plasminogen activity is fibrinolytic, preventing formation of a fibrin clot that would trap bacteria at the site of the flea bite, thus allowing the infection to spread. Scientists have also postulated other virulence roles for Pla, suggesting that by producing excess plasmin it causes ineffective structures between inflammatory cells and fibrin, and that by inhibiting production of interleukin-8 it reduces chemoattractants at the site of infection.[26]

The CD1 Plasmid The CD1 plasmid bears primary responsibility for encoding the antihost genome. The group of genes it encodes is commonly known as the low calcium response stimulon (LCRS), because expression and secretion of the gene products at 37°C are linked to one of two signals: the concentration of calcium ion in the environment or contact with a eukaryotic cell. The exact mechanism by which cell contact triggers gene activity is as yet unclear, but it is probably a response to the low-calcium environment inside the host cell, which is sensed by the bacillus when it punctures the cell and samples its cytoplasm upon contact.

The gene products of the LCRS include a variety of antigens, especially the virulence antigen (LcrV), *Yersinia* outer proteins (Yops), specific Yop chaperones (Syc), and the Yop secretion (Ysc) proteins.

The Ysc proteins are involved in the type III secretion system used by *Y. pestis* to inject pathogenicity proteins into the cytosol of eukaryotic host cells (figure 10.1). The type III secretion apparatus is a molecular machine that functions much like a hollow pushpin,[9] with Ysc proteins A–L and N–U forming a ring that passes through the inner and outer membranes of *Y. pestis*.

The processes by which the secretion apparatus is regulated are not yet clear. In the presence of calcium, or when the bacillus is not in contact with a eukaryotic cell, the Ysc secretion channel is blocked at both ends—by LcrG on the proximal end and LcrE and YopN on the distal end. YopN appears to be a surface-exposed sensor protein that acts in response to a signal on eukaryotic cells to facilitate the opening of the channel at the site of cell contact.[15] LcrE, also a surface-localized protein, may function as a calcium-ion sensor. In the ab-

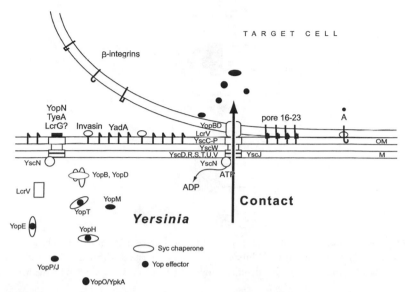

FIGURE 10.1. The *Yersinia* type III secretion system. Ysc proteins form a channel through the bacterial inner and outer membranes, with YscN acting as an ATPase. When there is no contact with a eukaryotic cell, the channel is blocked by YopN and LcrG. But on contact with a target cell, the bacterium attaches tightly by interaction between its YadA and Inv adhesions and β-integrins, and the secretion channel is opened. The Yop effectors B and D and LcrV then translocate across the plasma membrane, and the other Yops are transported through the Ysc channel.

sence of calcium or cell contact, LcrV antigen inhibits negative regulation of the secretion channel. This is partially achieved by a stable interaction with LcrG,[23] which prevents it from blocking the proximal end of the secretion channel, but the exact relationship between LcrV and the YopN/LcrE blockage is unknown (figure 10.2).

Once the Ysc secretion channel is open, a variety of proteins are sent through it. Some of these—YopB, YopD, YopK, and possibly YopR—are translocatory proteins that help to transfer pathogenic proteins from *Y. pestis* into the eukaryotic host cell. YopB and YopD act by forming a translocatory pore that breaches the target cell membrane. Other proteins, with direct antihost functions, are then injected into the host cell, including YopE, YopH, Yop T, YopJ, YpkA (or YopO), and YopM. YopE, YopH, and YopT all target the cytoskeleton, contributing to the strong resistance of *Y. pestis* to phagocytosis by macrophages. YopE and YopT

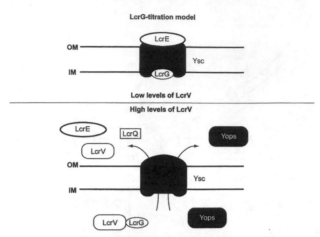

FIGURE 10.2. In the absence of LcrV, the Ysc secretion channel is blocked by LcrG and LcrE. However, when LcrV binds to LcrG and intersects with LcrE (in a way that is not yet clear), the Ysc secretion channel is unblocked and Yops are secreted by the bacteria.

do so by disrupting actin filaments, whereas YopH disrupts focal adhesions by dephosphorylating the proteins p130^{cas}, paxillin, and FAK. YopJ has been shown to counteract the normal inflammatory responses of several types of cells, including macrophages and epithelial and endothelial cells, by a variety of means. It also induces apoptosis in macrophages, although the mechanism by which this is done is not yet known. YpkA/YopO is a serine-threonine kinase that probably interferes with signal transduction in the host cell. YopM binds thrombin, but the significance of this binding is unclear. It may contribute to the spread of bacteria throughout the body by preventing blood-clot formation, or it may mute the local inflammatory response by inhibiting platelet activation.[15] Unlike the other Yops, YopM does not stay in the host cytoplasm but quickly migrates into the cell nucleus, where it may alter gene expression. The functions and effects of these proteins on the host cell are summarized in table 10.2.

The Syc proteins, which are encoded by a gene that is located close to the gene for the Yop it serves, are small (14–15 kd) and acidic (pH 4.4–5.2). They are categorized into two groups, one devoted to translocator Yops and the other to effectors. SycB and SycD, the translocator chaperones, bind to their Yops at several domains for a purpose not yet known. The effector chaperones—SycE,

TABLE 10.2 *YERSINIA* OUTER PROTEINS AND THEIR EFFECTS

PROTEIN	EFFECTS ON HOST CELL
YopE	Cytotoxic; disrupts actin filaments
YopH	Disrupts focal adhesion
YopT	Depolymerizes actin
YopJ	Downregulates inflammatory response; induces apoptosis in macrophages
YpkA/YopO	May interfere with signal transduction
YopM	Mutes local inflammatory response; prevents blood-clot formation; effects in nucleus unknown

SycH, SycT, and SycN—bind to their Yops at only one site. In the absence of a chaperone, secretion of its Yop is severely reduced, but if the binding site is removed from the Yop there is no reduction in secretion. This suggests that the binding site itself creates the need for the chaperone, which protects the site from premature interactions that would compromise the stability of the Yop. Other roles have been hypothesized for the Syc proteins as well, including activity as a pilot, guiding the Yop to the secretion apparatus or directing the order in which the Yops are secreted.[6]

The most enigmatic virulence factor is LcrV, the virulence antigen. In addition to its role in the regulation of secretion, described earlier, it also has a direct antihost function. The oldest known secreted *Yersinia* virulence protein, it downregulates the host immune system by inhibiting cytokine production and protects the bacillus from phagocytosis. The movement of LcrV is mysterious—although it sometimes leaves the bacteria with the Yops, through the type III secretion system, it can bypass secretion system entirely. Experiments have shown that the antibody against V antigen is protective for mice in immunizations against *Y. pestis*.[15]

ECOLOGY AND GEOGRAPHY

Plague is normally not a disease of humans. It is a zoonotic disease, occurring primarily by chronic, sustained circulation of *Y. pestis* in complex enzootic and epizootic transmission cycles involving rodents and their fleas.

Historically, the principal vector is *X. cheopis*, the Oriental rat flea, but any flea species may be capable of transmitting the plague. The infective potential of a given flea depends on a number of factors, including differences in size and structure of the proventriculus, a spiny, sphincter-like organ separating the stomach and esophagus (affecting gut-blockage potential); feeding frequencies; survival time after infection; and environmental factors.

Wild rodents, such as voles and deer mice, are the primary enzootic reservoir for the disease, as they generally experience milder symptoms than other mammals do, making them suitable maintenance hosts that ensure the long-term survival of *Y. pestis*. Other mammals—including the classic rodent host, the black rat *Rattus rattus*—are less resistant to plague infection, and serve as epizootic hosts. Susceptibility to plague prevents these animal species from effectively perpetuating the disease.

Humans are incidental hosts for *Y. pestis* (figure 10.3). Disease occurrence in humans depends on the frequency of infection in local rodent populations and the degree of contact between rodents and humans. Human outbreaks are usually preceded by epizootics with high mortality in rodent populations, forcing their infected fleas to find alternative hosts. Like humans, mammalian species other than rodents can serve as incidental hosts for *Y. pestis* but can also serve as sources of human exposure, either directly or through the flea vector. Domestic

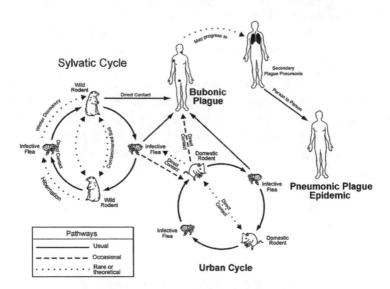

FIGURE 10.3. Usually, *Y. Pestis* is maintained in rural and urban rodent populations, cycling from flea to rodent and back to flea. Only occasionally will contact with infective animals transmit the plague to human beings.

cats are particularly susceptible to the plague, and commonly infect their owners in areas where plague is endemic.

Plague is currently endemic in every inhabited continent except Australia and Europe (figure 10.4a). The enzootic foci in North America—the largest in the world—are bounded geographically by Canada (British Columbia and Alberta) on the north, the western Great Plains on the east, Mexico on the south, and the Pacific coast on the west. Most human cases of plague in North America occur in the southwestern United States, specifically the regions that include northern New Mexico, northern Arizona, and southern Colorado, and California, southern Oregon, and far western Nevada (figure 10.4b).

Almost all plague cases are rural in origin, because the wild rodent reservoir is essential for the maintenance of the disease. Statistically, American Indians are more at risk than other populations in the United States, accounting for up to one-third of human plague cases in the country. This is because they, more than other populations, tend to live in areas of enzootic foci and engage in activities that put them in regular contact with wild animals, such as shepherding and hunting.

Recently, there has been an increase in domestic transmission of the plague, as suburban residential areas encroach on formerly rural enzootic foci, with a concomitant increase in cat-related cases. Twenty-eight percent of these human cases are of the more dangerous pneumonic form of the plague, to which cats are especially prone; *Y. pestis* is found in the oral cavities of 92% of cats killed by plague. Over 25% of these cat-related cases occur in veterinarians and their

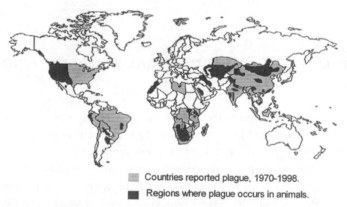

■ Countries reported plague, 1970-1998.
■ Regions where plague occurs in animals.

FIGURE 10.4A. World distribution of plague, 1970–1988. Plague is endemic on every inhabited continent except Europe and Australia. Lighter shading indicates coutries that reported plague from 1970 to 1998. Darker shading shows regions where plague occurs in animals.

Source: Centers for Disease Control and Prevention. Online: http://www.cdc.gov/ncidod/dvbid/plague/world98.htm.

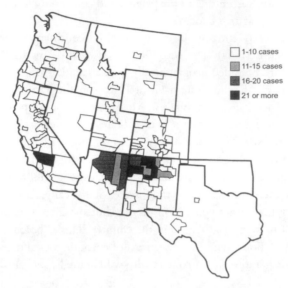

FIGURE 10.4B. Reported human plague cases in the United States by county, 1970–1997.

Source: Centers for Disease Control and Prevention. Online: http://www.cdc.gov/ncidod/dvbid/plague/plagwest.htm.

assistants, making plague a potential occupational risk for veterinary workers in areas where plague is endemic.

PATHOGENESIS AND CLINICAL MANIFESTATIONS

BUBONIC PLAGUE

Bubonic plague is the most common clinical manifestation of plague. In most cases, it begins with the bite of an infected flea, inoculating the patient's skin with thousands of bacteria. Because of the low temperature in the flea, the *Y. pestis* bacteria released into the skin do not express the F1 and LCRS genes and are therefore readily phagocytosed by macrophages and polymorphonuclear leukocytes (PMNs). The bacteria phagocytosed by PMNs are destroyed, but the macrophages, unable to kill *Y. pestis*, provide a protected environment in which the bacteria synthesize their capsule and virulence factors. Within 3 to 5 hours of entering the host, the bacteria are encapsulated and in possession of a complete set of virulence proteins. Thus armed, they lyse the macrophage and are released into the extracellular environment, where they resist phagocytosis by the PMNs.

Y. pestis migrates from the site of inoculation through the cutaneous lymphatics to the regional lymph nodes. There the bacteria multiply rapidly and are combated by many macrophages and PMNs. The resultant inflammatory response swells the lymph nodes, forming the characteristic buboes for which the plague is named. Buboes typically develop in the groin, axilla, or cervical region, and measure 1–10 cm in diameter. They are extremely tender, nonfluctuant, and warm, with erythematous overlying skin, and are often so painful as to prevent the patient from moving the affected body part. Other presenting symptoms include fever, headache, chills, and weakness; patients typically develop these and the buboes within 1–7 days of infection.

Despite the body's attempts to fight infection, the sheer numbers of bacteria overwhelm the immune system, and from the lymph nodes they go to the bloodstream, from which *Y. pestis* may colonize internal organs, especially the blood-filtration organs, the spleen and liver. Secondary septicemia is common, and can lead to disseminated intravascular coagulation (DIC), shock, and coma. In the terminal stages of bubonic plague, the bacteria may also spread to the lungs, causing secondary pneumonic plague in 5–15% of bubonic plague patients (table 10.3).

PRIMARY SEPTICEMIC PLAGUE

Primary septicemic plague occurs in only 10–25% of patients infected by fleabite. These patients have blood cultures that are positive for *Y. pestis* but no discernible buboes. Primary septicemic plague patients develop symptoms within 1–4 days of infection, including fever, chills, headache, malaise, and a higher incidence of gastrointestinal disturbances than in bubonic patients.

The septicemia may lead to intravascular coagulation, necrosis of small blood vessels, purpuric skin lesions, and, in the advanced disease, gangrene of acral regions such as the digits and nose. Secondary pneumonic plague develops in about 25% of primary septicemic patients, by hematogenous spread of *Y. pestis* to the lungs (table 10.4).

PNEUMONIC PLAGUE

Pneumonic plague is extremely rare. The initial symptoms of pneumonic plague are those of a febrile, flu-like illness, but rapidly progress to those of an overwhelming pneumonia, with coughing, chest pain, and production of bloody

TABLE 10.3 CLINICAL FEATURES OF BUBONIC PLAGUE

FEATURE	CHARACTERISTICS
Presenting features	Sudden onset of fever, chills, weakness
	Usually within 1 day, painful swollen lymph node or group of nodes (bubo) occurs in groin, axilla, or cervical region:
	1–10 cm, smooth, uniform, unfixed, egg-shaped mass or irregular cluster of several nodes
	Extremely tender
	Region may be erythematous, with surrounding edema
	Buboes usually occur in only one location, but multiple buboes may be seen
	Rarely, buboes may separate and rupture
	Skin lesions (papules, vesicles, pustules) may occur at site of flea bite but are absent in < 10% of cases
	Associated lymphagitis uncommonly occurs
	Presenting symptoms for 40 Vietnamese patients with bubonic plague
	Fever (100%) (mean temperature for 32 patients: 102.9°F [39.4°C])
	Chills (40%)
	Bubo (100%—groin, 88%; axilla, 15%; cervical, 5%; epitrochlear, 3%)
	Headache (85%)
	Prostration (75%)
	Altered mental status (38%) (lethargy, confusion, delirium, seizures)
	Anorexia (33%)
	Vomiting (25%)
	Abdominal pain (18%)
	Cough (25%)
	Chest pain (13%)
	Skin rash (23%) (petechiae, purpura, papular eruptions)
Complications	Secondary septicaemia (can lead to DIC, shock, multisystem involvement)
	Secondary pneumonic plague (5–15% of patients)
	Meningitis (may occur in patients with bubonic plague that was not adequately treated)
	Buboes may become infected with other bacterial pathogens

Source: Ref. 16.

sputum. Patients with pneumonic plague can spread the disease to others by respiratory droplet, and those within 2 m of an infected cough or sneeze are at risk of inhaling the bacteria and being infected through their mucous membranes. The infectious dose by inhalation is very small, estimated to be only 100–500 bacteria. Plague contracted by respiratory droplet is called primary pneumonic.

TABLE 10.4 CLINICAL FEATURES OF PRIMARY SEPTICEMIC PLAGUE

FEATURE	CHARACTERISTICS
Presenting features	10%–25% of US plague cases present with primary septicemic plague
	Presenting symptoms for 18 cases of septicemic plague in New Mexico
	Fever (100%)
	Chills (61%)
	Nausea (44%)
	Headache (44%)
	Vomiting (50%)
	Diarrhea (39%)
	Abdominal pain (39%)
	Any gastrointestinal symptom (72%)
	Presenting signs for 18 cases of primary septicemic plague in New Mexico:
	Mean temperature 38.5°C (range, 35.4–40.4°C)
	Mean pulse: 109 (range, 72-160)
	Mean respiratory rate: 31 (range, 16-60)
	Mean systolic BP: 104 (range, 80-130)
	Mean diastolic BP: 66 (range, 36-80)
	Mental status changes commonly occur (delirium, obtundation, coma)
Complications	Illness rapidly progresses to sepsis syndrome often with DIC, shock, and multisystem involvement
	Skin lesions reflect DIC (may be similar to meningococcemia)
	Purpura
	Petechiae
	Ecchymoses
	Gangrene of acral regions (caused by small artery thromboses)
	Ecthyma gangrenosum (rare)
	Meningitis
	Secondary plague pneumonia (about 25% of patients)

Source: Ref. 16.

Because *Y. pestis* transmitted by respiratory droplet comes from a mammalian host rather than a flea, they arrive in possession of a capsule and toxin delivery system. This gives the bacteria a head start in its race against the immune system, and the incubation period of pneumonic plague is 1–4 days, with symptoms usually appearing in only 1–2 days.

As in bubonic plague, the bacteria first encounter macrophages—in this case, alveolar macrophages, which defend the lungs. There is an influx of PMNs into the lungs, followed by inflammation and fluid buildup, which make breathing difficult. From the lungs, the bacteria may spread to the blood, so the initial pneumonic symptoms of coughing, chest pain, and bloody sputum may be fol-

TABLE 10.5 CLINICAL FEATURES OF PRIMARY PNEUMONIC PLAGUE

FEATURE	CHARACTERISTICS
Presenting features	Symptoms of primary plague pneumonia: 　Fever 　Chest pain 　Dyspnea 　Productive cough (sputum may be purulent or watery, frothy, blood-tinged) 　Hemoptysis 　Tachypnea (particularly in young children) 　Cyanosis 　Bubo not present (rarely, cervical bubo may be noted) Gastrointestinal symptoms (nausea, vomiting, abdominal pain, diarrhea) common
Complications	Septicemia with sepsis syndrome Meningitis

Source: Ref. 16.

lowed by septicemic plague symptoms, especially severe gastrointestinal symptoms, including nausea, vomiting, abdominal pain, and diarrhea (table 10.5).

MENINGITIS AND PHARYNGITIS

Plague syndromes that are even less common are meningitis and pharyngitis. Plague meningitis occurs as a complication of bacteremia, and may be the presenting clinical syndrome in some cases. Symptoms are typical of meningitis, and include fever, headache, meningismus, and changes in mental status. When meningitis occurs as a complication of bubonic plague, there is often a bubo in the axillary region.

Plague pharyngitis is a consequence of inhaling or ingesting *Y. pestis*, and its symptoms are similar to those of severe pharyngitis or tonsilitis. Inflamed cervical nodes with the characteristics of buboes are usually present. Secondary septicemia sometimes occurs.

DIAGNOSIS

According to the Centers for Disease Control and Prevention (CDC), certain conditions must be met for a definitive diagnosis of plague. For a presumptive diagnosis, the F1 antigen must be detected by fluorescent assay of a clinical specimen—bronchial wash or transtracheal aspirate in pneumonic cases, blood in septicemic cases, or aspirate of the involved tissue in bubonic cases. In theory, this can be done quickly—in less than 2 hours—but the test is available only at some state health departments, the CDC, and military laboratories, so there is a built-in delay in diagnosis for shipment of the samples. Moreover, samples that have been refrigerated during shipment or are from cultures that have been incubated at temperatures lower than 37°C would test negative.

To confirm a diagnosis of plague, it is necessary to isolate *Y. pestis* from the clinical specimen or to observe at least a fourfold elevation in serum antibody titer to the F1 antigen (smaller elevations are considered a presumptive diagnosis.) None of these techniques is fast: *Y. pestis* grows slowly in culture, and antibodies usually take several days or even weeks after disease onset to develop.

Only cases of plague defined as presumptive or confirmed are reported by the CDC to the WHO. However, suspected cases must be immediately reported to local and state health departments and laboratories, because of the possibility that a single case signals a larger epidemic.

It is fatal to delay treatment until the CDC conditions for diagnosis have been met. The prognoses for patients with untreated plague are extremely poor. Fifty to ninety percent of untreated bubonic plague patients die, and close to 100% of untreated septicemic and pneumonic plague patients die. Untreated pneumonic plague patients all die within 2 to 6 days of respiratory exposure, most within 48 hours of onset.

With appropriate medical intervention, the prognoses improve drastically. Among treated bubonic plague patients, the case fatality rate is less than 5%. The 30–50% mortality rate for septicemic plague patients is significantly higher, because the antibiotics generally used to treat undifferentiated sepsis are not effective against *Y. pestis*, minimizing the chance of an incidental cure. However, it still represents a major diminution of fatality from the numbers for untreated septicemic plague. To be effective, treatment of pneumonic plague must begin within 24 hours of symptom onset. It is therefore essential to act as quickly as possible, administering medical treatment as soon as plague is suspected.

The timely diagnosis of plague requires a high index of suspicion. Diagnosis based on clinical symptoms is difficult, especially in cases of primary septicemic and pneumonic plague, which lack the bubo that is the infamous sign of the disease. Septicemic plague symptoms resemble those of other gram-negative septicemias so closely as to be virtually indistinguishable, and pneumonic plague patients present with symptoms identical to those of other respiratory illnesses.

Microbiological studies are important, but not foolproof, evidence for a suspected diagnosis of plague. The rapid, automated microbiological identification systems commonly used by laboratories are often not programmed to identify *Y.*

pestis, instead identifying it as *Shigella*, HxS-negative *Salmonella*, Acinetobacter, or *Y. pseudotuberculosis*. In most biochemical or commercial identification systems, *Y. pestis* appears relatively inert, making further biochemical testing of little use. However, staining clinical specimens with Wright's, Giemsa's, or Wayson's stains to look for the characteristic bipolar staining pattern is a convenient and effective diagnostic method.[20] Also, a new rapid diagnostic test (RDT) for plague that uses monoclonal antibodies to the F1 antigen was recently tested by health workers in Madagascar with promising results. The RDT is a specific, sensitive, and reliable dipstick assay that can be used at a patient's bedside for diagnosis of bubonic and pneumonic plague.[4]

The most useful tool for rapid diagnosis of plague is the patient's history of exposure. Plague should be suspected in symptomatic patients who live in or have recently traveled to enzootic foci, especially in the summer, when incidence of plague tends to increase. Activities that put people in contact with wild rodents and other susceptible mammals, especially domestic cats, are also risk factors; hunters, campers, hikers, and veterinarians are among the groups at higher risk.

TREATMENT

Treatment with antibiotics should begin as soon as plague is suspected, and maintained for a minimum of 10 days, or 3–4 days after clinical recovery. *Y. pestis* has been considered universally susceptible to antibiotics, thanks to the wide variety of drugs that have been successfully used to treat plague. Historically, streptomycin, a treatment approved by the Food and Drug Administration (FDA) for plague, was preferred, but it is now infrequently used in the United States and in modest supply. Gentamicin, although not approved by the FDA, is widely available, inexpensive, and equivalent to streptomycin in its curative effects. Tetracycline and doxycycline are approved by the FDA for treatment and prophylaxis of plague, and anecdotal cases indicate that aminoglycosides are equally effective. Fluoroquinones, such as ciprofloxacin, levofloxacin, and ofloxacin, have demonstrated efficacy in combating plague in animal studies. Chloramphenicol may be used to treat plague, and is recommended for cases of plague meningitis because of its ability to cross the blood–brain barrier. Sulfonamides have also been used to successfully treat bubonic plague, but they are ineffective against pneumonic plague.[18]

The recommended therapies for adults, children, and pregnant women in both contained-casualty and mass-casualty settings, outlined in tables 10.6 and 10.7, include many of these antibiotics.

TABLE 10.6 TREATMENT FOR PLAGUE

PATIENT CATEGORY	RECOMMENDED THERAPY
	CONTAINED CASUALTY SETTING
Adults	Preferred choices Streptomycin, 1 g IM twice daily Gentamicin, 5 mg/kg IM or IV once daily or 2 mg/kg loading dose followed by 1.7 mg/kg IM or IV 3 times daily Alternative choices Doxycycline, 100 mg IV twice daily or 200 mg IV once daily Ciprofloxacin, 400 mg IV twice daily Chloramphenicol, 25 mg/kg IV 4 times daily
Children	Preferred choices Streptomycin, 15 mg/kg IM twice daily (maximum dose, 2 g) Gentamycin, 2.5 mg/kg IM or IV 3 times daily Alternative choices Doxycycline If ≥ 45 kg, give adult dosage If < 45 kg, give 2.2 mg/kg IV twice daily (maximum daily dose, 2 g) Ciprofloxacin, 15 mg/kg IV twice daily Chloramphenicol, 25 mg/kg IV 4 times daily
Pregnant women	Preferred choice Gentamicin, 5 mg/kg IM or IV once daily or 2 mg/kg loading dose followed by 1.7 mg/kg IM or IV 3 times daily Alternative choice Ciprofloxacin, 400 mg IV twice daily
	MASS CASUALTY SETTING
Adults	Preferred choices Doxycycline, 100 mg orally twice daily Ciprofloxacin, 500 mg orally twice daily lternative choice Chloramphenicol, 25 mg/kg orally 4 times daily
Children	Preferred choice Doxycycline If ≥ 45 kg, give adult dosage If < 45 kg, give 2.2 mg/kg orally twice daily Ciprofloxacin, 20 mg/kg orally twice daily Alternative choice Chloramphenicol, 25 mg/kg orally 4 times daily
Pregnant women	Preferred choice Doxycycline, 100 mg orally twice daily Ciprofloxacin, 500 mg orally twice daily Alternative choice Chloramphenicol, 25 mg/kg orally 4 times daily

Source: Ref. 13

TABLE 10.7 PROPHYLAXIS AFTER EXPOSURE TO *Y. PESTIS*

PATIENT CATEGORY	RECOMMENDED THERAPY
	POSTEXPOSURE PROPHYLAXIS
Adults	Preferred choices
	Doxycycline, 100 mg orally twice daily
	Ciprofloxacin, 500 mg orally twice daily
	Alternative choice
	Chloramphenicol, 25 mg/kg orally 4 times daily
Children	Preferred choice
	Doxycycline
	If ≥ 45 kg, give adult dosage
	If < 45 kg, give 2.2 mg/kg orally twice daily
	Ciprofloxacin, 20 mg/kg orally twice daily
	Alternative choice
	Chloramphenicol, 25 mg/kg orally 4 times daily
Pregnant women	Preferred choice
	Doxycycline, 100 mg orally twice daily
	Ciprofloxacin, 500 mg orally twice daily
	Alternative choices
	Chloramphenicol, 25 mg/kg orally 4 times daily

Source: Ref. 13

DRUG RESISTANCE

Drug resistance in *Y. pestis* is cause for concern. In 1995, a strain of *Y. pestis* called 17/95 was isolated from a bubonic plague patient in Madagascar. The isolate was found to have plasmid-mediated resistance to all the drugs recommended for prophylaxis and therapy, along with a few others. Specifically, it was resistant to ampicillin, chloramphenicol, kanamycin, streptomycin, spectinomycin, sulfonamides, tetracycline, and minocycline. The plasmid, named pIP1202, was highly transferable via conjugation in vitro to other strains of *Y. pestis*, where it remained stable. It is extremely likely that this process is replicable in the natural environment of *Y. pestis*, so pIP1202 may spread within the species. The plasmid is thought to have originated in an enterobacterium, but regardless of its origin the fact that *Y. pestis* was able to acquire a resistance plasmid under natural conditions is indicative that it may happen again, further threatening our ability to treat plague.[7]

Isolation is another important factor in the treatment of plague. All patients should be isolated for the first 48–72 hours after the start of antibiotic therapy, in case they develop pneumonic symptoms. If a patient has no pneumonia or draining lesions after 72 hours, he may be taken out of isolation. However, by law, patients with pneumonic plague must be isolated for their entire course of treatment, to prevent contagion. When caring for pneumonic plague patients, it is important to take droplet precautions and to wear eye protection, in addition to the usual standard precautions.

PREVENTION

Prevention is the best treatment. Attempts to eliminate fleas and rodents from the environment in areas where plague is endemic would be costly and futile, but preventive measures can be taken to reduce the threat of infection in humans. Controlling rodent and flea populations through the use of traps and insecticides can and does keep plague in check, as does advising people who live or engage in outdoor activities at enzootic foci to avoid rodent nests and to use insect repellents and insecticides. Hunters and people who work with potentially infected animals should always wear gloves. Veterinary workers in areas where plague is endemic should be informed of the particular risks of handling cats infected with *Y. pestis*, and should wear gloves, eye protection, and disposable surgical masks when handling suspect cats. *Y. pestis* organisms can be safely handled in laboratories using standard microbiological methods, but Biosafety Level 2 precautions and containment should be used when processing potentially infectious clinical specimens and cultures, and Biosafety Level 3 conditions should be used for work with large amounts of bacteria or when there is a high potential for aerosolization.

USE OF ANTIBIOTICS FOR PROPHYLAXIS

Antibiotics are sometimes recommended for prophylaxis, but only following high-risk exposure to pneumonic plague because of concerns about exacerbating drug resistance. The preferred treatment regimens for postexposure prophylaxis of adults, children, and pregnant women are outlined in table 10.7.[18]

Antibiotic prophylaxis should be maintained for 7days and patients monitored for fever and cough. Contacts who develop these symptoms should receive the antibiotic treatment appropriate for *Y. pestis* infection.

VACCINES

No vaccines are currently available in the United States for use against *Y. pestis*. Two vaccines were once used. The first, EV76, a live, attenuated mutant strain of *Y. pestis*, was introduced in 1908 and is no longer commonly available. EV76 had a number of flaws: it was not avirulent, offered no protection against pneu-

monic plague, and may have been unsafe. The second, USP, a formalin-inactivated, whole-cell vaccine, was manufactured from 1940 to 1999, when it was
discontinued; since then, all remaining stocks have expired.[24] USP was used for
the protection of people in high-risk groups—specifically, laboratory personnel
who routinely worked with *Y. pestis* and people who had regular contact with
rodents or fleas in enzootic foci. However, it was never tested in controlled studies: evidence of its efficacy against bubonic plague is available from retrospective
studies of U.S. military personnel who received the vaccine during World War
II and the Vietnam War, and there is no evidence of its efficacy against primary
pneumonic plague. Among the vaccine's other flaws, it caused adverse reactions
in a significant percentage of those vaccinated, including edema and induration
at the injection site, malaise, headache, fever, lymphadenopathy, and anaphylaxis. Also, the antibodies against the vaccine waned quickly, requiring booster
shots every 1–2 years, and adverse reactions tended to worsen with successive
doses of vaccine.[28]

New vaccines that address the problems of the two older ones are in development. A recombinant vaccine (F1+V) that includes both F1 and V antigen, in
a 2:1 ratio, adjuvanted with alhydrogel, offers protection against both bubonic
and pneumonic forms of plague. It has completed phase 1 clinical trials and is
beginning phase 2. Another recombinant vaccine (F1–V), composed of a single
fusion protein of F1 and V antigen, has been developed and is currently in preclinical trials in mice and nonhuman primates. This vaccine has been shown to
protect against both F1-positive and F1-negative strains of pneumonic plague,
as well as against bubonic plague.[11] The subunit vaccines F1+V and F1–V are so
similar that it is difficult to know which is better, but eventually one of the two
will be selected for use by the U.S. Department of Defense. A third promising
new vaccine, composed of V antigen in a plasmid DNA construct, is currently
in preclinical testing in mice, in which immunization boosted with recombinant antigens gave protective immunity against a lethal dose of *Y. pestis*.[24]

Other research is exploring nontraditional ways to deliver a vaccine, for example, nasal administration, which may provide better protection for against *Y.
pestis* challenge to mucosal surfaces.[22]

BIOTERRORISM POTENTIAL

The CDC has classified plague as a Category A, high-priority organism. This
means that plague poses a high risk to national security because of its suitability
for use as a biological weapon. *Y. pestis* is accessible, simple to produce, and,
because it can be delivered in aerosol form, is easily disseminated. It is also efficient: a dose of only 100–500 bacteria is sufficient to cause pneumonic plague,

which is a severe illness with a high mortality rate. Pneumonic plague is easily communicable from person to person, and large outbreaks have occurred in the past. An outbreak would cause widespread fear and panic that would be extremely difficult to contain, even with special action for public health preparedness.

CHARACTERISTICS OF AEROSOLIZED *Y. PESTIS* RELEASE

Y. pestis, which degenerates quickly in sunlight, is viable as an aerosol for only about 1 hour and a distance of up to 10 km. Environmental conditions at the location of release would affect the extent of the outbreak. Wind might spread the bacteria farther than expected; release of the aerosol at night, out of the sunlight, might lengthen the duration of its viability; and high humidity would increase the rate of infection.

No environmental warning system is in place to detect an aerosol of *Y. pestis* bacilli, but an autonomous pathogen-detection system (APDS) has recently been developed to continuously monitor the air for biological threat agents and to warn of a biological attack in critical or high-traffic facilities and at special events. In a research study, the system was challenged with a variety of pathogens, including *Y. pestis*, with promising results.[12]

CLINICAL FEATURES

Plague used as bioterrorist weapon would present with unusual clinical features. Previously healthy patients, probably in an urban area and without the usual risk factors for plague, would present with a severe and rapidly progressive pneumonia. Hemoptysis, gastrointestinal symptoms, and a fulminant course of symptoms would be suspicious signals of pneumonic plague. Many such cases would present over 1 to 6 days following the release of *Y. pestis*, with most occurring 2 to 4 days after release.

ESTIMATE OF THE DANGER OF A PLAGUE BIOWEAPON

In 1970, a WHO report estimated than an aerosol release of 50 kg of *Y. pestis* over a city of 5 million people in an economically developed country such as the United States would produce 150,000 incapacitating illnesses and 36,000 deaths. These estimates were modest, not taking into account the secondary pneumonic plague cases that would occur through person-to-person contact.

In 2001, a more thorough assessment of the dangers of a plague bioweapon to an American city was made, in TOPOFF, a 3-day drill testing government readiness to respond to terrorist attacks. TOPOFF theorized the covert release of an aerosol of *Y. pestis* in the Performing Arts Center in Denver, Colorado. By noon on the third day of the drill—the third day following release of the *Y. pestis*—there were an estimated 3060 cases of pneumonic plague, 795 of which resulted in death. The (theoretical) plague spread throughout the United States and to London and Tokyo, and travel in or out of state was prohibited in an attempt at limiting further spread of the disease. Colorado officials expressed concern about their ability to obtain food and supplies. Medical care in Denver began to shut down, as hospitals were understaffed, with insufficient antibiotics, ventilators, and beds to meet the enormous demand.[17]

DEFENSES

In event of attack with aerosolized *Y. pestis*, early treatment with antibiotics, coupled with the use of surgical masks to prevent further transmission of pneumonic plague, would be the primary means of defense.

Unfortunately, TOPOFF made it clear that, although straightforward, these defenses would not be easy to enact. By the second day of the drill, hospitals were full to capacity and running out of antibiotics. There were difficulties getting antibiotics from the National Pharmaceutical Stockpile (NPS) to distribution points where they could be given to people who needed them for treatment and prophylaxis. It took 3 days before word got out to the population of Denver that they should wear facemasks. Clearly, there is a need for better sources of information, communication, and supply in the medical and public health infrastructure in order to mount an effective defense against a plague bioweapon.[17]

The potential of *Y. pestis* for antibiotic resistance is also a major threat to defense against plague. The sequencing of the *Y. pestis* genome, which has made it easier for scientists to search for plague cures, might also make it easier for terrorists to create antibiotic resistant strains. A transfer of genes for antibiotic resistance in *Staphylococcus aureus*, an extraordinarily resistant bacterium, into the *Y. pestis* genome could result in an uncontrollably lethal strain. But terrorists looking for antibiotic-resistant *Y. pestis* may not need advanced degrees in microbiology; such strains may already be obtainable. The former Soviet Union is rumored to have developed a form of *Y. pestis* that is resistant to 16 antibiotics. The naturally resistant 17/95 strain also has ominous potential as a weapon, despite its known sensitivity to cephalosporins, other aminoglycosides, quinolones, and trimethoprim. In a bioweapon-induced epidemic, when a large, readily available supply of antibiotics is crucial, any limitation on the type of antibiotic that may be used would impose an insupportable restriction on the ability to treat the vast numbers of people affected.

The added threat of antibiotic resistance makes it all the more important that new therapies be found to supplement the drugs currently being used to combat plague. In addition to the new vaccine candidates, it would be useful to develop new drugs to impede *Y. pestis* pathogenesis. The yersiniabactin system, which is involved in iron acquisition during the earliest stages of plague, may be a good target for early intervention. Also showing promise is research to identify the molecules involved in the adhesion of *Y. pestis* to the lungs,[9] so that therapies might be developed to counteract them and prevent pneumonic plague infection from the start.

REFERENCES

1. Boccaccio, G. 1972. G. H. McWilliam, trans. *The Decameron*, 2nd ed., p. 23. London, Penguin Books.

2. Carmichael, A. G. 1986. *Plague and the Poor in Renaissance Florence*. Cambridge: Cambridge University Press.

3. Chandavarkar, R. 1992. Plague panic and epidemic politics in India, 1896–1914. In P. Slack, ed. *Epidemics and Ideas*, pp. 203–240. Cambridge: Cambridge University Press.

4. Chanteau, S., et al. 2003. Development and testing of a rapid diagnostic test for bubonic and pneumonic plague. *Lancet* 361: 211–216.

5. Cole, S. T. and C. Buchrieser. 2001. Bacterial genomics: A plague o' both your hosts. *Nature* 413: 467–470

6. Cornelis, G. R. 2000. Molecular and cell biology aspects of plague. *Proc Natl Acad Sci USA* 97(16): 8778–8783.

7. Galimand, M., et al. 1997. Multidrug resistance in *Yersinia pestis* mediated by a transferable plasmid. *N Engl J Med* 337(10): 677–680.

8. Gottfried, R. S. 1983. *The Black Death: Natural and Human Disaster in Medieval Europe*. New York: The Free Press.

9. Gregory, A. P. 2002. *Countering Bioterror: The New Threat of Plague*. University of Kentucky: Odyssey online, http://www.rgs.uky.edu/ca/odyssey/fall02/bioterrorism.html.

10. Guiyoule, A., et al. 1997. Recent emergence of new variants of *Yersinia pestis* in Madagascar. *J Clin Microbiol* 35: 2826–2833.

11. Heath, D. G., et al. 1998. Protection against experimental bubonic and pneumonic plague by a recombinant capsular F1-V antigen fusion protein vaccine. *Vaccine* 16(11–12): 1131–1137.

12. Hindsen, B. J., et al. 2005. Autonomous detection of aerosolized biological agents by multiplexed immunoassay with polymerase chain reaction confirmation. *Anal Chem* 77(1):284–289.

13. Hinnebusch, B. J., Perry, R. D., and Schwan, T. G. 1996. Role of the *Yersinia pestis* hemin storage (hms) locus in the transmission of plague by fleas. *Science* 273(5273): 367–370.

14. Hinnebusch, B. J., et al. 2002. Role of Yersinia murine toxin in survival of *Yersinia pestis* in the midgut of the flea vector. *Science* 296(5568): 733–735.

15. Hueck, C. J. 1998. Type III protein secretion systems in bacterial pathogens of animals and plants. *Microbiol Mol Biol Rev* 62: 379–433.

16. Infectious Diseases Society of America. *Bioterrorism Information and Resources.* Online: http://www.idsociety.org.

17. Inglesby, T. V., Grossman, R., and O'Toole, T. 2000. A plague on your city: Observations from TOPOFF. *Biodef Q* 2: 2.

18. Inglesby, T. V., et al. 2002. Plague as a biological weapon. In D. A. Henderson, T. V. Inglesby, and T. O'Toole, eds. *Bioterrorism: Guidelines for Medical and Public Health Management.* Chicago: AMA Press.

19. Kahn, J. 2002. Shouting the pain from Japan's germ attacks. *New York Times*, Nov. 23, p. A4.

20. Laboratory Response Network (LRN). 2001. *Level A Laboratory Procedures for Identification of* Yersinia pestis. Atlanta: Centers for Disease Control and Prevention.

21. Lillard, J. W., S. W. Bearden, J. D. Fetherston, and R. D. Perry. 1999. The haemin storage phenotype of *Yersinia pestis* is not essential for the pathogenesis of bubonic plague in mammals. Microbiology 145(Pt 1): 197–209.

22. Mandel, C. 2001. Sniffing out a plague vaccine. *Wired News.* Online: http://www.wired.com/news/conflict/0,2100,48432,00.html.

23. Matson, J. S. and M. L. Nilles. 2001. LcrG-LcrV interaction is required for control of Yops secretion in *Yersinia pestis. J Bacteriol* 183(17): 5082–5091.

24. Nierengarten, M. B. and L. I. Lutwick. 2002. Vaccine development for plague. *Medscape Infect Dis* 4:2.

25. Patt, H. A. and R. D. Feigin. 1997. Diagnosis and management of suspected cases of bioterrorism: A pediatric perspective. *Pediatrics* 109: 4.

26. Perry, R. D. and J. D. Fetherston. 1997. *Yersinia pestis*: Etiologic agent of plague. *Clin Microbiol Rev* 10:3–66.

27. Slack, P. 1985. *The Impact of Plague in Tudor and Stuart England.* London: Routledge and Kegan Paul.

28. U.S. Department of Health and Human Services. 1996. Prevention of plague: Recommendations of the Advisory Committee on Immunization Practices (ACIP). *MMWR Recomm Rep* 45(RR-14): 1–15.

CHAPTER 11

■ ■

SMALLPOX

(VARIOLA VIRUS)

Rohit Puskoor and Geoffrey Zubay

Prior to the mid-twentieth century, smallpox was one of the most dreaded diseases. Even if you recovered, your face and body were covered with pockmarks for the rest of your life. It was extremely infectious from one human to another and one of the most lethal diseases in circulation. Vaccines were first introduced against smallpox, and they were enormously effective almost from the very beginning. Vaccinia virus is currently used to make vaccine against variola virus, the virus that causes human smallpox. Except in very rare cases, the vaccinia-derived vaccine has no adverse reactions in humans. Because poxviruses tend to be very host-specific, there is no smallpox reservoir other than what exists in human hosts. Realizing this, immunologists organized a worldwide organization with the intention of completely eradicating smallpox. All they needed was a very large supply of effective vaccine and many field workers.

Area by area, smallpox was eliminated. If a case of smallpox was spotted, every effort was made to isolate the victim and vaccinate everyone who might have come into contact with him or her. By 1977 it appeared that smallpox had been eradicated worldwide, and no new cases have been reported since. The Soviet Union and the United States, however, held onto supplies of smallpox for military, defense, and research purposes. There is a fear that these stocks may have been pilfered and that there may be others—including terrorist groups—in possession of smallpox viruses. Because most people have not been vaccinated against smallpox for more than a quarter of a century, the possibility of uncontrolled possession makes smallpox an extremely serious threat, perhaps even more serious than it was when vaccination against smallpox was a common practice.

HISTORY

Virtually unstoppable smallpox epidemics date back many thousands of years (table 11.1). The search for a cure and the progress in our current understanding of immunity and vaccine development have gone hand in hand. Edward Jenner's development of the smallpox vaccine provided the means to control the spread of the pathogen and ultimately to eliminate it as a natural cause of disease.

In ancient times, smallpox ravaged Mesopotamia, China, and the Roman Empire.[18] Records from the Indian subcontinent from around 1500 B.C. de-

TABLE 11.1 THE HISTORY OF THE ERADICATION OF SMALLPOX

4000 B.C.	Smallpox orginiates in Asia or Africa
1157 B.C.	Pharaoh Ramses V dies of smallpox
910 A.D.	Clinical profile first described by Rhazes
1096–1291	Crusaders bring smallpox back to Europe
1400–1800	European fatalities routinely in excess of 500,000 per year
1520	Aztec empire collapses
1723	Variolation introduced to Europe
1763	First use of smallpox as a bioweapon (against native American Indians)
1796	Vaccination introduced by Jenner
1840	Variolation outlawed in Europe
1950	Freeze-dried vaccinia developed
1967	WHO-intensified eradication program initiated
1977	Last natural case of smallpox (Somalia)
1993	Variola DNA genome sequenced
1999	IOM report recommends further research into variola virus (United States)
2001	United States announces that the CDC's variola stock will be retained
2002	WHO recommends postponement of destruction of variola virus stock

Note: The advances in smallpox treatment and vaccines coincided.

scribe a smallpox-like disease. One hundred fifty years later, a smallpox epidemic spread from Egypt to the Hittite Empire in Syria. Around 250 B.C., smallpox was introduced to China during the Hun invasions from Central Asia. The pathogen swept throughout the entire Chinese empire, causing a major health crisis.

Only 200 years later, during the reign of the emperor Huang Wu Ti of the Han dynasty, another smallpox epidemic struck the nation. One of the emperor's best generals, Ma-Yuan, was sent with an army of 40,000 men to quell an uprising in the northern Hunan province. The general led his army into a district of the hostile province where an epidemic was in progress; the soldiers fell ill, and over half the troops, including Ma-Yuan himself, died of smallpox. The rest returned to southern China, bringing with them the disease.

The earliest known smallpox epidemic after the birth of Christ was the "Plague of Antoninus," which struck the Roman Empire during the reign of Marcus Aurelius. The contagion was described by the highly regarded Greek physician Galen as causing fever, thirst, vomiting, diarrhea, and lesions that covered the entire body. The epidemic started in a Roman army fighting under General Avidius Cassius in Mesopotamia. The soldiers brought the pathogen to Rome, where it killed nearly 2000 Romans a day for the next 15 years. Marcus Aurelius was among the fatalities. It is estimated that before the epidemic came to a close about 5 million Romans lost their lives.

EARLY METHODS OF DIFFERENTIAL DIAGNOSIS

By the tenth century, Islamic medical scholars and physicians began to focus on fighting deadly epidemics. In 911, the Islamic physician Rhazes published *A Treatise on Smallpox and Measles*, which explained how to differentiate between the two diseases. Rhazes believed that once the disease had passed the individual would be immune to smallpox, as pustules would not form once the "ferments had been permanently expelled." His theory on immunity would later prove to be true. He also hypothesized that the ideal treatment for a smallpox infection was bloodletting followed by exposure to high temperatures. Rhazes' paper became a landmark in the medical management of smallpox infection; it was translated into Greek and Latin and influenced the treatment of patients throughout the seventeenth century.

THE HISTORY OF SMALLPOX AS A WEAPON

In the Western world, the smallpox virus often had devastating effects; the native populations had never experienced the disease and thus had developed no immunity to it. In the fifteenth century, the Spanish explorer Camille Pizarro presented the Incas of Peru with variola-contaminated clothing in order to weaken opposition to Spanish conquest. Using similar techniques, Hernando Cortez was able to conquer the well-organized and militarily powerful Aztec empire in Mexico.

Smallpox virus was an effective weapon because it decimated the native population while leaving the European invaders—most of whom had developed immunity due to prior infection—relatively unscathed. This method of dealing with opposition became popular among European forces in the New World. During the French and Indian War, at the 1763 siege of Fort Pitt, the British gave items from their smallpox hospital to the local Indians, who were generally loyal to the French. Sir Jeffrey Amherst recommended this policy in a letter: "You will do well to try to inoculate the Indians by means of blankets, as well as to try every other method that can serve to extirpate this race."[12]

When the Continental Army surrounded and laid siege to Boston in 1775, General Howe, the British commander, ordered that fleeing civilians be inoculated with smallpox, hoping to spread the disease to the American troops surrounding the city. In the winter of 1775–1776, the British followed a similar policy, inoculating civilians and sending them to mingle with American troops. The ensuing devastation of the Continental Army led General Washington to order that all newly consigned troops be variolated (an early form of vaccination described below).

IMMUNITY THROUGH VARIOLATION

In response to the devastation of smallpox epidemics, several cultures developed methods of inoculating healthy individuals with pustule fluid or ground scabs from infected individuals. In Europe, the material was injected intradermally into the skin. The Chinese used a process called "insufflation" in which the inoculum was instilled into the nasal mucosa. European variolation usually produced an attenuated case of smallpox that eventually cleared up and left the individual with immunity against further infection. However, in about 1% of cases, this method led to severe infection and death. Also, variolated individuals could spread the smallpox virus to others, leading some to argue that variolation contributed to the epidemic spread of the disease.

Lady Mary Wortley Montague is believed to have been responsible for introducing a variant of the Chinese method to English society in the eighteenth century. She had her four children inoculated by this much safer Asian method and, on witnessing its safety and efficacy, used her influence in society to promote the technique. The method was used throughout Europe until Edward Jenner introduced the vaccination procedure in 1796.[18]

THE VACCINIA VIRUS-BASED VACCINE

The research of Edward Jenner yielded a successful and safer alternative to vari-olation that would become the foundation of the Global Eradication Program of the mid- to late 1900s. Noticing that milkmaids and dairy farm workers who had previously been infected by the cowpox virus (a homolog of the smallpox virus) were immune to the smallpox virus, Jenner investigated whether infection by a less lethal but related virus was responsible for mediating immunity. His article describing the technique marked the beginning of the vaccination era.

The idea of creating immunity in an individual by challenging his immune system with an altered, less virulent form of a pathogen, or an avirulent or less virulent infectious homolog, was a major milestone in smallpox vaccination methodology. Development of this method led to the use as a vaccine of the vaccinia virus, which shares over 90% DNA sequence homology with the small-pox virus but is avirulent in humans.

ERADICATION OF SMALLPOX

Despite the widespread administration of smallpox vaccine, little progress was made toward the goal of total eradication until 1950, when it became possible to preserve the vaccine by a freeze-drying process. The vaccine could be trans-ported and used in the field without refrigeration or any diminution of its effi-cacy. In 1959, the World Health Organization (WHO) gave formal support to a Soviet proposal to eradicate smallpox through a global vaccination program. The program did not begin to take root until 1967, when the WHO took the lead and commenced the Global Smallpox Eradication Program, which empha-sized surveillance and containment. In a process known as "ring vaccination," new cases of smallpox were identified and isolated rapidly. In addition, all indi-viduals who had had contact with those cases were vaccinated and quarantined, thus breaking the chain of human-to-human transmission.

The last known smallpox cases in the United States were observed in 1949. In 1967, when the global eradication program began, India had approximate-ly two-thirds of the world's smallpox cases. In 1974, one last epidemic swept through northeast India, killing over 11,000 residents in the Bihar state in one month. Shortly thereafter, the Indian subcontinent was cleansed of all smallpox; the last case in India occurred in 1975. The last naturally occurring case of smallpox occurred in Somalia.

In 1979, the WHO declared that the global eradication of smallpox was complete.[1] The success of this program was due to the fact that the smallpox virus is host-specific for humans and lacks an animal reservoir. Thus, once the disease had been eliminated from all human hosts, there was no risk of its being reintroduced to the human population through a mammalian or insect vector. Second, latent smallpox infections do not occur; individuals either die from the disease or recover and gain immunity. Finally, because the symptoms of smallpox infection are easily identifiable, infected persons were easily identified and quickly quarantined.

REMAINING STOCKS OF VARIOLA VIRUS

After the eradication of smallpox was confirmed in 1979 and certified in 1980, the WHO supported the destruction of smallpox virus stocks as the final step in the eradication program. Despite the recommendation, there are at least two known stocks of smallpox virus: one at the Centers for Disease Control and Prevention (CDC) in Atlanta, Georgia, and the other at the Russian State Center for Research on Virology and Biotechnology in Koltsovo. Arguments for and against total eradication have been raised. It seems unlikely that the existing stocks will be destroyed, but even if they were, because the total DNA sequence of the smallpox genome is known, reconstructing the virus is technically feasible.

MOLECULAR BIOLOGY

Ethical and public health concerns present major obstacles for in vivo smallpox research. Because of this, the molecular biology of the smallpox virus and its proteins has not been as well characterized as it has been for many related viruses. Most theories of the smallpox virus' molecular mechanism of action have been based on in vitro studies of molecularly engineered smallpox proteins or in vivo research performed on homologous viruses.

Smallpox belongs to the family *Orthopoxviridiae*, a family of viruses with highly specific host orientations. It is a human-specific orthopoxvirus, with a very narrow host range among laboratory animals. Macaques are among the few nonhuman mammals in which the virus is pathogenic. Cowpox, monkeypox, and vaccinia are the only other orthopoxviruses that can infect humans. Monkeypox causes a rash very similar to that of smallpox, whereas cowpox and vaccinia usually generate small lesions on the skin at the site of entry.

THE SMALLPOX GENOME

The brick-shaped virions of the smallpox virus are larger than many bacteria and can be visualized by light microscopy (figure 11.1). The pathogen becomes virulent once it has entered the cell and has lost enough of the viral envelope

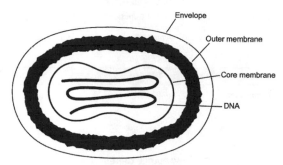

FIGURE 11.1. Cross-section of a smallpox virus. Smallpox is a very large virus, larger than many bacteria. The virion consists of four major parts: the core, lateral bodies (not shown), the outer membrane, and the envelope, the last of which is degraded once the virion enters the cell. Disulfide-linked proteins are believed to hold together the outer membrane. The dumbbell-shaped core contains the viral DNA and is surrounded by oval-shaped, trypsin-sensitive lateral bodies. Both are surrounded by the outer membrane, which is composed of large protein tubules and ribbed in texture. The viral DNA is associated with four proteins.

to allow its genetic material to be transcribed and translated. Infection of tissue cultures by orthopoxviruses results in cytopathic changes, changes in membrane permeability, and inhibition of host DNA, RNA, and protein synthesis.

The virion consists of four major parts: the core, the lateral bodies, the outer membrane, and the envelope, which is degraded once the virion has entered the cell. The dumbbell-shaped core containing the viral DNA is surrounded by oval, trypsin-sensitive lateral bodies. They are all surrounded by the outer membrane, which is composed of large protein tubules and ribbed in texture.

The 186-kb smallpox genome ranks as one of the longest viral genomes and consists of a single, linear, double-stranded piece of DNA (figure 11.2). All orthopoxviruses have *inverted terminal repeats* (ITRs)—identical but oppositely oriented sequences that occur near both ends of the genome.[14] Genes encoded within the terminal repeats are variable and possess host interaction functions. Genes encoded within the central region of the genome are usually highly conserved across the *Orthopoxviridiae* family[13] and perform essential functions, including transcription and replication.

THE INFECTIOUS AGENT

The existence of multiple infectious forms of the smallpox virion has made it difficult to identify the cellular receptors involved in smallpox entry into host

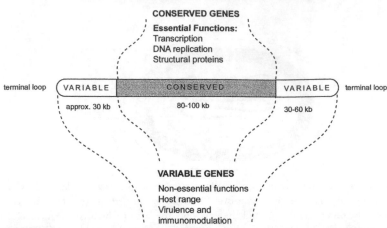

FIGURE 11.2. The orthopoxvirus genome. The average orthopoxvirus genome is 200 kb long. Near the termini of the viral genome, there are inverted terminal repeats, identical but oppositely oriented sequences that occur at both ends of the genome. The genome is divided into a highly conserved central region, which mainly encodes genes that are essential for virus replication and transcription and genes that code for structural proteins, as well as variable terminal regions that encode proteins that are nonessential for virus replication. These proteins govern the pathogen's virulence, ability to influence the host immune response, and the host range. The length of the central conserved region is fairly constant, but the terminal regions vary in length from 0.65 to 12 kb.

cells. Each form is believed to use a different receptor and interact with it using a variety of virion surface proteins. We restrict the discussion to the two most common forms.

Externally enveloped virions (EEVs) and cell-associated enveloped virions (CEVs) are particularly important for variola spread between the cells of the host organism.[15] EEVs are believed to enter cells by endocytosis; this is followed by disruption of the EEV outer membrane, which induces the release of a mature virus particle into the cytoplasm.

Cell-to-cell spread is best executed by CEVs, which possess the ability to attach to the plasma membrane of the host cell. CEVs are located at the tips of long extracellular microvilli that form by attachment of host actin tails to the virion. These actin tails are responsible for CEVs' greater efficiency in spreading,[11] because they allow the virions to be more effectively targeted to neighboring cells than if they were simply extruded into the extracellular matrix as EEVs are. CEVs are believed to have the same basic structure as EEVs and thus seem to follow a similar pathway of entry into and exit from the host cell. CEVs mediate efficient cell-to-cell spread between neighboring cells, whereas EEVs mediate long-range transmission.

THE TRANSCRIPTION MECHANISM

During the later stages of infection, the enzymes needed for early-stage transcription are synthesized and packaged with the core of the virion. Upon the core's entry to the host-cell cytoplasm and uncoating of the virion, the nucleoprotein complex passes through breaches in the core wall and enters the cytoplasm, where it is transcribed by the viral early-stage transcription machinery.

The primary enzyme of the smallpox transcriptional machinery is a viral DNA-dependent RNA polymerase (DdRp). Other enzymes involved in transcription of the smallpox genome include a 94-kd RNA polymerase–associated protein (RAP94), early transcription factor (ETF), capping and methylating enzymes, poly(A) polymerase, and nucleoside triphosphate phosphohydrolase I (NPH I). The RNA polymerase and its associated viral enzymes alone are sufficient to carry out mRNA synthesis, as long as they are provided with the necessary nucleotide building blocks. The mRNA products of the smallpox transcription machinery are polyadenylated, methylated and capped.

Because the early-stage machinery is packaged with the viral core, viral mRNA can be synthesized soon after infection of the host cell. Early-stage viral mRNA accumulates to maximum levels within 2 hours of infection.[17] Among the genes being transcribed at this stage are the encoding proteins involved in host interactions, viral DNA synthesis, and the regulation of intermediate gene expression.[16]

THE INFECTIOUS CYCLE

Following the expression of early-stage genes, orthopoxviruses replicate their genome, after which they direct resources toward the transcription of intermediate- and late-stage genes (figure 11.3). Intermediate-stage transcription begins about 2 hours after infection and reaches peak levels shortly thereafter. It is necessary for DNA replication to occur prior to intermediate- and late-stage transcription because the portions of the orthopoxvirus genome that encode intermediate- and late-stage genes are buried within the virion and are thus inaccessible to the early-stage transcription machinery. Replicated daughter DNA, which remains free-floating in the host cytoplasm until it is packaged in new infective particles, is free of any such hindrances and provides a suitable template for the transcription of intermediate- and late-stage genes.

Six orthopoxvirus genes have been formally classified as intermediate-stage genes, although additional genes are known to exist. Three of these encode activators of late gene expression, and another encodes an RNA helicase. The remaining identified genes encode a single-stranded DNA binding protein and a serine protease inhibitor (SERPIN).[5]

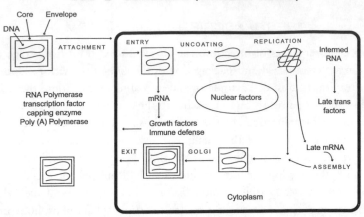

FIGURE 11.3. The smallpox life cycle. The virion consists of the viral DNA genome packaged with proteins necessary for the transcription of the early-stage genes. These proteins include RNA polymerase, transcription factor, 5′-capping enzyme, and a poly(A) tail polymerase. The virion attaches to and enters the host cell by endocytosis. While still packaged in the virion, the virus begins to transcribe its early genes, which encode growth factors and molecules for protecting the virus from the host immune response. After early-stage transcription is complete, the virus is activated by the uncoating of the viral DNA and packages proteins; a minor pH disruption across the virus particle's outer membrane triggers the release of viral DNA and proteins. The virus then begins to replicate its DNA genome, after which intermediate-stage transcription begins. Intermediate-stage genes generally encode late-stage transcription factors; the end of intermediate-stage transcription signals late-stage transcription to begin, during which genes encoding early transcription factors, virion structural proteins, and late-stage enzymes are expressed. Following late-stage transcription, the virion assembles and then receives its outer membrane through wrapping in the Golgi. After this, it is exocytosed by the cell.

The transcription of late-stage genes directly follows that of intermediate-stage genes. Synthesis of late-stage mRNA begins about 140 minutes after infection of the host cell and continues for roughly 48 hours. The proteins encoded by the late-stage genes include the major protein components of the virion, the early-stage transcription machinery (which is packaged with the orthopoxvirus virion), and other virion enzymes.[5]

DNA REPLICATION

Smallpox DNA replication, like transcription, takes place in the cytoplasm. The prereplication machinery, encoded by early-stage genes, includes thymidine kinase, thymidylate kinase, ribonucleotide reductase (an enzyme that converts ribonucleotides to deoxyribonucleotides), and dUTPase.

The presence of ribonucleotide reductase allows viral DNA synthesis to take place in the cytoplasm through the conversion of ribonucleotides, which are widely available in the cell cytoplasm, into deoxyribonucleotides, which are usually congregated in the host nucleus.

DNA replication begins a little over 1 hour following infection and results in the synthesis of approximately 11,000 copies of the orthopoxvirus genome per infected cell; half of these are eventually packed into virions. The terminal 200 bp of the viral genome have been shown to have an enhancing effect on DNA replication, suggesting that it acts as an origin of replication.

The most likely model for orthopoxvirus replication involves a nick near one or both of the hairpin termini. This is followed by addition of nucleotides at the now-freed 3′ end and gradual displacement of the mother strand. Replication of the hairpin termini forms a short piece of DNA that then serves as a primer for daughter-strand synthesis (a mechanism known as self-priming replication).

VIRION ASSEMBLY

When the components of the virion have been synthesized, they are ready to be assembled. This must occur before cell lysis, when the infectious virus particles are released.

Virion assembly begins in the host cytoplasm. The first visible virion structure formed is a crescent consisting of a membrane with a border of spicules on the bowed surface and granular material next to the hollow surface. The crescent structures then bubble out to form spherical intracellular immature virions. The nucleoprotein complex, made up of the newly synthesized genomic DNA and DNA-associated late-stage proteins (including ETFs), enters the crescent envelope just before it is sealed; the nucleoprotein complex is then embedded in the matrix of the mature, spherical membrane.

The movement of the immature virion to the periphery of the cell is mediated by host-cell microtubules. During this process, the immature virion is enveloped by additional membranes derived from the trans-Golgi and early endosomes that originate from the trans-Golgi. Fusion of the enveloped particles with the host-cell membrane results in the extrusion of the virus from the host cell and loss of the outermost membrane. Although a few of the extruded virions are found in the extracellular environment as fully independent EEVs, most adhere to the cell surface as CEVs.

THE IMMUNE RESPONSE

The smallpox virus produces several classes of defense proteins that protect it from the host immune system. These proteins fall into three major classes: virokines, which resemble cytokines and other secreted regulatory proteins and are secreted by infected cells; viroreceptors, which are homologs of cellular receptors lacking membrane-anchoring sequences that bind their ligands to form free-floating ligand–receptor complexes in the extracellular environment, and intracellular proteins, which interfere with the host's normal signaling and enzymatic pathways. The virulence of smallpox virus, despite its remarkable genetic similarity to the nonpathogenic vaccinia virus, suggests that smallpox proteins are uniquely suited to overcome the human immune response.[1,2,4,6–8,15–17]

CLINICAL DIAGNOSIS AND RESPONSE

SYMPTOMS

The smallpox virus results in characteristic bumpy lesions that cover a patient's face and body. Those who manage to survive the illness are usually left with disfiguring scars. Blindness occurs in about 10% of cases, especially when patients are malnourished or have a secondary infection. Smallpox lesions are concentrated on the face because of the high density of sebaceous glands and the deeper pitting associated with infection of such glands. The virus is relatively resistant to environmental conditions, persisting in aerosolized form for up to 1 week and in crusts for several years. Variola virus major causes four major clinical types of smallpox: the ordinary type, the modified type, the flat type, and the hemorrhagic type. The only known treatment for the disease is vaccination with the closely related vaccinia virus.

Ordinary-Type Smallpox For ordinary-type infections, the incubation period varies anywhere between 7 and 19 days, with most cases usually requiring 11–14 days after exposure before symptoms are manifest. The initial symptom is an abrupt fever of 38.5–40.5°C; in addition, most patients complain of malaise. Other early symptoms include a splitting headache and severe backaches. In rare cases, children experience convulsions and adults display delirium. Fifty percent of all patients present with nausea and vomiting, and diarrhea occurs in about 11% of infected individuals. During this early stage, known as the prodrome or prodromal phase, many clinicians misdiagnose the disease as appendicitis, because of the abdominal colic experienced by some patients. A transient rash sometimes occurs in the prodrome. By the second or third day of the early-stage symptoms, the fever abates.

When the initial fever has subsided, the eruptive phase begins. Lesions on the mucous membranes are the first to appear; minute red spots are visible on the tongue and palate 24 hours before the commencement of a macular rash on the skin. Lesions also occur along the pharynx, causing some patients to complain of sore throat. Although these pharyngeal lesions are not as visually impressive as skin lesions, they provide the major source of viral particles for airborne transmission to other individuals. The macular skin rash usually appears 2 to 4 days after the onset of fever as a few small "herald spots" on the face, especially the forehead. Lesions quickly spread first to the trunk and the proximal portions of the extremities and then to the distal portions of the extremities. Within 24 hours of the initial appearance of the rash, it has spread to all parts of the body.

By the second day of the rash, the lesions or macules have become raised because of the effusion of fluid into the tissue spaces and are now called *papules*. By the seventh day, these papules develop into pustules, which reach their maximum size on the tenth day of the eruptive phase. On the eleventh day, the pustules begin to flatten and the fluid is slowly reabsorbed. The central portion of the pustules harden and a scab or crust eventually forms. These scabs, which are highly infectious, flake off and leave a depigmented area within 3 weeks, although the lesions on the palms and soles persist much longer because of the thickness of the skin at those sites. Respiratory complications sometimes arise as a consequence of secondary bacterial or viral infections.

In fatal cases, death occurs between the eleventh and sixteenth days of the illness. Mortality depends on the extent of the rash. Three rash types are common with ordinary-type smallpox: confluent, semiconfluent, and discrete. Confluent ordinary-type smallpox describes cases in which the pustules on the extremities and face join together. Patients are febrile for a longer period of time than in semiconfluent and discrete infections, and toxemia sometimes does not end after the pustules have developed into scabs. The case mortality rate for unvaccinated individuals displaying confluent ordinary-type smallpox is 62%.

With semiconfluent ordinary-type smallpox, the rash is confluent only on the face. On the trunk and extremities, the pustules are discrete. A secondary fever sometimes develops during the pustular stage, but the temperature falls and toxemia abates as scabbing begins. In these cases, the fatality rate is about 37%. The most common form of ordinary-type smallpox is discrete, which occurs in 40–60% of patients infected with the variola major virus. In discrete ordinary-type smallpox, the patient has fewer lesions, which are separated by normal skin. The case fatality rate is much lower—approximately 9% of unvaccinated individuals.

Modified-Type Smallpox Modified-type smallpox is characterized by the same symptoms as seen in ordinary-type smallpox, but these symptoms present and conclude much more rapidly. Prodromal symptoms are far more severe and tend to continue after the appearance of a macular rash. There tend to be fewer lesions than in ordinary-type smallpox. Macules evolve more quickly, are more superficial, and are not as uniform as the lesions that develop in ordinary-type infections. Scabbing is complete within 11 days of initial appearance of symptoms. The most common reason for an accelerated course of smallpox infection is prior vaccination, although vaccination is not required for an individual to experience this form of smallpox infection. No fatalities have been observed among cases categorized as modified-type.

Flat-Type Smallpox Flat-type smallpox, which occasionally occurs in unvaccinated individuals, is so named because the lesions in this form of smallpox infection remain flush with the skin. The majority of cases occur in children. The prodromal phase lasts for 3–4 days and is similar to the prodromal phase for ordinary-type infections. However, toxemia is severe, the fever remains elevated throughout the prodromal phase, and the symptoms usually continue into the eruptive phase.

The enanthema (macular rash on the oropharynx and mucous membranes) is extensive and sometimes confluent on the tongue and palate. Unlike those in ordinary-type infections, the lesions evolve very slowly. When the macules develop into papules by the sixth day after the appearance of fever, a small depression in the papule is observed. By the seventh or eighth day, the lesions are flat and appear to be buried under the skin. Many patients experience hemorrhaging into the base of the lesion. The center of the papule turns purple or black and is surrounded by an erythematous areola. Because of the superficial nature of these lesions, the skin covering them often flakes off after slight trauma, exposing extensive raw flesh.

Throughout the progression of the disease, the patient is febrile and experiences toxemia. Respiratory complications such as pulmonary edema and pneumonia may set in by the seventh or eighth day after onset of fever. Unvaccinated children often experienced an acute expansion of the gut between the eighth and twelfth days after fever presentation; these children die within 48 hours of gut dilation. Before death, the flat-type lesions become ash gray. Blood and mucus often appear in the stool in the early stages of flat-type smallpox, and the rectal membrane is sometimes completely sloughed off before death.

In the few patients who survive, scabbing begins between the thirteenth and sixteenth days after fever presentation and ends on day 21. The scabs are thin and flake off rapidly, leaving superficial scarring. As a result of the hemorrhaging into the base of the lesions, the scabs are purplish before they dry and flake off.

Hemorrhagic-Type Smallpox Hemorrhagic smallpox is the most deadly form of the disease; the early subtype kills most of those infected with it. With respect to bioterrorist use of smallpox, this form is also the most worrisome, because it is the form that is believed would develop if individuals were infected with either of the two putative smallpox recombinant chimeras: Ebolapox and

Veepox (see "Weaponization" below). Hemorrhagic smallpox usually occurs in adults (88% of victims are 14 years old or more). Surprisingly, it is as common in vaccinated individuals as it is in unvaccinated individuals. Clearly a vaccination program featuring the vaccinia virus would provide little protection from hemorrhagic-type smallpox.

The early subtype of hemorrhagic smallpox is characterized by hemorrhages into the skin and mucous membranes in the early stages. Conjunctival hemorrhages are most prevalent, and bleeding from the gums, epistaxis, hematemesis, hemoptysis, hematuria, and vaginal bleeding often occur during the course of illness. No distinction between the prodrome and the eruptive phase is drawn, because death occurs before a macular rash fully develops. This subtype commences with the sudden onset of high fever, headache, and backache, which persist until the patient expires. On the second day of fever, generalized erythema (redness of the skin caused by dilation of the capillaries), petechiae, and localized ecchymosis appear. On the third day of fever, the skin feels velvety and has a matte appearance, and 24 hours later it is dark purple. Patients often complain of breathlessness and chest heaviness or pains. They usually die abruptly on the sixth day, maintaining consciousness until death. The cause of death is usually either heart failure or pulmonary edema rather than blood loss due to hemorrhaging.

The late subtype of hemorrhagic smallpox also features hemorrhages into the skin and mucous membranes, but it is further characterized by hemorrhaging into the base of the macules. In addition, hemorrhaging begins after the appearance of the rash instead of concurrently with macule development as is seen in the early subtype. The prodrome lasts for 3–4 days and features a fever of 40°C. Toxemic symptoms are similar in severity to those of the early subtype and continue through the eruptive phase. The macules quickly evolve into papules but mature very slowly from that point forward. They sometimes hemorrhage into their bases, giving them a flat appearance that may lead to misdiagnosis of late hemorrhagic-type smallpox as flat-type smallpox.

Bleeding occurs in the mucous membranes. Hemorrhaging papules flatten and become black; in about 15% of cases, they develop into pustules and then follow the course of ordinary-type smallpox. At that point, hemorrhaging from the pustules ceases and is seen only in the mucous membranes. The majority of patients with late hemorrhagic-type smallpox die within 8–11 days; the mortality rate in cases with flat lesions is slightly higher than in those with raised pustules. In those who survive, hemorrhaging gradually decreases, and patients then enter a long recovery stage.

Those who manage to recover from smallpox infection are physically scarred by permanent facial pockmarks. During the first months of recovery, the sites of lesions are abnormally pigmented. As the skin regains normal pigmentation, most patients develop pitted scars at these locations. These pockmarks, which are caused by fibrosis of the dermis, tend to congregate in the face because of the many sebaceous glands there. Secondary bacterial infections worsen the severity and irregularity of scarring.

Corneal ulceration and keratitis often lead to corneal scarring followed by blindness. Blindness affects only about 0.5% of survivors—most frequently malnourished patients, indicating that poor nutrition and hygiene are key factors. Another serious complication is limb deformity caused by bone lesions. As discussed earlier, bone lesions tend to lead to arthritis; in severe cases, the lesions can also cause acute skeletal manifestations including bone shortening, flail joints, and other bone deformities.

DIAGNOSIS FROM SKIN LESIONS

Clinicians rely on observations of the prodromal fever, the pattern of rash, and the evolution and feel of the skin lesions to diagnose ordinary-type smallpox. The distribution of skin lesions on the body is also a tool for diagnosis; concentration of macules on the extremities, soles, palms, and face, with relatively few on the trunk, indicates a smallpox infection. Vaccination complicates diagnosis because lesions appear only sparsely, if at all, and the course of the disease is accelerated. Flat-type smallpox is also easy to diagnose, owing to the unique flat shape of the pustules. Hemorrhagic-type smallpox is usually impossible to diagnose without laboratory tools, because of the speed with which the disease progresses and the absence of characteristic macules or papules.

It is impossible to clinically diagnose smallpox during the prodromal phase, but epidemiological factors such as exposure to a known smallpox case can suggest the diagnosis. The clinical diagnosis of smallpox during the eruptive phase is quite simple, although the disease is often confused with chickenpox despite the stark differences in character and progression of the rash—chickenpox lesions are far more superficial and appear in crops. In addition, they are concentrated on the trunk and extremities, whereas smallpox lesions are concentrated on the face. Smallpox is often misdiagnosed as human monkeypox, measles, or syphilis, even though the progression and distribution of the rash differentiate smallpox from these other pathogens. Diagnosis of all infections is necessary for infection-control purposes; individuals with partial immunity who develop mild cases of smallpox can serve as carriers of the disease.

ELECTRON MICROSCOPY AND PCR

Despite the relatively high rate of success associated with clinical methods of diagnosing smallpox, laboratory techniques are useful in confirming the initial diagnosis. A diagnosis of smallpox is generally confirmed using electron microscopy, in which Guarnieri bodies (clumps of variola virus particles) isolated from pustular scrapings are visualized. Another rapid technique involves treatment of pustular scrapings with Gilden's modified silver strain, which causes cytoplasmic inclusions such as the Guarnieri bodies to appear black.

These diagnostic tests do not differentiate between smallpox and other members of the orthopoxvirus family.[1] The traditional method of doing so required laboratory workers to isolate the virus from a patient's serum and observe its growth on a chorioallantoic egg membrane for certain smallpox-specific growth patterns. Currently, polymerase chain reaction (PCR)[19] and restriction fragment-length polymorphism techniques[10] are used to differentiate among members of the family, providing a simpler and more efficient method of confirming smallpox virus as the infectious agent.

MEDICAL MANAGEMENT

Treatment of a smallpox infection should involve rapid identification, recognition that the infection may be the product of a bioterrorist attack, and quarantine of all infected individuals. All confirmed cases of smallpox should immediately be reported to public health officials. The CDC guidelines *Droplet and Airborne Precautions* should be undertaken for a minimum of 17 days for all individuals who have had direct contact with the index case. Without vaccination, the likelihood of infection for close contacts ranges between 40 and 90%.[18] Any contacts of the index case with a fever of 38°C or higher within this period should be diagnosed as a smallpox patient. The patient should then be isolated, preferably within the home, and maintained in isolation until either the initial diagnosis has been ruled out or the patient's scabs flake off; smallpox patients are highly infectious until all scabs are gone. Although vaccination within 7 days of exposure can be of some utility in preventing disease or limiting the severity of symptoms, once the prodrome has begun, there are no proven drug or vaccine treatments. Supportive care, including analgesics and intravenous therapy, is highly recommended.[1] Patients should be isolated in a negative-pressure room with particulate air filtration. Masks and gloves should be worn by all health care providers and other individuals interacting with the patient, and waste and bed linens should be autoclaved before being incinerated or laundered.[16]

VACCINATION

The smallpox and vaccinia viruses share remarkable DNA homology (about 85%), but the smallpox virus is uniquely suited to evading the human immune response and causing infection. Vaccinia, on the other hand, causes no disease in nonimmunocompromised individuals, making it a relatively safe vaccine, and its antigenic and structural similarities to the smallpox virus make it an effective smallpox vaccine.

The vaccinia vaccine is believed to moderate toxemia, the density of skin lesions, and the development of the rash following a smallpox infection. Individual lesions tend to be more superficial and do not cause the pitted scars seen in unvaccinated smallpox victims. The progression from macule through papule to pustule occurs more rapidly, taking 3–4 days rather than the normal 7–8 days.

The smallpox vaccination process consists of intradermal administration of the vaccinia virus through a bifurcated needle.[1] The process is known as scarification, because of the permanent scar (known as a "take") that results. Within 7 days of inoculation, a pustule appears at the site, with surrounding erythema and induration; these are often misdiagnosed as the effects of a bacterial superinfection. The pustule forms a scab and slowly heals over the next 1–2 weeks.

Successful vaccination within 5 years of exposure provides strong resistance to smallpox. When symptoms do develop, they are generally very mild, except in cases of hemorrhagic smallpox. Side effects include low fever and axillary lymphodenopathy. Frequent complications of vaccination include secondary inoculation of the virus to other sites in the body, such as the eyelids or face, or generalized vaccinia, a systemic condition in which the virus produces mucocutaneous lesions distal from the vaccination site. In certain rare cases, vaccinated individuals develop neurological symptoms, most notably encephalitis.

Preexposure vaccination is contraindicated in immunocompromised and pregnant patients, because of the greater likelihood of complications, although postexposure prophylaxis using the vaccinia virus is recommended for all individuals who have been exposed to the smallpox virus. However, it is preferable that immunocompromised individuals be concurrently treated with vaccinia immune globulin (VIG) (see "Potential Defenses" below).

Vaccination performed more than 20 years before exposure usually has little or no effect on immunity. Even so, fatality rates for individuals vaccinated 20 or more years ago are significantly lower than those for unvaccinated individuals, perhaps because vaccination influences the frequency of different clinical types of smallpox. Modified-type smallpox is much more common among vaccinated individuals, and about 83.5% of ordinary-type cases are classified as discrete for vaccinees, compared with 47.4% of unvaccinated ordinary-type cases. Flat-type smallpox occurs five times more frequently among unvaccinated individuals as in those who have received the vaccine.

Immediate vaccination or revaccination should be administered to health care workers involved in countermeasures against weaponized or naturally occurring smallpox virus. In addition, it is highly recommended that all military personnel be revaccinated during wartime conditions if weaponized smallpox virus presents a serious threat.

WEAPONIZATION

Smallpox is a formidable bioterrorist weapon. History testifies to its use by militaries of many cultures, often with devastating and crippling results. The virus can be spread through airborne transmission and possesses an extremely stable genome capable of incorporating foreign virulence factors. While the wild-type smallpox virus alone is a strong candidate for use in bioterrorist attacks, modifications could increase the fatality rate of the disease caused by the virus to even more catastrophic proportions.

AEROSOLIZATION

Smallpox could be delivered by aerosolization or by infected "suicide carriers." Although it seems likely that infected carriers would be too ill and the rash too noticeable for terrorists to be able to infect a large number of people without detection, the disease is mild enough in its early stages that the rash could be disguised and an infected carrier could be given drugs to moderate symptoms. A few infected individuals in densely packed, cities distant from one another could infect enough people to cause major epidemics. The geographical separation would add to the straining of public health resources within the target nation.

ENHANCEMENT OF VIRULENCE

Researchers at the Cooperative Research Centre (CRC) for Biological Control of Pest Animals in Canberra, Australia, stumbled on a way to intensify the lethality of mousepox virus, a rodent analog of the smallpox virus. CRC scientists were hoping to sterilize female mice, and thus help control Australia's rodent problem, by inserting a gene for interleukin-4 (IL-4) into the mousepox genome. They hoped it would elicit an immune response against eggs in injected mice; however, the virus proved to be more virulent than intended, killing all test animals by suppressing the cell-mediated immune response to viral invasion. It is believed that a similar modification of the smallpox genome would prove equally lethal to humans, resulting in a "killer virus."

RECOMBINANT CHIMERA VIRUSES

Recombinant chimera viruses are genetically engineered viruses comprising genes from more than one virus. The discovery of a segment of the smallpox genome that plays no role in virulence or essential functions allowed researchers at Biopreparat, the Russian bioweapons program, to insert genes from other viruses into this site to create highly lethal hybrid viruses.

The first smallpox hybrid virus was created in the early 1990s by inserting genes from Venezuelan equine encephalitis (VEE) into the smallpox genome.[3] The resulting hybrid virus, called Veepox, looks like smallpox when visualized with electron microscopy. It is believed that the hybrid virus would cause, in addition to conventional smallpox symptoms, the significant neurological symptoms (including headache and coma) associated with VEE.

Ken Alibek, the former deputy director of Biopreparat, has maintained that the program developed, in addition to Veepox, a hybrid virus consisting of genes from the smallpox and Ebola viruses.[3] It is believed that Russian researchers used reverse transcriptase–PCR to create a DNA copy of the Ebola virus and then spliced the genes encoding key virulence factors into the unnecessary regions of the smallpox genome. Alibek believes that this chimera, which he has named Ebolapox, would be capable of causing hemorrhagic smallpox, the most fatal form of the disease, in all infected individuals. For more information about Ebolapox, see Chapter 5.

POTENTIAL DEFENSES

The main defense against a bioterrorist attack featuring smallpox is vaccination of the target population. Scientists continue to experiment with different formulations of the vaccinia-based vaccine, hoping to maintain its efficacy while moderating its occasionally severe side effects. Among the most promising new vaccines is a live-attenuated form of the vaccinia virus that appears to share the original Dryvax vaccine's immunogenicity while lacking its virulence. In addition, drug compounds for postexposure therapy are in development. The most promising of these include cidofovir and its analogs and a nucleoside analog believed to block virus replication. Many other antiviral drug candidates have displayed some efficacy against orthopoxvirus infection, indicating that further investigation is warranted.

The U.S. bioterrorism policy regarding smallpox is based on the ring-vaccination strategy, a method of vaccination developed during the 1960s that many scientists believe was responsible for the success of the Global Smallpox

Eradication Program. Recent evidence has cast doubt on the effectiveness of the program,[9] although D. A. Henderson and others associated with the original eradication program continue to maintain that the ring-vaccination strategy is the most effective method of combating a smallpox epidemic.

The aim of ring vaccination is to stop an epidemic following exposure by isolating and vaccinating infected individuals and all those who come into contact with them. Preemptive vaccination, on the other hand, would occur before exposure and involve immunizing all individuals in a society. The administration of George W. Bush plans to immunize 11 million individuals it believes are at high risk for contracting smallpox, including certain military personnel and health care workers. For the remaining individuals, the CDC calls for immediate ring vaccination at the first sign of a biological attack using smallpox.

Yale mathematician Edward Kaplan argues that the mass immunization procedure would be more effective than ring vaccination in preventing death and quickly limiting the spread of the disease. He notes that a famous graph from the eradication period (see figure 11.4) that shows a steep drop in smallpox cases following the introduction of the ring-vaccination procedure is misleading for several reasons. Most notably, William Foege, the author of the original graph, estimated the gains in overall vaccine coverage between January 1968 and March 1969 to be statistically insignificant, an assumption Kaplan said is invalid, given that coverage tripled during this period. Kaplan also charged that Foege skewed the steepness of the plunge by using a log curve. When Kaplan replotted the data using an arithmetic scale and correcting for the threefold increase in vaccinations between January 1968 and March 1969, he found that the ring-vaccination approach did not lead to a sharp decrease in smallpox cases. He concluded that the decrease in smallpox cases occurred gradually and was due to continuing mass vaccinations. At best, ring vaccination provided marginal help in the eradication program by increasing the total number of vaccinated individuals.

Foege disagrees with Kaplan, maintaining that ring vaccination was successful in eradicating smallpox from western and central Africa by 1969, when only 60% of the population had been vaccinated. He goes on to note that the disease continued to infect individuals in India, despite the fact that well over 90% of the population was vaccinated. Within a year of the introduction of the ring-vaccination method, however, the disease disappeared from India. Kaplan maintains that preliminary results from his study of the Indian epidemics actually support his view. In the United States, it seems that more analysis of the data from previous smallpox epidemics is necessary before the government can determine which vaccination program is truly superior.[16]

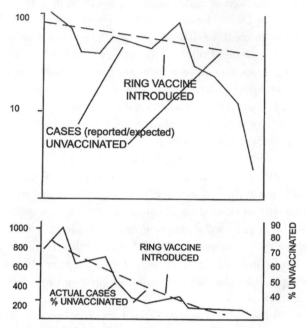

FIGURE 11.4. Two ways to look at the effectiveness of ring vaccination. (**Top**) William Foege used a log curve to depict the data and interpreted the shape of the curve to theorize that the introduction of ring vaccination in 1968 led to a rapid decrease in smallpox cases compared with the number of unvaccinated people. These data led the clinicians involved in the Global Eradication Program to choose ring vaccination (in which people who have come into contact with an index case, as well as the index case himself, are isolated and vaccinated to prevent further spread) over mass vaccination (in which an entire population is vaccinated). (**Bottom**) Mathematician Edward Kaplan has charged that Foege's disregard of the threefold increase in vaccine coverage between January 1968 and March of 1969 exaggerated the effects of ring vaccination. His graph shows that the number of smallpox cases decreased in tandem with the number of people not vaccinated, indicating that the number of smallpox cases decreased over time thanks to mass vaccination, and that ring vaccination aided the eradication effort only marginally by increasing the percentage of vaccinated individuals in the population. A linear time scale is used on the abscissa, which extends from January 1968 through March 1969.

REFERENCES

1. Alcami, A. and G. Smith. 1995. Vaccinia, cowpox, and camelpox viruses encode soluble gamma interferon receptors with novel broad species specificity. *J Virol* 69: 4633–4639.
2. Alcami, A. and G. Smith. 1996. A soluble receptor for interleukin-1 beta encoded by vaccinia virus: A novel mechanism of virus modulation of the host response to infection. *Cell* 71: 153–167.

3. Alibek, K. 1999. *Biohazard*. New York: Delta.

4. Beattie, E., J. Tartaglia, and E. Paoletti. 1991. Vaccinia virus encoded eIF-2a homolog abrogates the antiviral effect of interferon. *Virology* 183: 419–422.

5. Beaud, G. and A. Dru. 1980. Protein synthesis in vaccinia virus-infected cells in the presence of amino acid analogs: A translational control mechanism. *Virology* 100: 10–21.

6. Born, T., L. Morrison, and D. Esteban. 2000. A poxvirus protein that binds to and inactivates IL-18 and inhibits NK cell response. *J Immunol* 164: 3246–3254.

7. Chang, H., J. Watson, and B. Jacobs. 1992. The E3L gene of vaccinia virus enocdes an inhibitor of the interferon-induced, double-stranded RNA-dependent protein kinase. *Proc Natl Acad Sci USA* 89: 4825–4829.

8. Colamonici, O., Domanski, and S. Sweitzer. 1995. Vaccinia virus B18R gene encodes a type I interferon-binding protein that blocks interferon alpha transmembrane signalling. *J Biol Chem* 270: 15974–15978.

9. Enserink, M. 2003. New look at old data irks smallpox eradication experts. *Science* 299: 181.

10. Esposito, J., J. Obijeski, and J. Nakano. 1978. Orthopoxvirus DNA: strain differentiation by electrophoresis of restriction endonuclease fragmented DNA. *Virology* 89(1): 53–66.

11. Frischknecht, F., et al. 1999. Actin-based motility of vaccinia virus mimics receptor tyrosine kinase signalling. *Nature Lett* 401: 926–929.

12. Kawalek, A. and D. Rudikoff. 2002. A spotlight on smallpox. *Clin Dermatol* 20: 376–387.

13. Mackett, M. and L. Archard. 1979. Conservation and variation in Orthopoxvirus genome structure. *J Gen Virol* 45(3): 683–701.

14. Massung, R. F., et al. 1994. Analysis of the complete genome of smallpox variola major virus strain Bangladesh—1975. *Virology* 201: 215–240.

15. Moss, B. 2001. Poxviridae: The viruses and their replication. 2001. In D. M. Knipe and P. M. Howley, eds. *Fields Virology*, 4th ed., pp. 2849–2883. Philadelphia: Lippincott Williams and Wilkins.

16. Patt, H. A. and R. D. Feigin. 2002. Diagnosis and management of suspected cases of bioterrorism: A pediatric perspective. *Pediatrics* 109(4): 685–692.

17. Rosengard, A. M., et al. 2002. Variola virus immune evasion design: Expression of a highly efficient inhibitor of human complement. *Proc Natl Acad Sci USA* 99(13): 8808–8813.

18. Smith, C. A., et al. 1997. Poxvirus genomes encode a secreted, soluble protein that preferentially inhibits beta chemokine activity yet lacks sequence homology to known chemokine receptors. *Virology* 236: 316–327.

19. Smith, G. L. and G. McFadden. 2002. Smallpox: Anything to declare? *Nature Rev Immunol* 2: 521–527.

CHAPTER 12

■ ■

CHOLERA
(VIBRIO CHOLERAE)
Rohit Puskoor

Vibrio cholerae is the causative agent of the gastrointestinal disease cholera, noted for its characteristic dehydration and massive diarrhea. The disease devastated the world's populations in the nineteenth and early twentieth centuries, causing major pandemics. Although vaccines and therapeutic measures have since been developed, the disease is still a public health problem in parts of Africa, Asia, and Latin America, where health care is poor and extreme poverty precludes adequate medical care and preventive measures. An epidemic in central Africa in 1994 and 1995 presented case fatality rates as high as 30% in areas where medical facilities were limited.

Clinical symptoms appear soon after V. cholerae secretes the cholera toxin, an A-B-type toxin. The B subunit binds to galactose residues on the cell surface.[14] The A subunit triggers the irreversible activation of the cell's G proteins that causes efflux of ions and water into the intestinal lumen, leading to the watery diarrhea characteristic of cholera. Because of cholera's severe incapacitating symptoms, its high rate of fatality when untreated, and the possibility of genetically engineering a more potent strain, V. cholerae represents a significant threat as a biological weapon.

HISTORY

Cholera was first observed on the Indian subcontinent, with mentions in Sanskrit writings of a dehydrating diarrhea disease.[3] Epidemic cholera was first recorded in 1563. In 1854, John Snow hypothesized that water was the primary pathway of transmission for cholera following London's Broad Street outbreak, which led to 56 deaths within 2 days. In 1883, Robert Koch isolated V. cholerae from the feces of cholera patients and identified it as the infective agent. In an 1884 paper, Koch described the bacteria as "comma-shaped."[12]

CHOLERA-LIKE DISEASE IN ANCIENT TIMES

Although there is disagreement about whether true cholera (cholera whose causative agent is *Vibrio cholerae*) was present in Europe before the first pandemic in 1817, there is no debate that a cholera-like disease existed throughout most of the civilized world as early as the ancient period.[2] The *Sushruta Samhita*, an ancient Sanskrit text written around 500–400 B.C., makes mention of *visuchika*, an intestinal disease that closely resembles cholera. Indeed, the term is believed to be the Sanskrit word for cholera, indicating that the disease has been endemic to the Indian subcontinent for a very long time.

Both Hippocrates and Aretaeus of Capdadocia make mention of a cholera-like illness in their medical writings, evidence that cholera was present on the European continent prior to 1817. Hippocrates wrote:[2]

> At Athens, a man was seized with cholera. He vomited, and was purged and was in pain, and neither the vomiting nor the purging could be stopped; and his voice failed him, and he could not be moved from his bed, and his eyes were dark and hollow, and spasms from his stomach held him, and hiccup from his bowels. But the purging was much more than the vomiting. This man drank hellebore with juice of lentils; and he again drank juice of lentils, as much as he could, and after that he vomited. He was forced again to drink, and the [vomiting and purging] were stopped.

It is likely that lentil juice was an early, crude oral rehydration solution. The Islamic physician Rhazes also describes a cholera-like illness in 900 A.D., giving it the Arabic name *heyda*. A century later, Avicenna would adopt Rhazes's term for the milky, rice-water-like stool.

In 1643, the Belgian physician Herman Van der Heyden used the terms *trousse-galant* and *flux de ventre* to describe a severe diarrheal illness. Six years later, Rivirerus of Montpellier, France, wrote of a sporadic but very severe form of cholera plaguing the town. Cases of cholera or cholera-like illness were also noted in Brazil in 1658, Ghent in 1665, and England throughout the second half of the seventeenth century. Thomas Sydenham, the "English Hippocrates," noted that cholera occurred seasonally in the late summer and early autumn. He is also credited with developing the clinical term *cholera morbus* for the disease (the word "cholera" was also used at the time to describe a state of anger).

The British physician MacNamara, of the Indian Medical Services, cited the Bengali custom of worshipping the so-called goddess of cholera, Ola Beebee, as evidence of cholera's prior existence and its seriousness to the native peoples. He was told that the deity was initially located in a bamboo shed; a temple was provided around 1720 that was moved to a more accessible site in Calcutta in

1750. However, it was later discovered that the temple and the goddess were both fabricated by a Brahmin family as a scheme for making money.

The first outbreak following British occupation of India took place in 1781 among a division of Bengal troops headed toward Calcutta. The outbreak spread quickly through Calcutta, Sylhet, Hardwar, and other cities, leaving behind a trail of death and disease. Another outbreak took place in Vellore and Arcot from 1787 to 1789, in Ganjam in 1890, and in Malabar and Coromandel in 1796. In 1787, the Madras Medical Board complained that it could not make differential diagnosis of the epidemics. However, none of these minor outbreaks would prepare the world for the 1817 pandemic, the first of the six major cholera pandemics to spread from the Indian subcontinent through Europe and the United States.

PANDEMICS IN THE NINETEENTH AND TWENTIETH CENTURIES

In 1817, the British noted that cholera had begun to show an unusually potent virulence in India, and that the latest epidemic was sweeping from east to west. The outbreak, defined as the first cholera pandemic, swept through the Indian subcontinent from 1817 through 1823. Cholera traveled from India through the Middle East, reaching as far as Turkey and Persia, and through the Far East, invading Singapore, Japan, the Philippines, and other nations.[2]

The second pandemic took place from 1829 until 1851. It is believed that the first outbreak in this pandemic occurred in Astrakhan (part of the former Soviet Union). Some theorize that the disease spread to Astrakhan from India through Afghanistan and Persia to Orenburg, a city in Astrakhan. This pandemic reached the United States, entering through New York and New Orleans before proceeding to ravage the entire nation.

During this pandemic, a Moscow chemist conceived the idea of administering fluid intravenously to cholera patients as part of a treatment schedule. Although *The Lancet* highly approved of the technique, it did not gain widespread acceptance until the end of the nineteenth century, when a Calcutta physician used it to administer hypertonic saline solution. Alkali and potassium were added during the sixth pandemic, reducing case fatality rates from 60–70% to 30%. Normal saline replaced hypertonic saline in the intravenous fluid between 1958 to 1964, and antibiotics came to be used more frequently, reducing case fatality to around 1%. Orally administered glucose–electrolyte solutions were developed between 1964 and 1968, and were commonly used in the 1970s.

The third pandemic took place from 1852 to 1859; by 1853, the disease was rampant throughout the Middle East, northern Europe, and North and Central America. In 1854, Filippo Pacini examined the cadavers of cholera victims when the pandemic arrived in Tuscany and identified a curved bacterium, which he named *Vibrio cholerae*. His discovery of the causative agent, although a major turn of events in the fight against cholera, remained obscure for many years until Robert Koch's isolation of the bacteria in 1883.[12] The fourth pandemic stretched from 1863 to 1879. In 1865, a Mecca pilgrimage was severely affected, suffering fatalities of 33%. One of the more destructive periods of this pandemic occurred in 1866, perhaps exacerbated by the war between Germany and Austria.

The fifth pandemic, starting around 1881 and ending around 1896, primarily targeted the European continent. The first outbreaks were noted in India and Mecca. Pilgrims returning from Mecca and a fair in Egypt helped to spread the disease throughout the nation, where it claimed over 58,000 victims. From 1884 to 1887, the pandemic was confined mostly to France, Italy, and Spain. The disease made a brief appearance in 1887 and 1892 in the United States, although it was well controlled thanks to general improvement in living conditions and sanitation methods.

Between 1899 and 1923, the world was ravaged by the sixth pandemic. The first major epidemics of this pandemic struck Persia and Afghanistan in 1900. The last indigenous cases in the United States occurred during this pandemic, in Massachusetts and New York in 1912. These pandemics primarily affected continents in the Southern Hemisphere, but North America and Europe were also affected owing to increasing globalization. Although the early pandemics preceded Koch's discovery and specific strain variants could therefore not be identified, the infectious agent of the fifth and sixth pandemics was identified as serovar O1 biovar classical.

The bacterium then retreated to South Asia until 1961, when *serovar* O1 *biovar* El Tor, distinguished from previous forms by its ability to produce hemolysins, emerged in the Philippines and proceeded to cause a seventh pandemic. Stringent quarantine and vaccination measures were implemented to restrict spread, as there was no immunity to the new biovar in endemic areas. In spite of these measures, the spread continued. In 1970, about 9 years after the beginning of the seventh pandemic, cholera entered Africa and Europe. Case fatality rates for this pandemic were initially as high as 50% but fell to 7–10% upon organization of treatment procedures.

In 1972, 374 passengers and 19 crew members were exposed to cholera-infected food in a meal served on a flight from London to Sydney and New Zealand. Forty cases were diagnosed in Sydney and three were diagnosed in New Zealand, with one fatality. In 1978, Japan recorded a small cholera outbreak resulting from contaminated food at several wedding receptions at the same reception hall. These minor outbreaks following the seventh pandemic indicate how easily cholera can be spread via contaminated food and illustrate the importance of proper storage and preparation.

In December 1992, a large epidemic swept through Bangladesh. The infective strain was termed *V. cholerae O139*. Although the O139 serovar belongs to the biovar El Tor, there is currently no natural immunity to it because of the makeup of its cell-surface proteins.

THE USE OF CHOLERA AS A WEAPON

During World War I, there were allegations that the Germans attempted to spread cholera in Italy. If these allegations are true, this would be the first known use of cholera as a bioterror agent. Also, in the 1930s, Japan dropped bombs on the Chinese that released cholera and other biological pathogens.

MOLECULAR BIOLOGY

As marine organisms, *V. cholerae*, typically require 2–3% NaCl or seawater base for optimal growth. This may explain the reaction that the bacteria trigger within human hosts—the massive efflux of sodium and chlorine ions caused by toxin binding to the intestinal lining may be the bacteria's way of replicating its natural environment. The diarrhea and other symptoms of cholera are merely a side effect.

THE SEROVARS

Although over 200 serovars of *V. cholerae* have been identified, only O1 and O139 have been associated with pandemics.[4] The O1 serovar is further divided into classical and El Tor biovars and into three serosubtypes: Ogawa, Inaba, and Hikojima.

Intestinal and nonintestinal infections with non-O1 and non-O139 serotypes occur very rarely and do not cause clinically significant disease. *V. cholerae* O139 caused the first non-O1 outbreak in 1992 on the Indian subcontinent. Non-O1 and non-O139 strains are isolated from an aqueous environment more often than O1 and O139 strains, supporting the theory that strains become virulent through the acquisition of genetic mobile elements containing virulence genes from their environment. Toxic strains are extremely hardy, able to survive refrigeration and freezing in food samples. However, the bacterium is extremely sensitive to acidic environments.

V. cholerae O139 has much in common with the O1 El Tor biotype, but they differ in their cell-surface components. O1 alone is able to produce the lipopolysaccharide (LPS) somatic antigen, whereas O139 is unique in its ability to synthesize capsular polysaccharide. Because of its peripheral location in the cell,

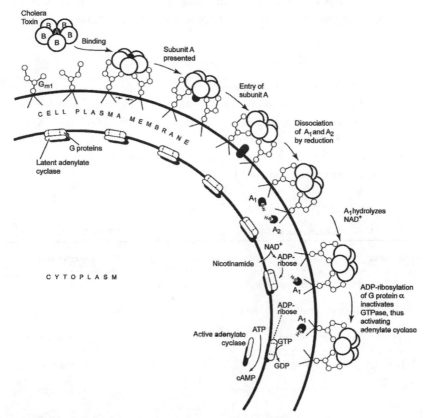

FIGURE 12.1. Cholera toxin approaches the target cell surface. B subunits bind to oligosaccharide fGM1 ganglioside. Conformational alteration of holotoxin occurs, allowing the presentation of the A subunit to the cell surface. The A subunit enters the cell and its disulfide bond is reduced, freeing A1 and A2. NAD is hydrolyzed by A1, yielding ADP ribose and nicotinamide. The α subunit of the G_s protein is ADP-ribosylated, preventing the G_s-bound GTP from being hydrolyzed to GDP and locking adenylate cyclase in the "on" mode.

the capsular polysaccharide is believed to play a pivotal role in the interaction of *V. cholerae* with the host and contribute to the O139 strain's virulence.[27]

Gram's stain of *V. cholerae* reveals that the bacterium forms curved, gramnegative rods (a negative result means that the bacterium does not retain the violet stain used in Gram's method) (figure 12.1). The bacteria in the *Vibrio* genus will grow on synthetic media with glucose as the sole carbon and energy source. In liquid media, vibrios are made motile by polar flagella that are enclosed in a sheath continuous with the outer membrane of the cell wall. On solid media, they may synthesize numerous lateral flagella that are not sheathed.

THE *V. CHOLERAE* GENOME

The vibrios are unusual in having a genome that consists of two circular chromosomes. The large chromosome, chromosome 1, is approximately 3 million bp long, whereas chromosome 2 is only a little over 1 million bp long. Together, they encode 3885 open reading frames.

The vast majority of recognizable genes for essential cell functions and pathogenicity are on chromosome 1 (table 12.1). Chromosome 2 has more hy-

TABLE 12.1 IMPORTANT VIRULENCE FACTORS

PROTEIN	FUNCTION	LOCATION
AcfABCD	Accessory colonization factors; function unknown	Chromosome 2
AldA	ToxT-activated aldehyde dehydrogenase	Chromosome 2
AphAB	Regulatory proteins	Chromosome 1
Chi group	Chitinase homologs	Chromosome 1
CtxAB	Cholera toxin; primary virulence factor	CTXφ bacteriophage
FlrC	Flagellar transcriptional activator	Chromosome 1
IrgA	Iron-regulated outer membrane protein	Chromosome 1
Msh	Mannose-sensitive hemagglutinin type IV pili	Chromosome 1
OmpU,T	Outer-membrane porins	Chromosome 1
Rst	Integration, replication, and regulation of CTXφ bacteriophage	CTXφ bacteriophage
RtxA	Repeats-in-toxin toxin; cross-links cellular actin	Chromosome 1
TcpA	Major component of toxin-coregulated pili (TCP); type IV pilus	Chromosome 2
TcpP, H	Transmembrane regulatory proteins	Chromosome 2
ToxR, S	Transmembrane regulatory proteins	Chromosome 1
ToxT	Transcriptional activator of TcpA	Chromosome 2
Wav proteins	LPS core polysaccharide synthesis	Chromosome 1

pothetical genes (genes that have been identified through molecular biology techniques but whose purpose and/or function are unknown) and more genes that come from a source other than the λ-proteobacterium, which is believed to the be the ancient parent of the vibrios. The small chromosome also carries a gene-capture system (the integron island) and host "addiction" genes that are typically found on plasmids.[9]

The fact that the two-chromosome structure is unique to the *Vibrio* genus, yet is shared across it, indicates that the gene content of the megaplasmid, as well as the two-chromosome structure, provides the bacterium with an evolutionary advantage. It is unknown why chromosome 2 has not been absorbed into chromosome1. One of the more popular theories is that, in response to environmental cues, one chromosome may partition to daughter cells in the absence of the other chromosome (aberrant segregation). Such single-chromosome-containing cells would be replication-defective but still maintain metabolic activity ("drone" cells) and therefore be a potential source of viable but nonculturable (VBNC) cells observed to occur in *V. cholerae*.[28] The drone cells produce extracellular chitinase, protease, and other degradative enzymes that enhance survival of cells retaining two chromosomes without directly competing with these cells for nutrients.

THE COLONIZATION PROCESS

Determinants of the colonization process include adhesins, neuraminidase, motility, chemotaxis, and toxin production. Chemotaxis is the characteristic movement or orientation of an organism along a chemical concentration gradient either toward or away from the chemical stimulus. *V. cholerae* is resistant to bile salts and can penetrate the mucous layer of the small intestine, possibly aided by secretion of neuraminidase and proteases. They withstand propulsive gut motility with their swimming ability and chemotaxis directed against the gut mucosa. Surface LPS and capsule structures protect *V. cholerae* against the entry of noxious compounds from the host environment.

Specific adherence of *V. cholerae* to the intestinal mucosa is probably mediated by long filamentous fimbriae that form bundles at the poles of the cells. These fimbriae have been called toxin coregulated pili (TCP) because expression of the pili genes is coregulated with expression of the cholera-toxin genes. Not much is known about the interaction of TCP with host cells, and the host-cell receptor for these fimbriae has not been identified. TCP is composed of a homopolymer of TcpA pilin, which is a 20.5-kd pilin peptide.

Other adhesion factors include hemagglutin and a group of outer-membrane proteins that are products of the accessory colonization factor (*acf*) genes. Hemagglutin is a surface protein that agglutinates red blood cells. *acf* mutants have been shown to be less able to colonize the intestinal tract. It has been suggested that *V. cholerae* use these nonfimbrial adhesins to mediate a tighter binding to host cells than is attainable with fimbriae alone.

To colonize the small intestine and cause diarrhea, *V. cholerae* must upregulate and coordinate the expression of many virulence factors, most notably the cholera toxin and TCP, which is a type IV pilus that is essential for intestinal colonization.[26]

TCP also serves as a receptor for the lysogenic bacteriophage CTXφ, which encodes the cholera toxin. During growth in vivo, ctxAB expression is dependent on TcpA. Full ctxAB expression requires prior TCP expression, indicating that expression of cholera toxin occurs after and is dependent on colonization. Expressing cholera toxin only after the bacterium is in close proximity to the target cells is consistent with the concept that cholera toxin should be efficiently targeted to host cells.

CHOLERA TOXIN

In the noninfected intestine, there is a net flow of ions from the lumen to intestinal tissue and a net uptake of water from the *lumen*. Ion-transport pumps for Na^+, Cl^-, HCO_3^-, and K^+ in mucosal cells normally maintain control over ion and water movement across the intestinal mucous membrane. Although cholera toxin plays no direct role in colonization of the host intestine, it has been shown to enhance growth of the bacteria after colonization, most likely owing to its destruction of epithelial cells, which allows the bacteria to penetrate further.

The cholera-toxin genes are located within CTXφ, which can both integrate into and self-excise from the *V. cholerae* chromosome, depending on environmental cues. The bacteriophage contains two regions, RS2 and the core. The former encodes genes involved in the integration, replication, and regulation of the bacteriophage, whereas the latter encodes the *ctxAB* operon and genes involved in CT morphogenesis. The bacteriophage is integrated into the large *V. cholerae* chromosome at the *attRS* element. Chromosomal restriction-fragment-length polymorphism studies performed on the O139 strains isolated in India and Bangladesh from 1992 to 1998 indicated 12 genotypes of the cholera toxin, suggesting that *V. cholerae* O139 mutates frequently.[17,29] Researchers have suggested that the toxin mutates in response to environmental factors.

Cholera toxin is an A-B type ADP-ribosylating toxin.[9] The genes encoding the A (*ctxA*) and B (*ctxB*) subunit form an operon. The A and B subunits are secreted to periplasm after expression, and assemble there in a 1:5 (A:B) ratio for excretion. The toxin contains five binding (B) subunits weighing 12.5 kd, an active (A1) subunit weighing 23.5 kd, and a bridging piece (A2) weighing 5.5 kd that links A1 to the five B subunits.[31] Normal transcription of the *ctxAB* operon is dependent on *V. cholerae* chemotaxis and motility, indicating that the bacterial cell must navigate to the correct location before cholera toxin can be expressed.

The mRNA transcript of the *ctxAB* operon has two ribosome binding sites (RBSs), one upstream of the A coding region and the other upstream of the B coding region. The RBS upstream of the B coding region is at least seven times stronger than the RBS of the A coding region. The organism can thus translate more B proteins than A proteins, enabling assembly of the toxin in the appropriate 1A:5B proportion. Any extra B subunits can be excreted by the cell, but A must be attached to five B subunits to exit the bacterium cell. Intact A subunit is not enzymatically active—the disulfide bond between the A1 and A2 components must be hydrolyzed, most likely by glutathione, to produce the enzymatically active A1 subunit.

Several other *V. cholerae* genes are coregulated with the *ctx* operon, including the *tcp* operon, which is concerned with fimbrial synthesis and assembly. The *ctx* and *tcp* operons are part of a regulon, the expression of which is controlled by the same environmental signals.

The B pentamer of cholera toxin binds to the G_M1 ganglioside on the host-cell surface.[15] Prior to cholera-toxin secretion, the invasin neuraminidase reduces all G_M1 gangliosides to their monosialosyl form, which is the substrate for the B subunit of cholera toxin. Once bound, the A subunit dissociates from the B pentamer and enters the host cell, where it separates into its A1 and A2 subunits.

When the disulfide bond between the A1 and A2 subunits has been nicked, the catalytic A1 subunit is free to carry out its function. The A1 subunit is a DP-ribosyltransferase that catalyzes the ADP-ribosylation of the α subunit of the G_s protein (for more information about G proteins, see box 12.1).[16]

The A1 subunit of cholera catalyzes the hydrolysis of cytosolic NAD to produce ADP-ribose and nicotinamide. This ADP-ribose is then transferred to an arginine residue of the 42-kd α subunit of the G_s protein, indicating that the cholera-toxin A1 subunit has both *NAD glycohydrolase* (NADase) and ADP-ribosyltransferase activity.[22] Once $G_{\alpha s}$ has been ADP-ribosylated, it can no longer hydrolyze its bound GTP to GDP and is thus permanently activated. ADP-ribosylation of $G_{\alpha s}$ is irreversible.[8] The permanently activated $G_{\alpha s}$ remains bound to adenylyl cyclase, keeping the enzyme continuously activated. The adenylyl cyclase thus continues to produce cAMP.[21] The net effect is that cAMP is produced at an abnormally high rate for a sustained period, which stimulates the mucosal cells to pump large quantities of Cl^- ions into the intestinal lumen.[8] Water, Na^+, and other electrolytes follow as a result of the osmotic and electrical gradients caused by the loss of Cl^-.[23] The lost water and electrolytes in mucosal cells are replaced from blood. The toxin-damaged cells effectively become pumps for water and electrolytes, causing the diarrhea, loss of electrolytes, and dehydration that are characteristic of cholera.

BOX 12.1. G PROTEINS

G proteins are membrane-bound proteins that mediate the effects of hormones once they have bound hormone-specific receptors embedded in the cell membrane. The G protein has three subunits: α, β, and γ. In its resting state, the α subunit binds a molecule of GDP. When the G protein interacts with the hormone-receptor complex, the affinity of the α subunit for GDP decreases, but its affinity for GTP is enhanced. The G protein therefore releases its bound GDP and binds a GTP molecule. The α subunit then dissociates from the $\beta\gamma$ complex. The GTP-charged α subunit binds to and activates adenylate cyclase, the enzyme responsible for synthesizing cyclic AMP (cAMP).

The hormone receptor, G protein, and adenylate cyclase are not linked together in the absence of the hormone. They find one another by lateral diffusion in the membrane after the hormone has bound the receptor. Each hormone receptor can modify the binding site for guanine nucleotides (GTP and GDP) for several G proteins. The GTP-α subunit can then activate several adenylate cyclase molecules. These events are halted as soon as the hormone has dissociated from the receptor, causing the receptor to revert to the inactive form. The molecule of GTP bound to the α subunit of the G protein is hydrolyzed to GDP. This reaction is catalyzed by the α subunit itself.

There are many different types of G proteins. G_s proteins have a stimulatory effect on adenylate cyclase because they contain a stimulatory α subunit known as $G_{\alpha s}$. G_i proteins have an inhibitory effect on adenylate cyclase because they contain an inhibitory α subunit known as $G_{\alpha i}$. The β and γ subunits of both G_s and G_i proteins are identical. Opiates and angiotensin, a liver hormone that regulates blood pressure, bind to receptors that interact with G_i proteins. Glucagon, parathyroid hormone, and many pituitary and gastrointestinal hormones bind receptors that work through G_s proteins.

Source: Adapted from G. L. Zubay. 1998. *Biochemistry*, 4th ed. Dubuque, Iowa: William C. Brown.

THE ROLE OF RtxA

The El Tor biovar of *V. cholerae* encodes a gene cluster of four genes that are physically linked to the *ctxAB* operon.[13] The toxin encoded by this cluster, RtxA, bears similarity to the bacterial RTX family of "repeats-in-toxin" toxin through its C-terminal-repeated Gly-Asp motif, necessary for target-cell binding. RtxA is activated by the acyltransferase RtxC and secreted by means of an ABC transporter system composed of the protein products of the other two gene-cluster components, RtxB and RtxD. RtxA comprises 4546 amino acid residues. The toxin features an N-terminal hydrophobic domain required for pore formation, centrally located activation sites, and C-terminal signal peptide necessary for secretion by the RtxB/RtxD transporter complex.

Although cholera toxin is the primary virulence factor of *V. cholerae*, cholera-toxin-deficient El Tor strains may elicit mild symptoms in infected individuals, indicating that *V. cholerae* produces additional toxins. RtxA has been shown to induce cell rounding by the depolymerization of actin stress fibers. Concurrently with F-actin disassembly, RtxA induces covalent cross-linking between two peptide fragments in the N-terminal of cellular G-actin, causing the formation of G-actin multimers.[7] It is believed that either these two differential functions are controlled by disparate domains of the protein or the protein activates two separate downstream pathways that perform these functions.

Most likely, G-actin is cross-linked by RtxA and removed from the G-actin pool, causing the actin stress fibers to depolymerize as F-actin is converted to G-actin to maintain equilibrium. An alternative model suggests that RtxA activates transglutaminases (TGases), a family of endogenous cross-linking proteins, which proceed to cross-link G-actin. G-actin has been shown to be a TGase substrate, giving further validity to this second model.

DNA ADENINE METHYLASE ACTIVITY

DNA adenine methylase (Dam) plays a significant role in many bacterial cells, most significantly in the replication, repair, transposition, and segregation of chromosomal DNA. In experiments conducted at the Mahan Laboratory at the University of California at Santa Barbara, it was shown that dam deletion mutants were unable to grow in culture unless dam was supplied externally. Mutation in the *dam* gene has been shown to attenuate the virulence of several disease-causing pathogens.

Interestingly, Dam overproduction attenuated the virulence of *V. cholerae*.[11] In mouse models, introduction of an *E. coli dam* gene attached to a strong promoter led to a 53-fold increase in intracellular Dam activity. Increased Dam activity caused a fivefold decrease in *V. cholerae* intestinal colonization through an unknown mechanism. The mechanism does not involve impairment of *V. cholerae*'s growth ability, as the bacteria grew at rates comparable to the wild-type bacteria in vitro, indicating that Dam overactivity somehow affects the ability of TCP to bind intestinal receptors and colonize the intestine.

ToxR, ToxS, AND ToxT

The proteins involved in control of expression of the regulon that include the *ctxAB* operon have been identified as ToxR, ToxS, and ToxT. ToxR is a transmembrane protein with about two-thirds of its N-terminal portion exposed to the cytoplasm. Only ToxR dimers bind to the operator region of the *ctxAB* operon and activate its transcription. ToxR is absolutely essential for induction of *ctxAB*. Miller and Mekalanos suggested in a 1984 paper that this is a consequence of the presence of ToxR binding domains within the *ctxAB* operator region.[19]

ToxS is a periplasmic protein. It is believed that ToxS can respond to environmental signals by changing its conformation and influencing dimerization of ToxR, which activates transcription of the *ctxAB* operon, through an unknown mechanism.

ToxT, a soluble protein, is a transcriptional activator of the *tcp* operon. It also serves as a transcriptional activator for expression of other virulence factors, most importantly the *ctxAB* operon. Expression of ToxT is activated by ToxR, and ToxT, in turn, directly activates transcription of *tcpA*.

Removal of the DNA-binding domain of ToxT will eradicate production of TCP. Induction of TCP is also abolished in *toxR* and *tcpP* mutants. *tcpP* has been shown to also activate ToxT production. During infection, only a *toxR* and *tcpP* double mutant will completely abolish TcpA induction; either mutant alone will fail to do so.

Loss of the TCP would prevent the double mutant from being exposed to intestinal signals necessary for activation of cholera-toxin expression. Thus, not only is the pathogen avirulent because of its inability to bind to the intestinal epithelium, but it is unable to produce its primary toxin.

Recently, TcpP and TcpH, which have significant homology to ToxR and ToxS, respectively, have been shown to regulate cholera-toxin and TcpA expression, possibly through ToxT. The current model states that ToxR and TcpP become activated in response to particular environmental signals by ToxS and TcpH, respectively. ToxR binds distally to the *toxT* gene and encourages binding of TcpP to a site adjacent to the RNA polymerase binding site, thus acting cooperatively to induce expression of ToxT. Null mutations in any of these regulatory genes prevent expression of cholera toxin and TCP and result in almost complete shutdown of *V. cholerae* virulence in animal models.

ToxR has been shown to regulate expression of the outer-membrane porins OmpU and OmpT. It activates expression of OmpU and inhibits production of OmpT. These porins mediate intestinal colonization and the resistance of *V. cholerae* to bile, both of which are important characteristics in a cholera infection. OmpU is much more protective against the bactericidal effects of bile salts and other anionic detergents, indicating why ToxR downregulates OmpT in favor of upregulation of OmpU. OmpT is more permeable to bile salts than OmpU. In addition, OmpU has been shown to mediate an increased resistance to organic acids and may act as an adhesin during the colonization process. Increased OmpT expression and decreased OmpU expression result in greater bile sensitivity, decreased TCP and cholera-toxin expression, and reduced colonization.[24] Thus, ToxR regulates the transcription and expression of the *ompU* and *ompT* genes in response to environmental signals.

CLINICAL DIAGNOSIS AND RESPONSE

Cholera results from ingesting an infectious dose of the bacterium. Usually, victims ingest food or water contaminated with *V. cholerae*. The infectious dose is fairly high—ranging from 10^6 to 10^{12} colonizing units—because the bacteria must pass through and survive the gastric acid barrier of the stomach. *V. cholerae* is especially sensitive to acidic conditions.

Once past the gastric acid barrier, the bacterium must penetrate the mucous lining of the intestinal epithelium. The bacterium then adheres to and colonizes the epithelial cells of the small intestine. The bacterium faces many challenges in the intestinal environment, including inhibitors such as bile salts and organic acids and immune-system factors such as complements secreted by intestinal epithelial cells and defensins secreted by intestinal Paneth's cells.

RISK FACTORS

In nonendemic populations, cholera can affect people of all ages equally. In endemic areas, however, cholera targets children, because of their lack of pre-existing immunity, and the elderly, because of their weakened immune systems and attenuated gastric acid production. Most adults in endemic areas have developed some form of memory immune response via previous symptomatic and asymptomatic infections. Lack of access to safe water supplies, poor hygiene, and dependence on unreliable sources of food are among the chief factors increasing the likelihood of contracting the disease, explaining why it is so prevalent in Third World countries but rare in more modernized nations.

Water from shallow wells, rivers, and streams is one of the leading sources of cholera infection. Also near the top of the list is seafood, especially raw or undercooked shellfish. Both the O1 and O139 serovars are killed by sunlight, drying, and acidity. However, they grow well on alkaline foods, as well as foods with neutral pH, such as cooked rice, lentils, millet, and other cooked grains.[3] Fruits and vegetables should be cooked or otherwise decontaminated if there is any doubt about their cleanliness. Freezing food and drinks does not kill the *V. cholerae* bacteria.

Because cholera is not transmissible person to person,[3] direct contact with infected individuals does not present a risk of contracting the infection. However, one can be infected by consuming food or water contaminated through handling by an infected individual.

SYMPTOMS

Cholera has a wide range of incubation periods. The appearance of symptoms takes between several hours to 5 days from exposure. The length of the incubation period depends on the size of the infecting dosage and whether the infection was contracted by infected food, which protects the bacteria from gastric acid and thus lowers incubation time.[26]

The best-known symptom of cholera infection is massive, watery diarrhea. In addition, cholera can cause hypotension due to water loss, hypoperfusion of critical organs, and, in some cases, death within 12 hours of infection.[26] The watery diarrhea is a result of the efflux of water and electrolytes caused by cholera toxin.

The watery diarrhea is accompanied by severe dehydration. Stool output can be as high as 0.5–1 L per hour. Unchecked, this massive efflux of water can cause hypotension, tachycardia, metabolic acidosis, potassium depletion, and vascular collapse.[3] In some cases, purging diarrhea can cause rapid loss of up to 10% of body weight, causing hypovolemic shock and death. Dehydration can also cause the patient to complain of malaise. Visible symptoms include sunken eyes and cheeks, decreased skin suppleness, and dry mucous membranes. Urine production declines sharply or stops altogether. Indeed, one of the more common complications seen in recent outbreaks is renal failure.[32]

Without treatment, death rates of 20% or more can be seen.[26] Death is usually due to circulatory collapse from the dehydrating effects of the pathogen. Untreated survivors generally recover within 4 to 6 days following presentation of symptoms. However, it appears that up to 75% of initial infections produce asymptomatic disease. Development of asymptomatic disease requires ingestion of a dosage of *V. cholerae* that is smaller than the minimum infectious dose.

CHOLERA GRAVIS

Cholera gravis refers to the severe, potentially fatal dehydration and diarrhea caused by *V. cholerae* infection in approximately 2% of infected individuals.[3] Patients with blood type O are more susceptible to development of cholera gravis. Around 1 million bacteria are necessary to cause cholera gravis, compared with the 100 to 1000 organisms needed to induce the mild form of the disease. It is characterized by voluminous expulsion of electrolyte-rich fluids in the stool, in amounts equal to or greater than the patient's blood volume. Cholera gravis responds well to adequate rehydration treatments, which can help keep fatalities well below 1%. However, when rehydration therapies are not readily or widely available, rates of death due to cholera gravis can be astronomical.

CONFIRMING THE PRESENCE OF *V. CHOLERAE*

Clinical diagnosis can be made through isolation of the bacteria from the extraintestinal environment or stool samples.[4,26] Isolation involves growth in an alkaline peptone water environment, blood- and chocolate-based solid media, or Muller-Hinton and MacConkey agar media. This is followed by growth on thiosulfate–citrate–bile salts–sucrose medium, which is highly selective for the vibrios. The bacterium's ability to ferment sucrose causes the formation of yellow colonies.

Biochemical screening tests including the oxidase test, the string reaction, the Kligler iron agar (KIA) test, the triple sugar iron (TSI) agar test, the lysine iron agar (LIA) test, Gram's stain, and the wet mount (which tests for motility) can also be performed to diagnose *V. cholerae* infection.

Molecular biology techniques have recently been applied to cholera diagnosis to allow more rapid diagnosis of clinical isolates. Polymerase chain reaction (PCR) protocols have been developed for *V. cholerae* identification, using primers for the *ctxA* gene, the *tcpA* gene, and the gene coding for the outer-membrane protein OmpW. PCR identification is highly sensitive, specific, and rapid—it can confirm the presence of *V. cholerae* within a few hours of the start of the reaction. Its high sensitivity enables the disease to be identified even if very little toxin is present in the serum. Immunoassays provide an alternative approach to the rapid detection of cholera.

Serovar identification is made by slide agglutination with O1 and/or O139 antibody, which identifies the serotype by specific binding with cell-surface antigens unique to the serovar. If the O1 serovar is present in the patient, the biovar (classical or El Tor) must be determined by physiological properties, such as polymyxin B resistance, number of genes encoding the toxin, hemolysin activity, and the presence of mannose-sensitive hemagglutinin.

TREATMENT

Rehydration Rapid and constant rehydration is the treatment most often used in response to cholera infection.[28] It is extremely important that glucose, salt, and water be administered to replenish the stores of nutrients and water depleted by cholera. Standard rehydration solutions called oral rehydration solutions (ORSs) are widely used to treat cholera victims. In mild cases, treatment with ORS alone is sufficient. The consumer beverage Gatorade, for example, has the necessary levels of electrolytes, water, and sugars to serve as an ORS in mild cases of cholera. In cases of cholera gravis, however, ORS may be supplemented with intravenous administration of fluids.

ORS—in volume of at least 1.5 times the loss through stool—can be administered in severe cases as soon as the patient is alert and can tolerate fluids taken orally. Stool volume and fluid intake should be monitored for the first 4–6 hours of treatment. Patients should continue to ingest ORS until the diarrhea has ended.

New ORSs have been developed with low osmolarity (low sodium concentration). These have proven to be extremely effective in children, decreasing stool output by 15–25% and lessening the need for supportive intravenous therapy by 33% compared with the standard ORS used by the World Health Organization.[1] Although the reduced need for intravenous intervention was ruled clinically significant and supported the use of low-osmolarity ORS in children, it was feared that the low sodium concentration might increase the risk of hyponatremia in adults.[1]

Subsequent studies have shown no significant difference in adult response (measured by stool output, duration of illness, and need for intravenous therapy) to low-osmolarity ORS versus standard ORS. The number of adult patients with hyponatremia was only slightly higher in individuals receiving the low-osmolarity ORS than in those receiving the standard ORS. These results indicate that low-osmolarity ORS could be used exclusively, simplifying treatment options and reducing the cost of therapeutic care.

Antibiotics Antibiotic agents can lessen the duration of illness and volume of liquid stool output.[3] However, given the short duration of the actual illness when treated with ORS and other supportive-care methods, antibiotics are not highly recommended in cholera treatment because of their high cost, potential for introducing antibiotic resistance to the bacterial population, and their limited benefit.

Tetracycline and doxycycline are the first-line drugs used against a cholera infection. For pregnant women and children under the age of 8, erythromycin, furazolidone, and cotrimoxazole are preferred. Quinolones can be used in cases of bacterial antibiotic resistance. Because of the possibility of antibiotic resistance, the susceptibility of *V. cholerae* strains to various antibiotics should be determined at the start of an epidemic. The strain should be monitored for indications of new resistance.

When used, antibiotics are generally administered in a single dosage and are used to ameliorate symptoms. Also, the shortened duration of illness helps prevent the spread of secondary infection.

WEAPONIZATION

Wild-type cholera is not likely to be used as a bioterror weapon because of the ease of treatment, the almost complete lack of fatality (less than 1%) if treated properly, and its inability to survive drying and temperatures above 70°C. Also, it is not contagious. However, it is entirely likely that the targets for cholera bio-terrorist attacks would be military rather than civilian populations. Although not lethal with proper treatment, cholera is a powerful incapacitating agent and could be used to knock out a military force for 24–72 hours.

Cholera could be an extremely potent bioweapon if the A-B operon encoding the cholera toxin were attached to a more powerful promoter. This could foil the traditional medical response of rehydration by expressing the cholera toxin to such a degree that moderate rehydration alone would not negate the dehydrating effects quickly enough to prevent death or severe disease.

Another possibility would be to determine the specific receptor(s) for TCP and use that information to genetically engineer a *tcpA* gene that would result in tighter binding between the bacterium cell and the small-intestine epithelium.

Also, one could enhance the adhesion strength of other adhesion factors such as the hemagglutinin and *acf*. As *acf* mutants have been shown to be less virulent, it stands to reason that *V. cholerae* with stronger adhesions as a result of a greater number of and/or stronger *acf* and stronger *tcp* pili fimbrial adhesions will be more virulent. One could also enhance the genes that mediate the chemotaxis ability of *V. cholerae* to make it more difficult to slough them off the epithelium and eject them from the gut.

Yet another option would be to attach a more powerful promoter to the gene that produces the invasin neuraminidase, so that all GM1 gangliosides are reduced to the monosialosyl form and can act as receptors for the cholera toxin. Expression of ToxR, ToxS, ToxT, and other virulence factors must also be enhanced—the greater the virulence, the more powerful the bioterror agent. ToxR is one way to indirectly increase the expression of *ctxAB*, and ToxT will lead to the increased expression of the *tcpA* gene, both of which are key to the creation of a more virulent and lethal form of *V. cholerae*.

It would be an interesting experiment in molecular biology to insert into the cholera genome the gene for the Ebola glycoprotein and manipulate it so that it is expressed properly. Specifically, we refer to the middle portion of the protein, which contains an amino acid moiety that has been shown to inhibit cytotoxic T-cell responses, monocyte chemotaxis, cytokine gene expression, and natural killer cell activity. It is believed that this motif is responsible for the lack of an inflammatory response to Ebola. Thus, if this protein could be co-opted by cholera, it could prevent the mucosal immune response for which it is well noted, which would make it far more dangerous.

DEFENSES

The previous section on weaponization illustrates how easily this seemingly innocuous and fairly mild pathogen could be genetically engineered to be far more lethal. However, steps can be taken to protect against cholera. Naturally, the biggest threat comes from contaminated water. In recent years, it has been suggested that houseflies may also play a role in the transmission of cholera. Simple protective measures such as filtering water, cooking food properly, washing utensils, and using fly netting in tropical climates will provide significant defense against cholera infection. For example, application of lime juice to foods stored in the home has been shown to inhibit the growth of *V. cholerae*, providing an effective and inexpensive way to protect against infection. However, for purposes of surviving a bioterrorist attack, drugs and vaccines are the best prophylactic measures.

ANTIBIOTICS

The use of appropriate antibiotics in cholera infection can reduce both the severity of diarrhea and the duration of the disease by speeding the excretion of *V. cholerae*. Tetracycline, furazolidane, and doxycycline have all been recommended for use against cholera; tetracycline is the most widely available and therefore most often used.

However, over the past decade, *V. cholerae* has steadily developed resistance to tetracycline and other oral antibiotics. Recent studies of cholera strains indicate the presence of multiple-resistance genes, especially among O1 strains. The O139 strains have been shown to be less resistant to antimicrobial agents. *V. cholerae* has developed resistance even to fluoroquinones, frustrating clinicians because fluoroquinones have not yet been used against cholera infection and resistance to them rules them out as potential treatment. Antibiotic-resistance genes are generally located on plasmids, indicating that the bacterial cells pick up the plasmids from resistant, avirulent gut coliforms in the organism.

OTHER DRUG CANDIDATES

X-ray crystallography is used as a tool for structure-based drug design targeting multiple sites on the toxins. The adhesins are one place to start, as it has been shown that *V. cholerae* that are unable to adhere to the intestinal epithelium are avirulent. A molecule could be synthesized that blocked binding of the TCP by preferentially binding with either it or its receptor (the latter requires specific knowledge of the receptor). The other adhesins, such as the hemagglutinins and the *acf*, could be targeted as well. Since modified DNA methylase (Dam) activity has also been shown to adversely affect intestinal colonization, this protein and its activity present novel drug targets.

The cholera toxin itself is susceptible at many points. One could devise a drug that would block one of the following: binding of the B_5 subunit with the

GM1 ganglioside, holotoxin assembly,[10] or the catalytic site on the A subunit.[6] Using structure-guided synthesis and combinatorial chemistry, one can start from galactose to find drug compounds that block binding of cholera toxin to GM1 gangliosides. The toxin recognizes the branched pentasaccharide portion of GM1 that protrudes from the exterior membrane surface. The structure of the cholera-toxin B-pentamer complexed with the complete GM1 pentasaccharide gives a conceptual view of toxin-receptor binding.[15,20]

The galactose-based compounds linked to form the latest drug candidate possess affinity for the B-pentamer that is many thousands of times greater than the B-pentamer's affinity for the GM1 ganglioside.[18] Increased receptor blocking occurs when the five such compounds are linked to form a pentapus whose symmetry matches that of the B-pentamer of the toxin. This pentapus structure is preferred because it is most like the original GM1 ganglioside receptor, and its symmetry perfectly matches the B-pentamer.[34] In short, it offers an alternative to the GM1 ganglioside for binding to the cholera toxin that is structurally the same but chemically far more attractive.

VACCINES

Two types of cholera vaccine are approved in the United States for use in humans. One is a killed-whole-cell formulation that includes killed bacterial cells from both biovars of serovar O1 and purified B subunit of the cholera toxin. The presence of purified B subunit causes production of antibody specific to this subunit of the toxin. Candidate vaccines expressing or containing only the B subunit are avirulent because of the absence of the A subunit and generate an immune response that builds antibody specific to the B subunit. Immunity to the B subunit prevents the toxin from entering the cell by preventing the cholera-toxin B subunit from binding to the cell's galactose residues. Multiple doses of this type of vaccine provide immunity to only 50% of adults and to less than 25% of children.

During the 1980s, work began in Vietnam on a killed-whole-cell oral vaccine that could be produced more inexpensively than European vaccines. A bivalent killed-whole-cell vaccine (including both O1 and O139 cells) was developed that displayed significant immunogenic properties.[25,30] Curiously, this vaccine elicited a more robust immune response against both the O1 and O139 serovars in children than in adults, which is of great significance because children are at higher risk for endemic cholera.

No serious side effects were observed. One problem observed in trials was that the response to the O139 serovar was not as pronounced as the response to the O1 serovar, a fact that should be addressed in future formulations of the vaccine. The cost of production is very low: 20 cents per dose. Thus, this vaccine holds great promise for use in developing countries, where cholera is most prevalent.

The other type of vaccine is a genetically engineered, live attenuated strain of *V. cholerae*. A single dose provides greater than 90% protection against the classical biovar and 65–80% protection against the El Tor biovar. Because doses of live attenuated vaccines resemble bacterial infection, they are able to stimulate both a rapid and a memory response from the immune system, inducing long-term protection against the bacteria.

However, these live vaccines present certain problems. First, they cause mild diarrhea, abdominal cramping, and a slight fever. Also, because the strains are made avirulent by deletion of the cholera-toxin genes carried on a bacteriophage, there is a danger that infection of vaccine strains by the bacteriophage could make them virulent, a major threat where cholera is endemic.

The live attenuated oral cholera vaccine Peru-15 is synthesized by a series of deletions and mutations made to the *V. cholerae* O1 El Tor Inaba strain.[5] The cholera-toxin operon is completely deleted. In addition, the *attRS1* insertion-like sequences are deleted. This deletion protects the bacteria from site-specific recombination with homologous DNA, preventing the bacteria from reacquiring the toxin operon. In addition, the binding B subunit of the cholera toxin has been attached to a heat-shock promoter and inserted into the *recA* gene, disabling that gene. *recA*-negative strains are unable to integrate foreign DNA into the bacterial genome, further protecting the mutant bacteria from being reactivated. The mutant strain is administered orally to individuals in the CeraVacx buffer, which was chosen for its ability to generate the strongest antivibrio response of tested buffers, including saline, bicarbonate–ascorbic acid, and Alka-Seltzer.

In a trial, 15% of volunteers experienced headache and 5% experienced diarrhea by the second day after receiving the vaccine. A statistically insignificant number also experienced abdominal cramps. No severe symptoms were reported, indicating a positive safety profile. As evidence of the vaccine's potent immunogenicity, 97% of volunteers experienced a fourfold increase in serum antibody levels, compared with a lack of immune response by candidates given placebo.

None of the vaccine recipients who were subsequently challenged by *V. cholerae* developed moderate or severe cholera, compared with 42% of the placebo recipients. In addition, only 7% of vaccine recipients experienced a statistically significant level of diarrhea. These data indicate that Peru-15 provides potent protection from the harmful effects of *V. cholerae* and that its effectiveness and ease of administration (the vaccine is taken orally) make it a strong preexposure prophylactic candidate.

No studies have yet been performed on the vaccine's postexposure prophy-lactic efficacy. In addition, although Peru-15 has proven ability to ward off infection in the human challenge model, it may not be as effective in areas where cholera is endemic. However, it is believed that Peru-15 will demonstrate strong efficacy in field trials as well, owing to its derivation from the El Tor strain, which is responsible for the most recent pandemic, and to its ability to colonize the intestine. These factors, as well as the vaccine genome's inability to recombine with homologous DNA, differentiate it from previous live atttenu-ated vaccines and suggest its promise as a prophylactic agent for endemic and nonendemic populations.

A new approach in vaccine design is to create a vaccine that targets a spe-cific subunit of the pathogen. As noted earlier, in "Molecular Biology," TCP is the primary factor mediating the *V. cholerae* bacterial cell's attachment to and colonization of the human intestine. Although TCP is not a dominant immu-nogen, with little antibody response seen in initial victims, more than 50% of individuals in endemic areas presented IgG and IgA TCP-specific responses, indicating that an antibody to TCP is synthesized in response to multiple *V. cholerae* infections.

TcpA, the peptide component of TCP, is very immunogenic when admin-istered alone. Anti-TcpA antibodies confer significant protection from *V. chol-erae* in infant mouse assays. In addition, TCP conjugated to class II–targeted monoclonal antibodies has been shown to induce an even more potent im-mune response, which can be enhanced by coimmunization with cholera toxin and reduced by coimmunization with anti-CD40 monoclonal antibodies. Bet-ter understanding of TCP interaction with B lymphocytes is needed to perfect TCP-conjugated monoclonal antibodies.

In a 2001 paper, Wu et al. described the preparation of synthetic peptides corresponding to portions of the C-terminus region of TcpA.[32] An experimental vaccine formulation composed of these peptides and polymer adjuvant elicited high levels of immunoglobin G (IgG), as well as lower levels of immunoglobin A (IgA), specific to the antigen TCP.

The use of polymer adjuvant ensures that the peptides robustly stimulate the immune system, despite their small size. In this case, Polydi(carboxylatopheno-xy)phosphazene (PCPP) from AVANT Immunotherapeutics (Needham, Mass.) was used. PCPP is a water-based ionically cross-linkable polymer adjuvant that promotes release of its coupled antigen for stimulation of a strong immune re-sponse while maintaining coupled-antigen integrity.

Another polymer adjuvant used in humans is a mixture of polyoxyethyl-ene (POE) and polyoxypropylene (POP). By adjusting the molecular weight and composition percentage of the POE and POP molecules, the copolymer mixtures can have different levels of adjuvant activity. One of these copolymer mixtures, CRL-1005, has been used in clinical trials to augment the immune response of mice and rhesus monkeys to the influenza vaccine.

Wu et al. formulated vaccines made up of synthetic peptides corresponding to portions of TcpA pilin and PCPP or CRL-1005.[33] The results showed that these vaccines can be used to immunize adult female mice and grant passive immunity to their offspring. Peptides emulsified in either polymer adjuvant cause antibody responses greater than peptide alone, supporting the use of polymer adjuvant in the construction of these targeted vaccines. The minimal immunogenic dosage and optimal immunization schedule and route of administration of these vaccines in humans have yet to be determined. Even so, these vaccines show great promise—they provide greater than 90% immunity against *V. cholerae* without the side effects of the live vaccine, and they may prove to have greater success with populations in endemic regions.

REFERENCES

1. Alam, N. H., R. N. Majumder, and G. L. Fuchs. 1999. Efficacy and safety of oral rehydration solution with reduced osmolarity in adults with cholera: A randomised double-blind clinical trial. *Lancet* 354: 296–299.

2. Barua, D. 1992. History of cholera. In D. Barua and W. B. Greenough, eds., *Cholera*, Chapter 1. New York: Plenum.

3. Bopp, C. A., A. A. Ries, and J. G. Wells, eds. 1999. Etiology and epidemiology of cholera. In *Laboratory Methods for the Diagnosis of Epidemic Dysentery and Cholera*. Atlanta: Centers for Disease Control and Prevention.

4. Bopp, C. A., A. A. Ries, and J. G. Wells, eds. 1999. Isolation and Identification of *Vibrio cholerae* Serogroups O1 and O139. In *Laboratory Methods for the Diagnosis of Epidemic Dysentery and Cholera*. Atlanta: Centers for Disease Control and Prevention.

5. Cohen, M. B., et al. 2002. Randomized, controlled human challenge study of the safety, immunogenicity, and protective efficacy of a single dose of Peru-15, a live attenuated oral cholera vaccine. *Infect Immun* 70(4): 1965–1970.

6. Fan, E., E. Merritt, and C. Verlinde. 2000. AB5 toxins: structures and inhibitor design. *Curr Opin Struct Biol* 10: 680–686.

7. Fullner, K. J. and J. J. Mekalanos. 2000. In vivo covalent cross-linking of cellular actin by the *Vibrio cholerae* RTX toxin. *EMBO J* 19(20): 5315–5323.

8. Ganguly, N. K. and T. Kaur. 1996. Mechanism of action of cholera toxin and other toxins. *Indian J Med Res* 104: 28–37.

9. Heidelberg, J.F. 2000. DNA sequence of both chromosomes of the cholera pathogen *Vibrio cholerae*. *Nature* 406: 477–483.

10. Hovey, T., C. Verlinde, and E. Merritt. 1999. Structure-based discovery of a pore-binding ligand: Towards assembly inhibitors for cholera and related AB5 toxins. *J Mol Biol* 285: 1169–1178.

11. Julio, S. M., et al. 2001. DNA adenine methylase is essential for viability and plays a role in the pathogenesis of *Yersinia pseudotuberculosis* and *Vibrio cholerae*. *Infect Immun* 69(12): 7610–7615.

12. Koch, R. 1884. An address on cholera and its bacillus. *Br Med J* 2: 403–407.

13. Lin, W., et al. 1999. Identification of a *Vibrio cholerae* RTX toxin gene cluster that is tightly linked to the cholera toxin prophage. *Proc Natl Acad Sci USA* 96: 1071–1076.

14. Merritt, E., et al. 1994. Galactose binding site in *E. coli* heat labile enterotoxin (LT) and cholera toxin (CT). *Mol Microbiol* 13: 745–753.

15. Merritt, E., et al. 1994. Crystal structure of cholera toxin B-pentamer bound to receptor GM1 pentasaccharide. *Protein Sci* 3: 166–175.

16. Merritt, E. and W. Hol. 1995. AB5 bacterial toxins. *Curr Opin Struct Biol* 5: 165–171.

17. Merritt, E., S. Sarfaty, and M. Jobling. 1997. Structural studies of receptor binding by cholera toxin mutants. *Protein Sci* 6: 1516–1528.

18. Merritt, E., et al. 2002. Characterization and crystal structure of a high-affinity pentavalent receptor-binding inhibitor for cholera toxin and *E. coli* heat-labile enterotoxin. *J Am Chem Soc* 124: 8818–8824.

19. Miller, V. and J. Mekalanos. 1984. Synthesis of cholera toxin is positively regulated at the transcriptional level by toxR. *Proc Natl Acad Sci USA* 81(11): 3471–3475.

20. Minke, W., C. Roach, and W. Hol. 1999. Structure-based exploration of the ganglioside GM1 binding sites of *Escherichia coli* heat-labile enterotoxin and cholera toxin for the discovery of receptor antagonists. *Biochemistry* 38: 5684–5692.

21. Moss J, Vaughan M. 1979. Activation of adenylate cyclase by choleragen. *Annu Rev Biochem* 48: 581-600.

22. Murayama, T., et al. 1993. Effects of temperature on ADP-ribosylation factor stimulation of cholera toxin activity. *Biochemistry* 32: 561–566.

23. Peterson, J. W. 1989. Role of prostaglandins and cAMP in the secretory effects of cholera toxin. *Science* 245: 857.

24. Provenzano, D. and K. E. Klose. 2000. Altered expression of the ToxR-regulated porins OmpU and OmpT diminishes *Vibrio cholerae* bile resistance, virulence factor expression, and intestinal colonization. *Proc Natl Acad Sci* 97(18): 10220–10224.

25. Quirk, M. 2002. Home-grown Vietnamese cholera vaccine "completely safe." *Lancet* ii: 198.

26. Reidl, J. and K. E. Klose. 2002. *Vibrio cholerae* and cholera: out of the water and into the host. *FEMS Microbiol Rev* 26: 125–139.

27. Sengupta, D. K., M. Boesman-Finkelstein, and R. A. Finkelstein. 1996. Antibody against the capsule of *Vibrio cholerae* 0139 protects against experimental challenge. *Infect Immun* 64(1): 343–345.

28. Shears, P. 2001. Recent developments in cholera. *Curr Opin Infect Dis* 14: 553–558.

29. Shoham, M., T. Scherf, and J. Anglister. 1995. Structural diversity in a conserved cholera toxin epitope involved in ganglioside binding. *Protein Sci* 4: 841–848.

30. Trach, D. D., et al. 2002. Investigations into the safety and immunogenicity of a killed oral cholera vaccine developed in Viet Nam. *Bull World Health Organ* 80(1): 2–8.

31. van den Akker, F., E. A. Merritt, and W. G. J. Hol. 2000. Structure and function of cholera toxin and related enterotoxins. In *Handbook of Experimental Pharmacology*, pp. 109–131. Berlin: Springer-Verlag.

32. Wu, J.-Y., R. K. Taylor, and W. F. Wade. 2001. Anti-class II monoclonal antibody targeted vibrio cholerae TcpA Pilin: Modulation of serologic response, epitope specificity, and isotype. *Infect Immun* 69(12): 7679–7686.

33. Wu, J.-Y., W. F. Wade, and R. K. Taylor. 2001. Evaluation of cholera vaccines formulated with toxin-coregulated pilin peptide plus polymer adjuvant in mice. *Infect Immun* 69(12): 7695–7702.

34. Zhang, Z., E. Merritt, and M. Ahn. 2002. Solution and crystallographic studies of branched multivalent ligands that inhibit the receptor binding of cholera toxin. *J Am Chem Soc* 124: 12991–12998.

CHAPTER 13

■ ■

SALMONELLA

Kira Morser, Rohit Puskoor, and Geoffrey Zubay

Bacteria of the genus *Salmonella* account for the majority of food poisoning cases in the United States, with the Centers for Disease Control and Prevention (CDC) estimating 1.4 million cases annually. It is a rod-shaped, gram-negative, facultative anaerobe that infects a wide variety of organisms, from lizards to humans. The main feature of salmonella pathogenesis is the type III secretion system (TTSS), a needle-like multiprotein complex that delivers toxic proteins to host cells. One of these proteins, SopB, causes the diarrhea characteristic of salmonella disease.

Salmonella infection results in either typhoid fever (if the individual is infected by the typhi or paratyphi serovars) or gastroenteritis (if infected by the enterditis serovar). Typhoid fever was a potent killer prior to the development of modern antibiotics. The pathogen's weaponization potential and the severe gastrointestinal symptoms it inflicts on the majority of infected individuals have led to its classification as a Category B biological threat by the CDC.

HISTORY

Salmonella is believed to have killed many famous historical figures, most notably Alexander the Great and Prince Albert, the husband of Queen Victoria. The pathogen was identified in 1885 by the American veterinarian Dr. Daniel E. Salmon, for whom the pathogen is named.

In 1898—a little over a decade after Salmon's isolation of the choleraesuis strain from the intestine of a pig—British surgeon Almroth E. Wright developed an antityphoid inoculation. The first salmonella vaccine consisted of a heat-denatured bacterium, a rudimentary killed-whole-cell vaccine, which was successfully used during World War I to reduce the number of soldiers who died from enteric fever.[3] Despite this, the disease was not deemed to have a cure until the discovery of antibiotics in the twentieth century.

Some individuals have natural immunity to salmonella. Although most infections result in a period of illness followed by the flushing of bacteria out of the victim's system, in some cases the bacteria remain in the body. Such individuals, known as carriers, normally shed large amounts of bacteria in their stools.[21] These people, in essence, serve as a natural reservoir for the disease, able to infect others susceptible to the pathogen although they themselves contract only mild or asymptomatic disease. The best-known carrier of the disease was Mary Mallon, more familiar as "Typhoid Mary"; the epidemic she caused in the New York City area in the early 1900s led to marked improvements in public health approaches to dealing with the pathogen.

During the summer of 1906, Charles Henry Warren, a New York City banker, rented a summer home for his family in Oyster Bay, Long Island, and hired Mallon as his family's cook. During the summer, one of the Warren daughters fell ill with typhoid fever. Mrs. Warren, several maids, and a gardener contracted the illness soon thereafter.

Various investigators were called in to determine the source of the outbreak, all without success. Then, George Soper, a civil engineer familiar with typhoid fever, was hired. Soper suspected that Mallon, who had abruptly resigned 3 weeks before he was hired, had been the source. Tracing her recent employment history, Soper found that typhoid outbreaks had often followed in her wake. By the time of his investigation, she had worked for seven families, who collectively represented 22 cases of typhoid fever, with one death.

When Soper located Mallon in 1907, she vehemently refused to provide stool, blood, and urine samples and chased him away with a kitchen knife. It was not until Soper presented his evidence to New York City health officials that he and five policemen were able to get her to cooperate. After her stool tested positive for salmonella, she was diagnosed as a carrier for the pathogen. Soon thereafter, she was tried and found guilty of endangering the public health. Mallon was remanded to an isolated cottage on North Brother Island, an island in the East River, and warned never to work in a position that involved food preparation. Mallon's infamy was solidified by her blatant disregard of this warning—because cooks were usually better paid than most of their fellow servants, she continued to defy the warnings of health officials. She was certain that, since she herself had never manifested typhoid fever, she could not possibly be a reservoir for the disease. After being released from the island in 1910, she resurfaced in Newfoundland, New Jersey, as a cook for a sanatorium, where an ensuing outbreak forced her to flee yet again. After officials caught her in Westchester County, New York, Mallon was returned to North Brother Island, where she lived until her death in 1938.

As late as 1947, there was no specific treatment for dealing with a salmonella infection, leading to mortality rates ranging from 10 to 30%. Then, in

1948, Woodward and Phillip published a report detailing the salutary effects of chloramphenicol (an antibiotic extracted from a mold) on typhoid fever.[22] The antibiotic reduced the length of the illness from 35 days to 4 days, with a corresponding reduction in mortality.

In 1990, the FDA approved irradiation of poultry as a method for controlling salmonella and other foodborne bacteria, in an attempt to curtail the number of natural cases that resulted each year from improper preparation. These more stringent safety measures do not ensure complete protection against salmonella; food-poisoning cases continue to be attributable to undercooked or improperly prepared food.

In late 1984, local health authorities shut down 10 area restaurants in The Dalles, Oregon, that were linked to a series of salmonella outbreaks affecting 751 people. It was eventually discovered that in September and October members of a cult had used *Salmonella typhimurium* to contaminate salad dressings, fruit, vegetables, and coffee creamers at these restaurants.

At first, federal and state health officials believed that restaurant workers' poor hygiene was responsible for the incident, until cult leader Bhagwan Shree Rajneesh admitted responsibility for the outbreak. Federal agents confiscated from his compound a vial containing a strain of *S. typhimurium* whose genetic sequence was identical to the one responsible for the outbreak. The agents also found vials of hepatitis B virus, *Francisella tularemia*, and *Salmonella typhi*. Cult members had intended to sabotage a local election by poisoning the community with salmonella so that residents would be too ill to vote. Interestingly, only 11% of those infected contracted the virus through secondary sources, indicating the rarity of person-to-person spread.[19]

MOLECULAR BIOLOGY

Salmonella infections usually begin with the ingestion of contaminated food or water. After entering the small intestine, the bacterium passes through the mucosa of the intestine to reach the intestinal epithelial cells. There, the salmonella bacterial cell adheres to epithelial cells, aided by several fimbriae expressed by the bacterium. The bacterium then releases through its TTSS effector proteins that mediate the endocytosis of the bacterium.

SURFACE STRUCTURES

The genes responsible for the synthesis and organization of sugars into the polysaccharide side chains of surface lipopolysaccharide (LPS) molecules are clustered within the *rfb* locus on the salmonella chromosome. Although the lipid core of the LPS is conserved across serotypes, the polysaccharides composing the sugar side chains are highly polymorphic owing to the polymorphic nature of the *rfb* genes.[5]

Whereas a few serotypes, such as *S. typhi*, possess an outer capsular layer, the LPS forms the protective outer layer of most salmonella strains.[6] Salmonella mutants lacking the protective LPS layer are killed by complement proteins, indicating that LPS mediates resistance to complement-directed killing of bacteria. Most antibodies, especially those of the memory immune response, are directed against the LPS, indicating that the bacteria should benefit from altering the polysaccharide component of the LPS during the course of infection to avoid detection by the memory immune response. Salmonella is able to cycle expression of different fimbriae during the infectious cycle, which helps it evade immunological recognition of its cell-surface components.[13]

The flagella are another potent antigen on the bacterial cell surface. Indeed, the host cytotoxic T-cell response is directed largely against salmonella flagellar epitopes. The middle domains of the flagella vary greatly in different serovars, while the N- and C-termini are highly conserved. The variation seen in the middle domains is likely to influence the virulence of different serovars.[4] The operon encoding phase I flagella also encodes a protein that represses expression of phase II flagella. The switch from phase I flagella expression to phase II flagella expression is mediated by an enzyme that inverts a segment of the phase I operon, preventing transcription of the entire operon. Thus, no phase I flagella are expressed and the phase II repressor is inhibited, allowing the transcription and expression of phase II flagella to proceed. This flip-flop switching provides a potent mechanism for escaping the cell-mediated immune response.

Flagella from both *S. typhi* and *S. typhimurium* have been shown to induce expression of tumor necrosis factor–α (TNF-α). Phase II flagella have been shown to be less potent inducers of TNF-α, indicating that the switching mechanism may also provide the bacteria with a way to downregulate the inflammatory response within the host.

GENES FOR VIRULENCE FACTORS

Genes for the virulence factors of the salmonella bacterium tend to cluster in the salmonella chromosome in regions called pathogenicity islands. Bacteriophage and/or transposon insertion sequences often flank pathogenicity islands, suggesting that they once served as the vehicles for the transfer of pathogenicity islands to salmonella.

Acquisition of a pathogenicity island greatly enhances the virulence of the bacterium. There are several known pathogenicity islands in salmonella; each can present new mechanisms of attack and new potential hosts for the bacterium. For example, salmonella pathogenicity island 1 (SPI-1) encodes the genes necessary for bacterial entry into the cells of the intestinal epithelium whereas pathogenicity island 2 (SPI-2) carries genes necessary for intracellular bacterial replication and the initiation of systemic infection in the host. The systemic infection promotes enteric fever symptoms seen with *S. typhi* and *S. paratyphi*.[4] Of the three additional pathogenicity islands that have been clearly identified within the Salmonella genome, SPI-5 is the most significant because it encodes SopB (the primary virulence factor of salmonella), which causes the efflux of water into the gut that is responsible for diarrheal symptoms.

SECRETION OF EFFECTOR PROTEINS

The TTSS complex of proteins found in salmonella mediates the transfer of virulence factors from the bacterium into the host cell. The complex is highly regulated during and after its transcription, ensuring controlled secretion and delivery of virulence factors to the host cell. The TTSS complex is present in many bacteria besides salmonella, including *E. coli* and *Y. pestis* (the causative agent of plague).

The structural proteins of TTSS form a long, thin, pointed structure called the needle complex (see figure 13.1). The base of the needle complex is made up of PrgH, PrgK, and InvG. PrgH is located in the inner membrane, InvG in the outer membrane. PrgK connects PrgH and InvG to form the base of the TTSS complex. The PrgI needle structure is secreted by the needle-complex base into the extracellular environment; its formation and final length are highly regulated by another protein called InvJ (see table 13.1).[6,7] PrgI then makes contact with the eukaryotic host cell, forming a channel between the bacteria and the host cell that can then be traversed by secreted effector proteins.

Proteins targeted for secretion carry multiple signal peptides that route them to the TTSS complex. No conserved code has been discovered for these signal peptides; it appears that each secreted protein bears a unique signal. In addition to regulating PrgI formation, InvJ also controls the ability of the TTSS complex to recognize the different signal peptides of proteins targeted for secretion at different time points in the infectious cycle. This recognition system is the mechanism for the TTSS complex's delivery of proteins at the "right" time and place.

Salmonella contains two sets of genes encoding two forms of TTSS. One set is located on SPI-1, the other on SPI-2. Mutants lacking working copies of these sets of genes are avirulent. It is believed that SPI-1 TTSS mediates the entry of salmonella into host cells by initiating actin rearrangements. Entry into the host cell activates the production of the SPI-2 TTSS.[15]

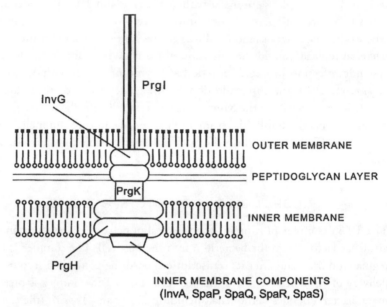

FIGURE 13.1. Structure of the type III secretion system (TTSS). Salmonella pathogenicity islands 1 and 2 encode TTSS and are expressed at different points in the infectious process. The TTSS allows the bacteria to inject virulence factors into the host cell. The structural proteins of the TTSS adopt a long, thin, pointy structure called the needle complex. PrgH is located in the inner membrane; InvG, a homolog of the secretin family, is located in the outer membrane; and PrgK connects the two proteins. PrgI is anchored in the outer membrane, bound to InvG; it has the needle-like shape characteristic of the comlex. At least 17 polypeptides are secreted by SPI-1 TTSS, and four of them (SpaO, InvJ, PrgI, and PrgJ) are involved in the assembly of the needle complex and protein secretion. Inner-membrane components InvA, SpaP, SpaQ, SpaR, and SpaS help regulate the secretion complex.

SPI-1 TTSS is expressed in the intestinal lumen during the initial contact between salmonella and the intestinal epithelium. Numerous polypeptides are secreted by SPI-1 TTSS (see table 13.1), three of which (SpaO, InvJ, and PrgJ) regulate assembly of the needle complex and secretion of effector proteins. Three proteins—SipB, SipC, and SipD—are required for entry of secreted proteins into the eukaryotic host cell.

All other secreted proteins, referred to as effector proteins, carry out specialized virulence functions in the host cell. The genes encoding the effector proteins are located on mobile genetic elements and other pathogenicity islands. Different salmonella strains encode different effector proteins, explaining the varying virulence and epidemiology between strains.[1]

TABLE 13.1 MAJOR EFFECTOR PROTEINS OF SALMONELLA
PATHOGENICITY ISLAND–1 TYPE III SECRETION SYSTEM

SECRETORY PROTEIN	FUNCTION
PrgH	Component of the base of the TTSS complex; located in the inner membrane
InvG	Component of the base of the TTSS complex; located in the outer membrane
Prgk	Component of the base of the TTSS complex; connects PrgH and InvG
InvJ	Required for needle complex assembly and protein secretion; controls length of needle component (PprgI) of TTSS
SpaO	Required for the needle complex assembly and protein secretion
PrgI	Required for the assembly of the needle portion of the TTSS complex
SipA	Binds actin, stabilized actin filaments
SipB	Translocation of virulence factors secreted by TTSS through needle complex and eukaryotic cell membranes
SipC	Translocation of virulence factors secreted by TTSS through needle complex and eukaryotic cell membranes
SipD	Translocation of virulence factors secreted by TTSS through needle complex and eukaryotic cell membranes
SopB	Inositol phosphatase; stimulates actin cytoskeleton rearrangements, nuclear responses, and cloride secretion

Salmonella's ability to enter the host cell aids the bacteria in avoiding the host immune response, as the immune response will not target host cells unless those cells present foreign antigen on their surface. Furthermore, the cytosolic environment is more favorable than the extracellular environment for bacterial replication. Finally, by breaching the intestinal epithelium (epithelial cells are the first to be attacked by the bacteria), the bacteria gain access to deeper, more vulnerable tissues (figure 13.2).

HOST CYTOSKELETAL REARRANGEMENTS

The actin cytoskeleton is a dynamic support matrix. It is highly regulated by actin-binding proteins that mediate the transformation from monomeric actin, known as G-actin, to polymeric F-actin. The filament structures formed by F-actin are also controlled and stabilized by actin filament-binding proteins. The

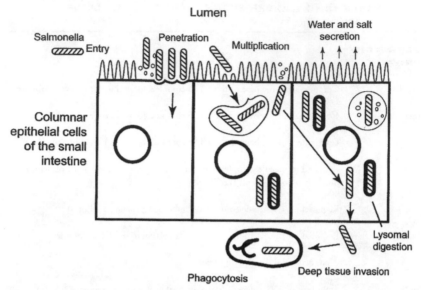

FIGURE 13.2. Overview of salmonella pathogenesis. As salmonella passes through the digestive tract, it adheres to the intestinal wall. It then uses what can be characterized as a molecular syringe mechanism to inject virulence proteins into the cell wall. It multiplies, modifying the host-cell environment to suit its needs. The release of electrolytes and water from the intestinal cell causes diarrhea. Although the bacteria can eventually be flushed, they can also penetrate more deeply into the tissue by spreading through the blood (septicemia).

level of G-actin concentration in host cells is normally kept below the amount required for polymerization, thus making the synthesis of functional G-actin a requirement for filament formation.

SipA is an SPI-1 effector protein that is secreted into the host-cell cytoplasm, where it binds actin and further stabilizes actin filaments, thus decreasing the minimum cellular concentration of G-actin necessary for actin polymerization. SipC, in addition to aiding the entry of other SPI-1 effector proteins, is believed to activate actin monomers to polymerize and stimulate actin-filament bundling. Through this activity, SipA and SipC mediate the highly localized nature of cytoskeletal rearrangements that allow for optimal bacteria-mediated endocytosis.

Upon contact of the salmonella bacterium with the host cell, cytoskeleton-associated proteins relocate to the site of bacterial entry. The apical surface of the infected cell undergoes structural changes in response to interaction with salmonella effector proteins, causing its brush border to resemble the membrane ruffles that appear in response to growth-factor release or oncogene activation.[9] By a pathway called bacteria-mediated endocytosis, the membrane ruffles elongate and enclose the bacterium in membrane-bound vesicles.[17] After internalization, these vesicles translocate to the basolateral surface of the infected cell, allowing the apical membrane to reassume its normal, nonruffled appearance.

THE ROLE OF SOPB

The enterotoxin SopB causes the fluid secretion and neutrophil accumulation in the gut characteristic of gastrointestinal salmonella infections.[17,21] This physiological response is the result of a many-fold increase in the concentration of inositol-1,4,5,6-tetrakisphosphate [Ins $(1,4,5,6)P_4$] induced by SopB. The increase triggers a rise in chloride secretion from the intestinal epithelium into the lumen, which alters the charge equilibrium across the epithelium, causing the secretion of sodium ions and water. The secreted fluid causes the watery stool characteristic of salmonella gastroenteritis. It is believed that the increased levels of Ins $(1,4,5,6)P_4$ also cause intestinal inflammation by attracting neutrophils to the local site of infection.

M CELLS

Membranous epithelial cells (M cells) comprise the follicle-associated epithelium (FAE) overlying Peyer's patches and other intestinal mucosa-associated lymphoid tissues. These cells function as antigen-sampling cells, transporting material across the FAE to the lymphoid tissues beneath, thereby stimulating an immune response.

Invasion of M cells by salmonella induces extensive membrane ruffling and results in M-cell destruction and sloughing off of the FAE.[19] Even though the FAE comprises only a minor portion of the intestinal epithelium, damage at these sites creates epithelial perforations (or breaches) through which bacteria can freely enter.[14] This epithelial damage also results in intestinal ulcers.

THE HOST-CELL ORGANELLE

In the host cell, salmonella resides in an organelle known as the salmonella-containing vacuole (SCV). Very little is known about the characteristics of the SCV or the mechanisms by which salmonella manages to survive within it. After crossing the epithelial lining of the intestine, salmonella colonizes Peyer's patches and recruits macrophages to the area by stimulating the production of interleukin-8 (IL-8). Macrophages, which are very susceptible to salmonella infection, carry these bacteria to the spleen and liver via the circulatory system.[12]

After the salmonella bacteria have gained access to host macrophages, SPI-2 TTSS expresses itself. The SPI-2 TTSS complex is required for a thorough infection; mice infected with SPI-2 TTSS(–) mutants do not develop systemic infection. The complex secretes its effector proteins from the phagosome into the macrophage cytosol. The SPI-2 TTSS effector proteins interfere with the maturation of the phagosomes and dismantle macrophage attack mechanisms. Immature phagosomes are unable to fuse with lysosomes to form the phagolysosome, allowing salmonella to avoid degradation by the phagolysosome. The phagosome eventually develops into a protective vacuole encapsulating the SCV and providing it with a suitable environment for survival.

The bacterium produces several enzymes that directly inactivate toxic compounds in the macrophages. For example, salmonella synthesis of homocysteine, an antagonist of nitric oxide (NO), is responsible for its ability to resist attack by NO and other related reactive nitrogen compounds. Salmonella also produces at least one superoxide dismutase that inactivates reactive peroxides. To survive the limited nutrient base of the macrophages, salmonella must synthesize additional factors that upregulate the enzymes necessary for de novo biosynthesis of protein building blocks, including aromatic amino acids and purines.

PhoP AND PhoQ

Salmonella strains that cause typhoid fever are spread to distal sites by the macrophages; this allows them to produce systemic infections. Fever-producing strains must possess a mechanism by which they can survive the antimicrobial responses of the macrophage. A two-component system known as PhoP/PhoQ is responsible for mediating salmonella's resistance to destruction in the macrophages. The PhoP/PhoQ system modulates salmonella gene expression in response to cues from the extracellular environment. PhoQ is a bacterial membrane–spanning regulatory protein that phosphorylates the transcription factor PhoP.

PhoP then induces transcription of genes called PhoP-activated genes (*pag*) and represses transcription of genes known as PhoP-repressed genes (*prg*). *pag* genes are transcribed by salmonella in the macrophage and encode proteins, such as superoxide dismutase, that are necessary for survival in the macrophages.[11] By contrast, transcription of *prg* genes is inhibited in the macrophages. The fact that *prg* genes mostly encode effector proteins indicates that salmonella directs its production resources away from factors necessary for growth and virulence and toward factors necessary for survival while in the macrophages.

The genes of the *spv* operon have been shown to influence the spread of systemic infection as well. This operon comprises five genes: a positive regulatory gene (*spvR*) and four genes encoding structural proteins (*spvA, spvB, spvC,* and *spvD*).[4] Mutations in *spvR* abolish expression of all four structural *spv* genes. In human macrophages, the main effect of the *spv* genes appears to be detachment and apoptosis of infected cells.[16,21] Once the cells detach from the extracellular matrix, they can move to distal sites and spread infection. By inducing cell death, salmonella bacteria can escape the host cell and spread infection when apoptotic bodies are phagocytosed by uninfected cells.

DNA ADENINE METHYLASE MUTANTS

The Dam enzyme methylates specific adenine residues in the salmonella genome, thereby disrupting regulation of DNA replication, mismatch repair, and transposition. Most important, Dam regulates the expression of 20 or more bacterial genes that are active during infection. As a result, Dam(–) mutants are avirulent. Dam inhibitors are likely to have broad antimicrobial action and Dam(–) strains have proven useful for making live attenuated vaccines.[10]

CLINICAL DIAGNOSIS AND RESPONSE

An infection by *S. enteriditis* or *S. typhimurium* causes the primarily gastrointestinal symptoms that most people associate with food poisoning, including diarrhea, nausea, and cramps. *S. typhi* and *S. paratyphi* cause symptoms resembling those of influenza, making differential diagnosis difficult. The clinical response is fairly simple, usually consisting solely of supportive care. Although antimicrobial drugs are available, they are not recommended for most salmonella-infected patients unless complications or long-term illness seem likely. This is because the drugs kill most of the enterobacteria that may be beneficial to the host.

SYMPTOMS

Nontyphoidal Infections The most common symptom of salmonella infection is diarrhea accompanied by stomach cramps. Occasionally, vomiting, bloody stool, and a fever of 100–102°F are also manifest. The young and the elderly must be monitored because the dehydration caused by diarrhea and/or vomiting can lead to a life-threatening situation.[20]

For most cases, supportive treatment to counter the dehydrative effects of the infection is sufficient. If dehydration is unusually severe, intravenous replenishment of lost fluids may be called for. Initial symptoms appear 12 to 36 hours after infection and generally last 5 days, although the exact length of the infection fluctuates with individual susceptibility, amount of bacteria ingested, and the virulence of the strain. The estimated number of bacteria needed to initiate infection is 10^4, but some suggest that as few as 100 bacteria can suffice.

Typhoidal Infections *S. typhi* and *S. paratyphi* cause a set of symptoms commonly referred to as typhoid, or enteric, fever. The symptoms of infection with *S. typhi* are slightly more severe than those induced by *S. paratyphi*.

The inoculum size of typhoidal strains is much larger than that of nontyphoidal strains, with an LD_{50} of 1000 to 10^8 bacteria. Typhoidal infections manifest prolonged fever as their main symptom. Diarrhea and signs of general malaise (such as chills, sweating, headache, loss of appetite, weakness, sore throat, and muscle pains) are also sometimes exhibited. The majority of typhoid fever patients complain of abdominal tenderness. "Rose spots" (faint salmon-colored spots on the trunk) are seen in 30% of cases; they disappear after 2–5 days, with no permanent scarring. Enlargement of the spleen and liver, nosebleeds, and bradycardia are also seen. In 5–10% of typhoidal infections, there are neurological symptoms, notably psychosis and confusion.

Late-stage complications appear in the third or fourth week. The most common of these—intestinal perforation and gastrointestinal hemorrhage—can be fatal and must be treated immediately with antibiotics (to control peritonitis) and surgical resection of the bowel (to control hemorrhaging). Other, rare complications include infection of the pancreas, endocardium, pericardium, myocardium, testis, liver, meninges, kidneys, joints, bone, lungs, and parotid gland. Spleen and liver abscesses may also form.

In more than 90% of patients, the bacteria is cleared from the intestinal tract within 8 weeks. However, 1–5% will become asymptomatic chronic carriers (and shed the bacteria for at least a year in their urine and stool); in these cases, the gall bladder is the primary source of bacteria. Women and individuals with biliary and gastrointestinal malignancies are at greater risk of becoming carriers.

As with nontyphoidal bacteremia and localized infections, quinolones and third-generation cephalosporins are usually prescribed for typhoidal salmonella infections. Ceftriaxone (1–2 g intravenously or intramuscularly) is also recommended; a 5–7-day treatment schedule is used for cases without complication and a 10–14-day treatment schedule for more severe cases. For infections by multiple-drug-resistant (MDR) strains of *S. typhi* or *S. paratyphi*, quinolones are the only effective oral treatment. Dexamethasone is used to treat very severe cases of typhoid fever (marked by neurological symptoms and septicemia). Patients designated as chronic carriers are treated with a 6-week schedule of oral amoxicillin, ciprofloxacin, or norfloxacin. Surgical intervention may also be required to remove damaged cells,[14] depending on the scale of necrosis.

DIAGNOSIS

Biochemical tests performed on stool samples provide a quick and definitive method of diagnosing patients with salmonella. Blood can also be assayed if septicemia is suspected.

The typical confirmation involves stool culture on an agar plate, which is an inexpensive method to determine bacterial presence[1] (a blood sample may also be used). The color of the colonies identifies the bacteria in the stool; salmonella colonies are pink. The enzyme-linked immunosorbent assay (ELISA) test can also be used to detect the presence of bacterial antigens and make differential diagnosis of the particular serovar, although it is expensive and not recommended given the lack of a pressing need for precise diagnosis of salmonella infections.

WEAPONIZATION

The ease with which it can be isolated and cultured makes salmonella a unique bioterrorist threat. Although death due to infection is rare, the pathogen can cause death or severe complications in immunocompromised individuals.

CONTAMINATION OF FOOD

Salmonella is generally not considered a serious bioterrorist threat to life, because of the relative transience of symptoms and low mortality. However, for rogue groups wishing to spread panic, causing people to doubt the safety of anything they might eat could be a very powerful psychological tool.

ENHANCEMENT OF VIRULENCE

The virulence of salmonella has evolved through lateral gene transfer.[2] Rather than develop new virulence strategies for host infection, all members of the genus *Salmonella* share certain characteristics, such as their ability to invade intestinal epithelial cells and multiply within gut-associated lymphoid tissue (GALT). The differential virulence of various strains is due to the adoption by individual strains of various virulence factors located on nomadic plasmids through horizontal gene transfer. Acquisition of virulence genes by horizontal gene transfer mediates the formation of new strains and a broadening of the host range.

This is an ongoing process; genetic exchange continues to occur at a high frequency for the salmonella serovars. It is all but certain that new strains will arise naturally via horizontal gene transfer. However, the bacterium could also be weaponized by such genetic exchange mechanisms. For example, if a plasmid encoding anthrax lethal factor (LF) or other toxic protein attached to a salmonella-specific promoter were transfected into the salmonella bacterium, it is possible that the gene would be transcribed and that the protein product would be added to salmonella's already large arsenal of effector proteins.

Salmonella's reliance on lateral gene transfer for the evolution of virulence indicates that the pathogen provides a receptive environment for transcription and translation of nonself genes, leading to the conclusion that more toxic proteins could be produced by this bacterium. This, combined with the bacteria's noted ease of spread and growth and the effortlessness with which food can be contaminated by the bacteria, could present a serious threat.

CULTIVATION OF ANTIBIOTIC RESISTANCE

Antibiotics provide a method of treating salmonella victims, but the range of antibiotics that can be used for successful treatment has narrowed because of the rise of antibiotic-resistant strains. In spite of this, an antibiotic can usually be found for which the infecting agent has a high susceptibility.

Salmonella transfected with plasmids that carry genes for MDR genes could theoretically resist treatment by a wide range of antibiotics. Bacteria so reinforced could induce disease unchecked by modern medicine. The only available therapy would be supportive care to alleviate the symptoms until the patient had either recovered or perished.

To give an idea of the potential devastation of such a situation: before the advent of antibiotics, salmonella had fatality rates of 30%—greater when the infection developed into a systemic one. Although mortality rates would probably not be as high now owing to improved supportive measures, we can only imagine how dangerous salmonella might be if multiple-drug resistance were incorporated into the *S. typhi* or *S. paratyphi* strains, both of which cause systemic illness.

POTENTIAL DEFENSES

Salmonella has become increasingly resistant to traditional antibiotic treatment. A new research area—the development of live attenuated vaccine candidates—

may soon lead to the availability of safe and effective preprophylactic treatments for salmonella infections.

The *S. typhi*–derived Vi vaccine, a type of killed-whole-cell vaccine, prevents the formation of the bacterium's lipopolysaccharide layer, which is essential for virulence. However, salmonella vaccines also cause a variety of side effects, including a high incidence of adverse systemic reactions. Scientists are now studying attenuated strains of salmonella to find a less reactive alternative.

Live-attenuated vaccines present the risk that the attenuated strain may somehow revert to virulence and cause infection.[8] With this in mind, two approaches have been used to design the strains for these vaccines. One relies on mutations in genes controlling essential metabolic functions, and the second utilizes mutations in genes encoding virulence factors. The strategy behind a metabolic mutation is to create a bacterial auxotroph that has specific nutrient needs not provided by the host environment, thus interfering with the vaccine strain's ability to proliferate and cause disease while maintaining the strain's ability to stimulate the immune system.[18]

One such auxotroph provides insight into the process of creating a safe and efficient vaccine. The *aroA* gene codes for 5-enolpyruvylshikimate-3-phosphate synthetase in *S. typhimurium*. *aroA*(–) strains require a concentrated amount of *para*-aminobenzoic acid (PABA) for growth in host tissues. Because the tissues in which salmonella bacteria normally localize synthesize PABA in very minute amounts, the attenuated status of the *aroA*(–) auxotroph is assured. However, a mutant with only one defect is likely to revert to wild-type, so bacteriologists have tried combining *aroA* mutations with others, such as the mutation in the *purA* gene. *purA* encodes an enzyme called adenylosuccinate synthetase that converts IMP to AMP. A double-mutant strain would seem to provide the solution to reversion. However, studies with mice indicate that this particular combination is poorly immunogenic.

Combinations of double mutations within the genes controlling synthesis of aromatic amino acids (*aroC* codes for chorismate synthase, and *aroD* encodes 3-dehydroquinase) with a third mutation in the *htrA* gene resulted in a successful vaccine candidate. *htrA* is a component of the stress-induced response that protects the bacteria from oxidative killing by macrophages. The vaccine, known as CVD 908-*htr*, has developed into a leading oral vaccine candidate. It has already passed phase I and phase II clinical trials. Although CVD 908-*htr* must pass many more tests, it has been shown to possess strong immunogenicity.[8]

As described above in "Molecular Biology," Dam methylates specific adenines in the bacterial DNA. These methylations influence expression of many of the genes required for a full-blown bacterial infection. In addition to being avirulent, a Dam(–) mutant constitutively expresses about 20 of the genes that are induced during the normal infectious process. These same genes are expressed in wild-type Dam(+) cells but only at precise times during the infectious process. Because Dam is a pleiotropic regulator of genes expressed during infection, it is not surprising that vaccines prepared with Dam(–) mutants of *S. typhimurium* have elicited a cross-protective immune response against other pathogenic strains of salmonella.

REFERENCES

1 Baumler, A. J. 1997. The record of horizontal gene transfer in Salmonella. *Trends Microbiol* 5(8): 318–322.

2. Burrows, W. D. and S. E. Renner. 1999. Biological warfare as threats to potable water. Environ Health Perspect 107: 995–984.

3. Christopher, G., et al. 1997. Biological warfare: A historical perspective. *JAMA* 278: 412–417.

4. Fierer, J. and D. G. Guiney. 2001. Diverse virulence traits underlying different clinical outcomes of Salmonella infection. *J Clin Invest* 107(7): 775–780.

5. Freudenberg, M. A., et al. 2001. Role of lipopolysaccharide susceptibility in the innate immune response to *Salmonella typhimurium* infection: LPS, a primary target for recognition of Gram-negative bacteria. *Microbes Infect* 3: 1213–1222.

6. Galan, J. E. and A. Collmer. 1999. Type III secretion machines: bacterial devices for protein delivery into host cells. *Science* 284: 1322–1328.

7. Galan, J. 2001. Salmonella interactions with host cells: Type III secretion system at work. *Annu Rev Cell Dev Biol* 17: 53–86.

8. Garmory, H., K. Brown, and R. Titball. 2002. Salmonella vaccines for use in humans: present and future perspectives. *FEMS Microbiol Rev* 26(4): 339–353.

9. Goosney, D. L., D. G. Knoechel, and B. B. Finlay. 1999. Enteropathogenic *E. coli*, Salmonella, and Shigella: masters of host cell cytoskeletal exploitation. *Emerg Infect Dis* 5(2): 216–223.

10. Heithoff, D. M., et al. 1999. An essential role for DNA adenine methylation in bacterial virulence. *Science* 284: 967–971.

11. Hohmann, E., et al. 1995. Macrophage-inducible expression of a model antigen in *Salmonella typhimurium* enhances immunogenicity. *Proc Natl Acad Sci USA* 92: 2904–2908.

12. Holden, D. W. 2002. Trafficking of the salmonella vacuole in macrophages. *Traffic* 3(3): 161–169.

13. Hueck, C. J. 1998. Type III protein secretion systems in bacterial pathogens of animal and plants. *Microbiol Mol Biol Rev* 62: 379–433.

14. Jepson, M. A. and M. A. Clark. 2001. The role of M cells in Salmonella infection. *Microbes Infect* 3: 1183–1190.

15. Leavitt, J. 1995. "Typhoid Mary" strikes back: Bacteriological theory and practice in early twentieth-century public health. *Isis* 86(4): 617–618.

16. Monack, D. M., W. W. Navarre, and S. Falkow. 2001. Salmonella-induced macrophage death: The role of caspase-1 in death and inflammation. *Microbes Infect* 3: 1201–1212.

17. Ohl, M. E. and S. I. Miller. 2001. Salmonella: A model for bacterial pathogenesis. *Annu Rev Med* 52: 259–274.
18. Raupach, B. and S. Kaufmann. 2001. Bacterial virulence, proinflammatory cytokines, and host immunity: How to choose the appropriate Salmonella vaccine strain. *Microbes Infect* 3(14–15): 1261–1269.
19. Torok, T., R. Tauxe, and R. Wise. 1997. A large outbreak of salmonellosis caused by intentional contamination of restaurant salad bars. *JAMA* 278(5): 389–395.
20. U.S. Food and Drug Administration Center for Food Safety and Applied Nutrition. 1992. *Foodborne Pathogenic Microorganisms and Natural Toxins Handbook*. Rockville, Maryland.
21. Wallis, T. S. and E. E. Galyov. 2000. Molecular basis of salmonella-induced enteritis. *Mol Microbiol* 36(5): 997–1005.
22. Woodward, T. and C. B. Phillip. 1948. Chloromycetin in the chemoprophylaxis of scrub (tsutsugamush disease) *Rickettsiasis orientalis*: Control and prevention. In *Quarterly Progress Report of the Committee on Immunization*, Vol. 6(1), pp. 1–32. Washington, DC: U.S. Army.

APPENDIX 1

■ ■

DRUG DISCOVERY AND BIODEFENSE

William Edstrom and Geoffrey Zubay

The search for a new medicine to combat the pestilence of a pathogen requires major research effort and significant financial investment. Furthermore, it involves a calculated risk—of several thousand compounds with antitoxin properties, typically only a handful will reach the point where they are considered suitable for testing on human patients, and maybe only one of these will ever be approved for patient use. Furthermore, it is not uncommon for a drug that has gained general acceptance and is in widespread use to be pulled off the market because of harmful side effects on a significant minority of users.[5]

For infectious diseases that could be spread by bioterrorists, the medicinal needs could be very pressing. The threat of a high death toll might induce the U.S. Food and Drug Administration (FDA) to tolerate a less than perfect drug. Nevertheless, the time it takes to develop a medicine to counteract a microorganism for which no treatment is known must still be measured in years. It seems clear that many toxins of bacterial, viral, or chemical origin that are available to bioterrorists can be confronted only with new drugs. If we are to survive in these tumultuous times, a way must be found to develop and distribute new medicines more rapidly, especially to counter the class of toxins that are most likely to serve as weapons for bioterrorists.

IDENTIFYING THE RESPONSIBLE AGENT

An investigation invariably starts with tissue samples obtained from victims in an attack zone with brief details of the victims' ages and sex. The disease-causing agent in each sample must be identified, and the effective drugs must be found to neutralize or destroy it without harming the infected individual. The next step is to find a means for mass production of an effective counteragent.

The goal is to have all this happen in the shortest possible time to minimize casualties.

Patient samples should be tested for growth of the suspected pathogen in different media. Most pathogens will probably grow, or at least survive, in at least one of the many kinds of media. The cultures in which the suspected agent grows can be inoculated into several species of animals in the hope of identifying animals whose response to the agent is similar to that in humans. Animals that become ill after inoculation can serve as model systems for in vivo testing in a drug-screening program. Moreover, such animals may supply investigators with badly needed samples for characterization of the pathogen under investigation.[2,45]

Samples to be examined typically include blood or lung tissue. Additional tissue would be available if victims have already begun to pass away. Such tissues could first be characterized by analysis of their protein components on two-dimensional (2D) gels (see box A1.1). Gels[10,19,33,34,65] could be compared for tissue extracts from normal individuals of both sexes and a broad spectrum of agents. A broad distribution of normal tissues is needed for comparison purposes because protein expression is a function of many factors including age and gender. A robust database on normal individuals should facilitate programmed computer analyses of 2D gels.

When anthrax spores were delivered by mail to various people in the United States during the fall of 2001, identification times ranged from a few minutes to several days. Methods now exist for making most such identifications in minutes.

Bacterial spores are like seeds and cannot be quickly grown into rapidly dividing bacteria. The only readily identifiable material in anthrax spores is the chromosomal DNA of the bacterium. Bacterial chromosomes can be readily

---

BOX A1.1. PROTEIN ANALYSIS BY 2D GELS

Two-dimensional (2D) gels separate proteins based on their isoelectric points (first dimension) and molecular weights (second dimension). A complex mixture of proteins from a bacterial cell such as *E. coli* can be resolved by 2D gels into a 2D pattern of hundreds of spots, each spot indicative of a specific protein. At the end of 2D gel analysis, each spot, or a select group of spots, can be cut out of the gel and subjected to amino acid sequencing. Mass spectrometry (MS) is the preferred tool for determining the identity of the proteins.[20,21,32,44,49] There are two key principles of MS: all proteins can be sorted on a mass-to-charge ratio, and proteins can be broken into peptide fragments, facilitating their identification.

isolated, but single-copy chromosomal amounts of raw DNA are insufficient for precise genetic characterization. Polymerase chain reaction (PCR) is a robust procedure for DNA amplification,[55] but it typically takes 2–3 hours. The PCR products then need to be analyzed by gel electrophoresis, which takes additional time.

In 1998, Andreas Manz described a fast method of PCR called chemical-amplification continuous-flow PCR, performed on a chip.[29] DNA extracted from a biological sample is mixed with PCR reagents, and, once combined, they rapidly travel through several temperature zones to produce PCR products for later analysis. The Manz laboratory had provided an excellent tool to speed up the PCR process, but identification of the products was still suboptimal.[70] What was needed was a self-contained unit that would perform the PCR and do gel analysis in one step. David Burke developed an integrated nanoliter DNA-analysis device that solves the problem.[35,63] The entire chip is produced as a single piece, and the user supplies the standard PCR reagents and template DNA. The PCR product can be placed directly onto a miniature gel that takes about 30 minutes to resolve into specific bands. This device presents several advantages. First, the person in the field only needs to load DNA from a biological sample; the chip does everything else, including recording the results. The chips are small and their operation does not require much training. Phillip Belgrader helped to develop a more efficient, self-contained, and commercially available machine that can identify samples in as little as 7 minutes.[4] With such machines in the field, pathogens that contain nucleic acids can be rapidly identified.[26,54,61,67,69] It seems likely that other devices that operate by identifying protein products[31,38,58,71] or chemical products will be designed in the near future.

RECENT ADVANCES IN DRUG DEVELOPMENT

Once the pathogen is known, the search begins for a new drug. Whereas a few years ago the entire range of pharmaceutical products was based on only a few hundred targets, the number of targets now available for investigation has soared into the thousands and is ever-growing. To match this, the number of available drug candidates measures in the millions. Recent advances in our knowledge of molecular structure and drug-screening procedures have increased the speed and efficiency of developing drugs that neutralize toxin-related targets.

The availability of complete genome-sequence data for many important human pathogens provides a wealth of fundamental information on which a drug-screening program can be based. The combination of genome-sequence data and new technologies makes it possible to systematically explore the function of each open reading frame in a genome and identify any potential molecular targets for drugs.

Once an assay has been developed, researchers may screen thousands of compounds daily to find new leads. Robots[7,17,59] that can be programmed to carry out routine tasks make this possible. Interactions between potentially usable compounds and their targets are then classified and retested while chemists synthesize similar compounds that may also show activity against the target. A series of assays and screens should result in a collection of lead compounds for further, more refined analyses.

Methodologies for drug screening with robots are in common use by most large pharmaceutical companies. They usually involve plating picogram amounts of protein per well in 1536-well plates, then adding a different chemical (drug candidate) to each well, washing, and scanning plates in a reader for binding.[3,28,53] Ideally, the screening is coupled to an activity assay to demonstrate in vitro efficacy as well as binding of the drug candidate to the target.[12]

NEW APPROACHES TO LEAD GENERATION

Once a target has been identified, compounds can be synthesized that interact with it. Such a compound is called a lead, and the discovery process is called lead generation. Two complementary methodologies in quest of new drugs should be conducted in parallel with screening of known drug candidates: structure-based drug design and phage-display technology.

STRUCTURE-BASED DRUG DESIGN

Structure-based drug design[6,41,48,64] is most likely to be successful when the structure of the toxin is known. In the absence of a known structure with 100% sequence identity, homology modeling based on significant sequence homology of the pathogenic target to an already known structures can be utilized.[68] Antitoxins that have structures complementary to the toxin are chemically synthesized. It is not uncommon to make many derivatives with structures that show promise for making good antitoxins because their structures are complementary to critical regions of the toxin. The arms of structural biology that have proven useful in structure-based drug design are nuclear magnetic resonance (NMR) and x-ray crystallography.[42,50]

After the target has been identified, the objective of drug design is to develop a molecule (drug) that can bind to and inhibit it. If the target's function is essential to the life processes of the pathogen, inhibition of the target will stop the growth of the pathogen or even destroy it. An understanding of the structure and function of proteins is crucial in drug development because proteins are by far the most common drug targets.

The first step in the discovery of an inhibitor for a particular target protein usually involves identification of one or more lead compounds that bind to it and that might therefore bind to the active site(s) on the protein. This is traditionally a trial-and-error process in which numerous compounds are tested using various assays until enough compounds with inhibitory effect have been found. Recently, high-throughput screening (HTS) methods have made the procedure much more efficient, but the underlying process remains essentially the same.

Frequently, the three-dimensional structure of a protein and its ligand is known but the structure of the complex they form is unknown. The objective of "computational docking" is to determine how two molecules of known structure are likely to interact. In the cases of drug design, "molecular docking" is most often employed to aid in determining how a particular drug lead will bind to the target or how two proteins can interact to form a binding pocket.[30,60] Another important consideration in drug design is how to deal with flexibility in both the protein and the putative ligand.

PHAGE-DISPLAY TECHNOLOGY

Antibody fragments can be displayed on the surface of phage particles by fusion of the antibody-variable genes to one of the phage-coat protein genes. Antigen-specific phage antibodies can subsequently be enriched by multiple rounds of affinity selection because the phage particle carries the gene encoding the displayed antibody.[11,16,22,24,52,66] From such findings, it has become possible to make phage-antibody libraries by PCR cloning of large collections of variable-region genes, each expressing the binding sites on the surface of a different phage particle and harvesting the antigen-specific binding sites by in vitro selection of the phage mixture on a chosen antigen.

Before building a library of antigen-binding sites, the source of the sites should be considered. Selection is based on the future applications of the library. The source of the variable-domain genes and the type of complementary determining regions included will determine the specificity and frequency of antigen-specific clones in a binding-site library. High-affinity antibody-binding sites are found in immunogenic or disease-related libraries, whereas natural or synthetic libraries are a reliable source of antibodies to a more diverse collection of antigens. The antibody genes may be amplified from their natural sources by PCR. Finally, the antibody may be displayed on a major coat protein present in many copies on the phage or a minor coat protein with only a few copies.

Similarly, screening a large repertoire of degenerate peptide sequences could elucidate which peptide(s) bind with highest affinity to an essential pathogen protein. This peptide sequence can be readily subcloned into the variable region of a humanized antibody clone,[24] and this recombinant clone can then be expressed and purified for use as a humanized antibody drug that will be an antidote to the pathogen.

LEAD OPTIMIZATION

Having identified a number of leads, investigators focus on narrowing the field to a set of lead compounds that show the most promising interactions with the target.[47,57,62] Optimizing any particular compound may enhance its potency, selectivity, and availability. The goal of lead optimization is to fine-tune the lead series of compounds and enhance their potential for testing on humans. The lead candidates selected for further development should possess characteristics required for clinical success. Multidisciplinary teams of investigators that select candidates must address three questions: 1) which compounds are likely to be the most effective? 2) what are the projected dose range and easiest method of delivery for each compound? 3) which compounds are feasible to produce on a large scale?

In vitro tests[9,51] are performed to evaluate potency and selectivity of promising leads. These tests are complemented by in vivo tests on animals[39] to determine dose response, duration of action, and how the body modifies the drug and vice versa. Other questions asked during lead candidate selection are whether the candidates are part of a known toxic group, whether they are soluble, and whether they are effective in animal tests.

The molecular structures of leads are carefully scrutinized for their potential to be chemically altered (i.e., by combinatorial chemistry) to produce a range of modified small-molecule structures that could have superior medicinal properties.[1]

PRECLINICAL TRIALS

Before a new drug is tested on humans, a number of preclinical tasks are completed, including the development of dosage forms and assays as well as animal testing. New drugs may be screened out at this stage if they are toxic or not well absorbed in the body. When these data have been analyzed, a protocol is compiled for the first human studies, which are almost always conducted on healthy volunteers.

As a promising drug moves through a clinical trial program, much of the scientific support work continues. This includes toxicology studies,[13,27,40] which have two critical aims to be met before the drug is used in humans: to determine toxicity levels and to identify the target organs for toxic effects. The latter facilitates the early identification of adverse effects.

Metabolism studies are required to learn how the body responds to the drug. Scientists look at how the drug is absorbed, distributed, metabolized, and excreted (ADME).

The fastest possible development of a drug requires that it be available in sufficient quantity for clinical trials. The purity of the drug needs to be assessed, and the shelf-life and requisite storage conditions must be determined. As the drug is tested, more and more of it is required. Chemical procedures for production must be adapted so that the drug can be made on a larger scale. At this point, manufacturing experts should advise scientists about the compounds that can be mass-produced most readily as well as about any problems related to future scaling-up.

TESTS OF NEW DRUGS ON HUMANS

Testing of new drugs on humans is usually divided into three phases.[36] In phase I, the effects of a new drug on healthy volunteers are tested (these trials may be extended to desperately ill patients). Phase I determines safety and tolerability, how the body interacts with the drug, and what the drug does to the body. The results of these studies should underpin the design of subsequent phases, when the drug is tested on human patients. If the drug appears to be safe and to provide therapeutic benefit, it may be taken further in the clinical trial program.

In phase II, the drug is usually tested on 100 to 300 patients, with the primary emphasis on assessing safety and developing side-effect profiles. Evidence of efficacy is gathered as well, but obtaining absolute proof of efficacy is unlikely until large-scale phase III trials are well underway. Phase II studies also clarify the design and dosage regimen to be used in phase III trials. At this point, significant resources are needed to mold the drug molecule into a medicine.

Phase III studies typically involve a thousand or more patients in several centers and may require many months or even years. Additional data on safety are compiled, and it is usually these studies that establish efficacy.

If the population of human patients is limited, much of the in vivo testing may need to be carried out on animal-model systems.

ACCELERATING THE DRUG-DISCOVERY PROCESS

Existing procedures for drug discovery take too long to be useful in the event of an attack by bioterrorists.[46] There are several ways in which the drug-discovery process might be accelerated to produce an effective biodefense.[18]

GOVERNMENT-SPONSORED DRUG DISCOVERY UNITS

Just as the National Institutes of Health (NIH) has played a pivotal role in cancer research, government-sponsored facilities are needed to function as rapid discovery units. Such units should be organized to discover, produce, and oversee the distribution of bioweapon counteragents—for postexposure treatment, preexposure prophylaxis to protect uninfected individuals, and postexposure prophylaxis to prevent illness in recently exposed individuals.[18,23,25] In the event of an attack involving biological weapons for which no drugs exist, a rapid discovery unit would initially focus its efforts on expeditious discovery and production of the best possible medicines. Increases in the number of specialized personnel and machines (e.g., mass spectrometers, robots, and synthesizers) will facilitate accelerated discovery of antidotes.[18] For example, a facility that has 10 scientists working with one robot, one plate reader, and a small-molecule library of 50,000 drug candidates will be slower and less effective than a team of a thousand professionals working with 100 drug-screening robots, 100 plate readers, and a library with more than a million small molecules. In addition, implementation of gene synthesis[8] and de novo protein synthesis[14,15,37,43,56] as well as other newer technologies described earlier may help to speed the drug discovery and development process.

To some extent, this role is being played by the Centers for Disease Control and Prevention (CDC). Although the original purpose of the CDC (established in 1946) did not include dealing with bioterrorism, its scope and responsibility could be expanded to inaugurate and oversee rapid-drug-discovery facilities for biodefense.

In implementing established drug-discovery procedures for use against bioterrorism, further improvements in most phases of such investigations must be made and test times must be shortened. A concerted effort on many fronts should be initiated now so that in a few years we will be able to say that we have a solid biodefense system in place.

THE ROLE OF PHARMACEUTICAL COMPANIES

It is likely that the vast facilities and technical knowledge of the U.S. pharmaceutical industry would be harnessed to the task of finding new drugs for a wide variety of pathogens. Because pharmaceutical companies are, understandably, motivated by profit considerations, the inducement to engage in life-saving drug-discovery research must be provided in the form of monetary compensation by the federal government.

A radio network should be in place to communicate (e.g., in the event of a crisis) the findings on lead drugs so that one or more pharmacological units can join in the validation and manufacturing process. Collaborations of this sort between pharmaceutical companies and rapid discovery units are crucial to the success of the process in times of crisis.

RAPID SCREENING

If cultures grown from original patient samples and animal models are both positive, the animal models should be used to test with different dosage levels to make a rapid assessment of effective dosages and toxicology. Simultaneously, drugs can be tested for optimal dosage range and toxicology in human cells in culture as well as by proteomic studies in vitro. If consenting human volunteers are available, small amounts of the drug can be given to them for an initial assessment; otherwise, it may become necessary to move forward but without the support of certain information. This risk factor must be weighed against the potential seriousness of the bioattack.

REFERENCES

1. Adang, A. E. and P. H. Hermkens. 2001. The contribution of combinatorial chemistry to lead generation: An interim analysis. *Curr Med Chem* 8(9): 985–998.
2. Balk, M. W. 1987. Emerging models in the U.S.A.: Swine, woodchucks, and the hairless guinea pig. *Prog Clin Biol Res* 229: 311–326.
3. Battersby, B. J. and M. Trau. 2002. Novel miniaturized systems in high-throughput screening. *Trends Biotechnol* 20(4): 167–173.
4. Belgrader, P., C. J. Elkin, S. B. Brown, S. N. Nasarabadi, R. G. Langlois, F. P. Milanovich, B. W. Colston Jr., and G. D. Marshall. 2003. A reusable flow-through polymerase chain reaction instrument for the continuous monitoring of infectious biological agents. *Anal Chem* 75(14): 3114–3118.
5. Bensoussan, A., S. P. Myers, A. K. Drew, I. M. Whyte, and A. H. Dawson. 2002. Development of a Chinese herbal medicine toxicology database. *J Toxicol Clin Toxicol* 40(2): 159–167.
6. Bertelli, M., E. El-Bastawissy, M. H. Knaggs, M. P. Barrett, S. Hanau, and I. H. Gilbert. 2001. Selective inhibition of 6-phosphogluconate dehydrogenase from *Trypanosoma brucei. J Comput Aided Mol Des* 15(5): 465–475.
7. Beydon, M. H., A. Fournier, L. Drugeault, and J. Becquart. 2000. Microbiological high throughput screening: An opportunity for the lead discovery process. *J Biomol Screen* 5(1): 13–22.
8. Cello, J., A. V. Paul, and E. Wimmer. 2002. Chemical synthesis of poliovirus cDNA: Generation of infectious virus in the absence of natural template. *Science* 297(5583): 1016–1018.
9. Clarysse, C., J. Fevery, and S. H. Yap. 2000. In vitro assays for drug testing: Continuous cell lines. *Acta Gastroenterol Belg* 63(2): 213–215.
10. Collins, P. J., C. Juhl, and J. L. Lognonne. 1994. Image analysis of 2D gels: Considerations and insights. *Cell Mol Biol* (Noisy-le-Grand, France) 40(1): 77–83.

11. Coomber, D. W. 2002. Panning of antibody phage-display libraries: Standard protocols. *Methods Mol Biol* 178: 133–145.

12. Croston, G. E. 2002. Functional cell-based uHTS in chemical genomic drug discovery. *Trends Biotechnol* 20(3): 110–115.

13. Davila, J. C., R. J. Rodriguez, R. B. Melchert, and D. Acosta, Jr. 1998. Predictive value of in vitro model systems in toxicology. *Annu Rev Pharmacol Toxicol* 38: 63–96.

14. Dawson, P. E. and S. B. Kent. 2000. Synthesis of native proteins by chemical ligation. *Annu Rev Biochem* 69: 923–960.

15. Dawson, P. E., T. W. Muir, I. Clark-Lewis, and S. B. Kent. 1994. Synthesis of proteins by native chemical ligation. *Science* 266(5186): 776–779.

16. Dower, W. J. and L. C. Mattheakis. 2002. In vitro selection as a powerful tool for the applied evolution of proteins and peptides. *Curr Opin Chem Biol* 6(3): 390–398.

17. Dunn, D. A. and I. Feygin. 2000. Challenges and solutions to ultra-high-throughput screening assay miniaturization: Submicroliter fluid handling. *Drug Discov Today* 5(12 suppl 1): 84–91.

18. Edstrom W. 2003. *Bright summer: Rapid drug discovery biodefenses*. Unpublished master's thesis. New York: Columbia University, New York.

19. Fey, S. J. and P. M. Larsen. 2001. Two-dimensional gel electrophoresis: 2D or not 2D. *Curr Opin Chem Biol* 5(1): 26–33.

20. Gevaert, K. and J. Vandekerckhove. 2000. Protein identification methods in proteomics. *Electrophoresis* 21(6): 1145–1154.

21. Hess, S., F. J. Cassels, and L. K. Pannell. 2002. Identification and characterization of hydrophobic *Escherichia coli* virulence proteins by liquid chromatography-electrospray ionization mass spectrometry. *Anal Biochem* 302(1): 123–130.

22. Hoogenboom, H. R. 2002. Overview of antibody phage-display technology and its applications. *Methods Mol Biol* 178: 1–37.

23. Hupert, N., A. I. Mushlin, and M. A. Callahan. 2002. Modeling the public health response to bioterrorism: Using discrete event simulation to design antibiotic distribution centers. *Med Decis Making* 22(5 suppl): S17–25.

24. Huston, J. S. and A. J. George. 2001. Engineered antibodies take center stage. *Hum Antibodies* 10(3–4): 127–142.

25. Kahn, J. O., J. N. Martin, M. E. Roland, J. D. Bamberger, M. Chesney, D. Chambers, K. Franses, T. J. Coates, and M. H. Katz. 2001. Feasibility of postexposure prophylaxis (PEP) against human immunodeficiency virus infection after sexual or injection drug use exposure: The San Francisco PEP Study. *J Infect Dis* 183(5): 707–714.

26. Kato-Maeda, M., Q. Gao, and P. M. Small. 2001. Microarray analysis of pathogens and their interaction with hosts. *Cell Microbiol* 3(11): 713–719.

27. Kennedy, S. 2002. The role of proteomics in toxicology: Identification of biomarkers of toxicity by protein expression analysis. *Biomarkers* 7(4): 269–290.

28. Keusgen, M. 2002. Biosensors: New approaches in drug discovery. *Die Naturwissenschaften* 89(10): 433–444.

29. Kopp, M. U., A. J. Mello, and A. Manz. 1998. Chemical amplification: Continuous-flow PCR on a chip. *Science* 280(5366): 1046–1048.

30. Krumrine, J., F. Raubacher, N. Brooijmans, and I. Kuntz. 2003. Principles and methods of docking and ligand design. *Methods Biochem Anal* 44: 443–476.

31. Kumble, K. D. 2003. Protein microarrays: New tools for pharmaceutical development. *Anal Bioanal Chem* 377(5): 812–819.

32. Landreau, A., Y. F. Pouchus, C. Sallenave-Namont, J. F. Biard, M. C. Boumard, T. Robiou du Pont, F. Mondeguer, C. Goulard, and J. F. Verbist. 2002. Combined use of LC/MS and a biological test for rapid identification of marine mycotoxins produced by *Trichoderma koningii*. *J Microbiol Methods* 48(2–3): 181–194.

33. Lemkin, P. F. and G. Thornwall. 1999. Flicker image comparison of 2-D gel images for putative protein identification using the 2DWG meta-database. *Mol Biotechnol* 12(2): 159–172.

34. Lilley, K. S., A. Razzaq, and P. Dupree. 2002. Two-dimensional gel electrophoresis: Recent advances in sample preparation, detection and quantitation. *Curr Opin Chem Biol* 6(1): 46–50.

35. Lin, R., D. T. Burke, and M. A. Burns. 2003. Selective extraction of size-fractioned DNA samples in microfabricated electrophoresis devices. *J Chromat A*. 1010(2): 255–268.

36. Lipsky, M. S. and L. K. Sharp. 2001. From idea to market: The drug approval process. *J Am Board Fam Pract* 14(5): 362–367.

37. Low, D. W., M. G. Hill, M. R. Carrasco, S. B. Kent, and P. Botti. 2001. Total synthesis of cytochrome b562 by native chemical ligation using a removable auxiliary. *Proc Natl Acad Sci USA* 98(12): 6554–6559.

38. MacBeath, G. and S. L. Schreiber. 2000. Printing proteins as microarrays for high-throughput function determination. *Science* 289(5485): 1760–1763.

39. Mahmood, I. and J. D. Balian. 1996. Interspecies scaling: A comparative study for the prediction of clearance and volume using two or more than two species. *Life Sci* 59(7): 579–585.

40. Meador, V., W. Jordan, and J. Zimmermann. 2002. Increasing throughput in lead optimization in vivo toxicity screens. *Curr Opin Drug Discov Devel* 5(1): 72–78.

41. Molteni, V., J. Greenwald, D. Rhodes, Y. Hwang, W. Kwiatkowski, F. D. Bushman, J. S. Siegel, and S. Choe. 2001. Identification of a small-molecule binding site at the dimer interface of the HIV integrase catalytic domain. *Acta Crystallogr D Biol Crystallogr* 57(Pt 4): 536–544.

42. Muchmore, S. W. and P. J. Hajduk. 2003. Crystallography, NMR and virtual screening: Integrated tools for drug discovery. *Curr Opin Drug Discov Devel* 6(4): 544–549.

43. Muir, T. W., P. E. Dawson, and S. B. Kent. 1997. Protein synthesis by chemical ligation of unprotected peptides in aqueous solution. *Methods Enzymol* 289: 266–298.

44. Nedelkov, D., A. Rasooly, and R. W. Nelson. 2000. Multitoxin biosensor-mass spectrometry analysis: A new approach for rapid, real-time, sensitive analysis of staphylococcal toxins in food. *Int J Food Microbiol* 60(1): 1–13.

45. Niewiesk, S. and G. Prince. 2002. Diversifying animal models: The use of hispid cotton rats (*Sigmodon hispidus*) in infectious diseases. *Lab Anim* 36(4): 357–372.

46. O'Toole, T., M. Mair, and T. V. Inglesby. 2002. Shining light on "Dark Winter." *Clin Infect Dis* 34(7): 972–983.

47. Pagel, M. 2000. Phylogenetic-evolutionary approaches to bioinformatics. *Brief Inform* 1(2): 117–130.

48. Pattabiraman, N. 2002. Analysis of ligand macromolecule contacts: Computational methods. *Curr Med Chem* 9(5): 609–621.

49. Peng, J. and S. P. Gygi. 2001. Proteomics: The move to mixtures. *J Mass Spectrom* 36(10): 1083–1091.

50. Powers, R. 2002. Applications of NMR to structure-based drug design in structural genomics. *J Struct Funct Genomics* 2(2): 113–123.

51. Pritchard, J. F., M. Jurima-Romet, M. L. Reimer, E. Mortimer, B. Rolfe, and M. N. Cayen. 2003. Making better drugs: Decision gates in non-clinical drug development. *Nat Rev Drug Discov* 2(7): 542–553.

52. Rader, C. and C. F. Barbas 3rd. 1997. Phage display of combinatorial antibody libraries. *Curr Opin Biotechnol* 8(4): 503–508.

53. Ramstrom, O. and J. M. Lehn. 2002. Drug discovery by dynamic combinatorial libraries. *Nat Rev Drug Discov* 1(1): 26–36.

54. Relman, D. A. 1999. The search for unrecognized pathogens. *Science* 284(5418): 1308–1310.

55. Saiki, R. K., T. L. Bugawan, G. T. Horn, K. B. Mullis, and H. A. Erlich. 1986. Analysis of enzymatically amplified beta-globin and HLA-DQ alpha DNA with allele-specific oligonucleotide probes. *Nature* 324(6093): 163–166.

56. Schnolzer, M., H. R. Rackwitz, A. Gustchina, G. S. Laco, A. Wlodawer, J. H. Elder, and S. B. Kent. 1996. Comparative properties of feline immunodeficiency virus (FIV) and human immunodeficiency virus type 1 (HIV-1) proteinases prepared by total chemical synthesis. *Virology* 224(1): 268–275.

57. Selzer, P. M., S. Brutsche, P. Wiesner, P. Schmid, and H. Mullner. 2000. Target-based drug discovery for the development of novel antiinfectives. *Int J Med Microbiol* 290(2): 191–201.

58. Seong, S. Y. 2002. Microimmunoassay using a protein chip: Optimizing conditions for protein immobilization. *Clin Diagn Lab Immunol* 9(4): 927–930.

59. Taylor, P. B., F. P. Stewart, D. J. Dunnington, S. T. Quinn, C. K. Schulz, K. S. Vaidya, E. Kurali, T. R. Lane, W. C. Xiong, T. P. Sherrill, J. S. Snider, N. D. Terpstra, and R. P. Hertzberg. 2000. Automated assay optimization with integrated statistics and smart robotics. *J Biomol Screen* 5(4): 213–226.

60. Taylor, R. D., P. J. Jewsbury, and J. W. Essex. 2002. A review of protein-small molecule docking methods. *J Comput Aided Mol Design* 16(3): 151–166.

61. Tefferi, A., M. E. Bolander, S. M. Ansell, E. D. Wieben, and T. C. Spelsberg. 2002. Primer on medical genomics. Part III: Microarray experiments and data analysis. *Mayo Clin Proc* 77(9): 927–940.

62. Tong, A. H., M. Evangelista, A. B. Parsons, H. Xu, G. D. Bader, N. Page, M. Robinson, S. Raghibizadeh, C. W. Hogue, H. Bussey, B. Andrews, M. Tyers, and C. Boone. 2001. Systematic genetic analysis with ordered arrays of yeast deletion mutants. *Science* 294(5550): 2364–2368.

63. Ugaz, V. M., S. N. Brahmasandra, D. T. Burke, and M. A. Burns. 2002. Cross-linked polyacrylamide gel electrophoresis of single-stranded DNA for microfabricated genomic analysis systems. *Electrophoresis* 23(10): 1450–1459.

64. Verlinde, C. L., E. A. Merritt, F. Van den Akker, H. Kim, I. Feil, L. F. Delboni, S. C. Mande, S. Sarfaty, P. H. Petra, and W. G. Hol. 1994. Protein crystallography and infectious diseases. *Protein Sci* 3(10): 1670–1686.

65. Vijg, J. 1995. Two-dimensional DNA typing: A cost-effective way of analyzing complex mixtures of DNA fragments for sequence variations. *Mol Biotechnol* 4(3): 275–295.

66. Walter, G., Z. Konthur, H. Lehrach, and A. S. Biorchard. 2001. High-throughput screening of surface displayed gene products. *Comb Chem High Throughput Screen* 4(2): 193–205.

67. Wilgenbus, K. K. and P. Lichter. 1999. DNA chip technology ante portas. *J Mol Med* 77(11): 761–768.
68. Williams, M. G., H. Shirai, J. Shi, H. G. Nagendra, J. Mueller, K. Mizuguchi, R. N. Miguel, S. C. Lovell, C. A. Innis, C. M. Deane, L. Chen, N. Campillo, D. F. Burke, T. L. Blundell, and P. I. de Bakker. 2001. Sequence-structure homology recognition by iterative alignment refinement and comparative modeling. *Proteins* suppl 5: 92–97.
69. Yowe, D., W. J. Cook, and J. C. Gutierrez-Ramos. 2001. Microarrays for studying the host transcriptional response to microbial infection and for the identification of host drug targets. *Microbes Infect* 3(10): 813–821.
70. Zhang, C. X. and A. Manz. 2003. High-speed free-flow electrophoresis on chip. *Anal Chem* 75(21): 5759–5766.
71. Zhu, H., M. Bilgin, R. Bangham, D. Hall, A. Casamayor, P. Bertone, N. Lan, R. Jansen, S. Bidlingmaier, T. Houfek, T. Mitchell, P. Miller, R. A. Dean, M. Gerstein, and M. Snyder. 2001. Global analysis of protein activities using proteome chips. *Science* 293(5537): 2101–2105.

APPENDIX 2

■ ■

THE SEARCH FOR VACCINES

William Edstrom

The fountain of scientific successes includes drugs, medical devices, and prophylactic modalities such as vaccines. The following overview examines the history and formulation of vaccines, along with case studies not only of the successes but also of failures to develop safe and effective vaccines despite decades of intensive efforts.

It is good, of course, for people to have access to drugs and devices when disease strikes, but it is even better for people not to become sick in the first place, hence the importance of preventive medicine and biodefenses. Safe and effective vaccines enable people to stay healthy and avoid the pain, suffering, and sometimes even premature death that often accompany disease.

Given a hypothetical choice of becoming sick and then taking therapeutics to treat the illness or taking a safe and effective vaccine and not getting sick, most people would prefer the latter. There are pathogens for which no treatment exists and vaccines offer the only pharmaceutical protection available. For the many pathogens for which neither drugs nor vaccines exist, simultaneous research and development of both are desirable.

HISTORY

The two public health interventions that have had the greatest impact on public health and saved the most lives have been efforts to ensure the availability of potable water and vaccines. A handful of vaccines prevent illness and death for millions of people each year.[45] Vaccination, which is synonymous with immunization, has been defined as "protection of susceptible individuals from disease by the administration of a living modified agent, a suspension of killed organisms, or an inactivated toxin."[38] The word *vaccine* comes from *vaccinia*, the microorganism used by Edward Jenner in effectively vaccinating humans against smallpox by preexposure to cowpox, also known as vaccinia.[22] As a country physician, Jenner had heard of milkmaids who had been exposed to cowpox and were then apparently protected against infection by smallpox. To test the theory, he inoculated humans with cowpox, beginning in May 1796. It worked: prior exposure of an individual to cowpox resulted in immunity against the killer smallpox.[4,5]

Although Jenner's vaccinia vaccine was widely believed to be cowpox, it is now thought that the vaccinia virus used in smallpox vaccines might actually have been horsepox. Because horsepox was eradicated years ago, it cannot now be ascertained whether vaccinia was horsepox. The discovery that vaccinia is genetically distinct from both cowpox and smallpox led to speculation that vaccinia might have been horsepox. Possibly vaccinia is a hybrid or, more likely, a close relative of both smallpox and cowpox. What is known is that the current stockpile of smallpox vaccine consists of a live-animal poxvirus (vaccinia virus) that was grown on the skin of calves. As time and scientific prowess progress, we have learned more and more.

Since the early success—over 200 years ago—of the smallpox vaccine, vaccines have been developed for a host of other diseases. In 1885, Louis Pasteur developed a vaccine to protect against rabies by inoculation with an emulsion made from the spinal cord of a rabbit that had died of rabies.[42] Waldemar Haffkine developed a first-generation killed-whole-cell vaccine that was introduced in 1897 for protection against *Yersinia pestis*, the causative agent of plague.[36] Vaccines for typhoid[19] and cholera[25] were also developed in the late 1800s. In the first half of the twentieth century, vaccines became available for diphtheria,[63] pertussis, tuberculosis, tetanus,[63] and yellow fever. Since 1950, more vaccines have been developed, tested, and introduced for protection against polio, measles, mumps, rubella, and hepatitis A and B. Table A2.1 is a more complete list of vaccines approved for use in humans in the United States. These vaccines have saved millions of lives and prevented needless suffering for millions of people. However, several microorganisms (e.g., HIV-1 and *Plasmodium falciparum*, the causative agent of malaria) have remained refractory to the development of successful vaccines despite decades of intensive efforts.

TYPES OF VACCINES

An effective vaccine must contain an antigen—or, more specifically, one or more antigenic epitopes—that evokes an immune response. An antigen is any substance that stimulates the immune system. The earliest vaccines (e.g., for smallpox and rabies) as well as some of those developed more recently (e.g., for measles, mumps, rubella, and adenovirus) are live attenuated vaccines. Other

TABLE A2.1 APPROVED VACCINES

APPROVED VACCINES	VENDOR
Acellular pertussis multivalent[49]	Sanofi Pasteur
Anthrax[14]	BioPort Corporation
Diphtheria and tetanus toxoids	GlaxoSmithKline
Hepatitis A[6]	GlaxoSmithKline and Merck and Co.
Hepatitis B[32]	GlaxoSmithKline and Merck and Co.
Influenza A and B[40]	Chiron, MedImmune, and Sanofi Pasteur
Influenza B conjugate[58]	Sanofi Pasteur and Merck and Co.
Japanese encephalitis virus[60]	Sanofi Pasteur
Lyme disease (recombinant OspA)[21]	Sanofi Pasteur
Measles, mumps, and rubella[31]	Merck and Co. and GlaxoSmithKline
Neisseria meningitides strains A, C, Y, and W-135 (polysaccharide vaccine)[53]	Sanofi Pasteur
Tuberculosis/BCG[10]	Sanofi Pasteur
Polyvalent pneumococcal vaccine[59]	Merck and Co.
Poliovirus vaccine inactivated[61]	Sanofi Pasteur
Rabies[29]	Sanofi Pasteur
Tetanus/*Clostridium tetani*[27]	Chiron Corporation
Typhoid Vi polysaccharide[2]	Sanofi Pasteur
Varicella virus live[24]	Merck & Co
Yellow fever[47]	Chiron and Sanofi Pasteur

types of approved vaccines are killed-whole-organism, purified protein/toxoid, and genetically engineered. Live attenuated vaccines contain live organisms that are similar enough to the organism of interest to confer postadministration immunity but are not capable of causing disease. For example, cowpox is similar enough to smallpox (they are both closely related members of the orthopoxvirus family)[52] that administration of cowpox to humans confers immunity against smallpox. Cowpox causes disease in cows, whereas smallpox causes disease, and often death, in humans.

Yellow fever is another example of a live attenuated vaccine. In 1937, Theiler and Smith at Rockefeller University (then known as the Rockefeller Foundation Laboratories) passaged the Asibi strain of yellow fever virus 176 times in mouse embryo and chicken tissue. This resulted in attenuation of the yellow fever virus to such an extent that human administration of it (notated as the 17D yellow fever virus) resulted in the introduction of a form of the virus that is incapable of causing disease in humans. But it is similar enough to the wild-type, disease-causing yellow fever virus that this attenuated form confers immunity to it in humans.[35]

To make killed-whole-organism vaccines, the offending agent is killed by heat treatment and then further inactivated by chemical treatment. The killed-whole-microorganism is then used as the vaccine. The cholera vaccine is an example. Developed in 1896, it was produced by growing the virus (*Vibrio cholerae*), then heating the culture at 56°C for 1 hour and adding 0.5% phenol.[8] Another example is the vaccine for pertussis, also known as whooping cough. *Bordetella pertussis* bacteria are heat-killed and then treated with formalin to denature toxic protein components.[15] In 1996, the U.S. Food and Drug Administration (FDA) approved an acellular pertussis vaccine without trace amounts of formalin. It contains two important antigens of *B. pertussis*: pertussis toxin and filamentous hemagglutinin. These components are included in a vaccine marketed under the name Tripedia*.

Purified protein vaccines are made from a key antigenic protein of the offending organism. *Corynebacterium diphtheriae*, the causative agent of diphtheria, produces a toxin that is a polypeptide of approximately 62 kd. Isolation of this protein, followed by formalin and heat treatment of the diphtheria toxin protein, results in production of the diphtheria "toxoid," which is approximately 1000-fold less toxic than the toxin. The heat and formalin treatment results in the loss of the diphtheria toxin's attachment and enzymatic activities but not of its immunogenic properties. Inoculation of humans with the diphtheria toxoid therefore results in immunity against the disease-causing element of *C. diphtheriae*—the diphtheria toxin—by priming the immune system to produce antibodies to it. The lingering memory cells are then able to rapidly produce antitoxin if needed.

Tetanus vaccine provides another example of a manufacturing strategy that entails isolation and treatment of a protein from the offending organism. *Clostridium tetani*—the causative agent of tetanus—produces tetanospasmin, a potent exotoxin. Tetanospasmin is a neurotoxin protein of approximately 150 kd. As little as 2.5 ng of toxin per kilogram of body weight will result in spasms, rigidity, central nervous system failure, and death, usually by respiratory failure.[44] The tetanus vaccine is produced by isolating and purifying tetanospasmin from cultures of *C. tetani*, then detoxifying it into a toxoid by treating the purified tetanospasmin with 40% formaldehyde at 37°C. The toxoid no longer has enzymatic or pathogenic activity but does retain immunogenic activity to prime the immune system into producing both antibodies against tetanospasmin and memory cells that keep the instructions for producing tetanospasmin antibodies in the event of a challenge.[17] Vaccines based on only a portion (e.g., a protein or a toxin) of the offending organism are referred to as subunit vaccines.

Genetically engineered vaccines are composed of recombinant materials such as a protein or nucleic acid. Recombivax HB, a hepatitis B vaccine produced by Merck and Co. (West Point, Pennsylvania), is made by recombinant technology. The 226-amino-acid hepatitis B surface-antigen protein (HBsAg) is expressed in yeast cells. These yeast cells are then harvested and HBsAg protein is purified from the harvested cells; the purified HBsAg protein is the vaccine.[23]

A new frontier in vaccine development is nucleic-acid-based vaccines. Several such vaccines—for example, against HIV-1, malaria, and HSV—have progressed to human clinical trials.

THE IMMUNE RESPONSE

The human immune system can be divided into the innate and the acquired (also known as adaptive) immune systems. The innate immune system is nonspecific and nonadaptive. Its components include specialized cells such as neutrophils and natural killer cells. The innate system can be activated within minutes of an infection.[45] The acquired immune system, in contrast, is specific and adaptive. Its three major components are two classes of lymphocytes and cells that specialize in antigen presentation (e.g., dendritic cells and macrophages). Upon infection of a type not previously encountered, the acquired immune system takes several days to be activated. The acquired immune system has specificity and memory for specific epitopes on specific pathogens and immunogens.[45]

Vaccines work via the acquired immune system. B lymphocytes (also known as B cells) recognize nonnative biological macromolecules (e.g., pathogens and nonnative peptides and proteins), internalize them, and, through recombination mechanisms, generate novel DNA sequences through somatic hypermutation.[48] This then translates into antibodies that can bind to epitopes on nonnative macromolecules with high specificity.[62] B lymphocytes are small white blood cells derived from bone marrow; they mature into plasma cells that produce antibodies and secrete antibodies into the circulatory system. If the existence of a pathogen is sensed by B lymphocytes, then the specific B lymphocytes—with

DNA-sequence combinations that code for antibodies to that pathogen—will proliferate, leading to more plasma cells capable of secreting antibodies with specificity to the pathogen. This leads to higher levels of antibodies in the circulatory system capable of binding to the offending pathogen. If the pathogen is eliminated, the population of memory B lymphocytes specific for that pathogen will dwindle to only a few cells. Memory B lymphocytes are often referred to as simply memory cells or memory B cells.

These memory B lymphocytes reside in secondary lymphoid organs and remain quiescent unless there is a reinfection by the pathogen for which they have specificity. In that event, the memory B lymphocytes proliferate and once again secrete antibodies specific to the pathogen.[34] This is known as a secondary response, and it is much faster than a primary response. This is the essence of immunity.

Thymus-derived lymphocytes (T cells) express a broader repertoire of receptors on their cell surface. T cells are subdivided into CD4+ and CD8+ T cells. CD4+ T cells express the CD4 transmembrane receptor and CD8+ T cells the CD8 transmembrane receptor. CD4+ T cells are also known as helper T cells because they promote maturation of B cells to make antibody. CD8+ T cells are also known as cytotoxic T lymphocytes (CTLs) or killer T cells because one of their major functions is to recognize and lyse target cells infected with a specific antigen (e.g., a virus, a foreign protein, or a pathogen).

The sequence of events involving antigen-processing cells, helper T cells, and antibody-forming B cells is as follows. Typically, an antigen is phagocytosed by a macrophage. The phagocyte partially digests the antigen and presents the processed antigen on its outer plasma membrane. A specific helper T cell binds the antigen to a receptor bound to its outer plasma membrane. The T cell usually has many of the same types of receptors and binds many copies of the antigen in a similar way. This stimulates proliferation of T cells, which interact with B cells that display similar processed antigen. Several B-cell receptors involved in a similar way become focused in a local region of the outer plasma membrane. This focusing triggers the proliferation and differentiation of the B cells to become antibody-secreting or memory B cells.

One aspect of immunity for which there still is not complete consensus within the scientific community is exactly how long memory B cells persist and what the factors are that determine duration of persistence. This question is important in assessing how long a vaccination will be effective. Concerns in this regard have been voiced recently about smallpox. Because this scourge was officially declared eradicated in 1979, people who have been vaccinated received the vaccine decades ago. This raises the question of whether someone vaccinated against smallpox in the 1950s or 1960s still has effective immunity against smallpox. Knowing how long specific memory B cells persist would help answer questions about the durability of immunity provided by vaccines.

ASSESSING THE EFFICACY OF VACCINATION

A way to test whether immunity still exists is to test for the presence of antibodies. Such tests are available for hepatitis A and B. Typically a "booster shot" is needed after a period of time; for example, if antibodies to hepatitis B fall below 10 mIU/mL of blood, hepatitis B vaccine booster immunization is recommended.[13] The longevity of vaccine efficacy varies from person to person, and diagnostic assays for the presence of antibodies should be conducted at regular intervals (e.g., every 4 years). To illustrate, one study found that of infants vaccinated against diphtheria and tetanus, 89% still had antibodies against diphtheria greater than or equal to 0.01 IU/mL of serum 5 years later, and 93% had such levels of antibodies against tetanus 5 years later.[11]

Another factor in assessing efficacy is the type of assay used to determine antibody persistence. For example, in one study of infants evaluated 5 years after vaccination against pertussis, only 44% of infants had antibodies against the pertussis toxoid detected by enzyme-linked immunosorbent assay (ELISA), but 99% had the antibodies according to the Chinese hamster ovary cell neutralization test.[11] A Medline search yields few citations that directly addresses longevity of vaccine efficacy, but one article states that vaccines are "generally thought to be effective for three to five years, and in some cases as long as 10 years or more."[50] There does not appear to be a medical or scientific consensus regarding the longevity of vaccine efficacy. Immunity seems to diminish over the years following vaccination, and regular booster shots or revaccination may be advisable (e.g., every 4 or 8 years). Studies can be constructed and conducted to assess, on a vaccine-by-vaccine basis, for how long each vaccine confers immunity and how often booster shots should be given.

DOSING REGIMEN

Another aspect of assessment is efficacy of the dosage schedule. A clinical vaccine trial of more than 10,000 people with a three-dose schedule at 0, 1, and 12 months for Lyme disease (the LYMErix vaccine) resulted in 76% efficacy after the third dose. Additional booster shots 12 and 24 months later resulted in 96% efficacy.[46] With three inoculations of LYMErix, approximately three of four people were protected; with five inoculations, 96% were protected. How many inoculations to give, and how often, are clearly important considerations.

BENEFITS VERSUS THE RISKS OF VACCINATION

In another study of LYMErix, 905 patients had experienced adverse events by a point when more than 1,400,000 doses had been distributed. Fifty-nine of the most serious events reported were arthritis, 34 were arthrosis, nine were rheumatoid arthritis, and 12 were facial paralysis.[30] According to these figures, less than 1% of people vaccinated had adverse reactions and less than 0.1% had serious adverse reactions. A question that arises when one is vaccinated is the degree of risk that one may actually be exposed to the pathogen against which the vaccine is directed. Even when the risk of a serious adverse reaction is less than 0.1%, if the risk of being exposed to a particular pathogen without vaccine protection is also less than 0.1%, then it may be advantageous for an individual not to take that vaccine.

The risks of taking versus not taking certain vaccines are exceedingly difficult to assess. Although smallpox has been eradicated, concern has recently arisen that it may be used as a bioweapon. People who would be first-line providers (e.g., nurses and physicians) in the event of a serious bioattack are being encouraged to receive the smallpox vaccine. It is estimated that vaccinating all Americans aged 1 to 65 would result in 4600 serious adverse reactions and 285 deaths;[26] these figures represent less than 0.01% of the American population.

For many vaccines, the rate of serious adverse events, including death, due to vaccination is less than 1%. There have been notable incidents of vaccine-related adverse events. For instance, following the 1976–1977 swine flu vaccination program, Guillain-Barré syndrome developed in some people who had taken the vaccine.[18]

REFRACTORY PATHOGENS

There are still many pathogens for which vaccines do not exist. For example, many researchers have attempted for a long time to discover a safe and effective vaccine against *Plasmodium falciparum* (the causative agent of malaria in humans). A more recently discovered pathogen is the HIV-1 virus, which causes acquired immunodeficiency syndrome (AIDS) and has infected tens of millions of people and killed millions. Despite decades of intensive work by many researchers, there is still no viable vaccine against the HIV-1 virus.

Tuberculosis is one of the 10 leading killers of humans. The vaccine currently used for tuberculosis—the BCG vaccine—is of limited efficacy, at best.[9] A vaccine that is both safe and has significantly more efficacy is needed but has eluded researchers. The development of vaccines against countless other pathogens would be highly advantageous for the protection of human life and health, but unfortunately there are still no vaccines to protect us from most pathogens. The successes to date (vaccines for smallpox, polio, pertussis, etc.) are outnumbered by the failures.

The life cycle of *P. falciparum* has multiple stages, and a vaccine that immunizes against this parasite in one of its stages may not immunize against it in another.[37] Also, because *P. falciparum* lives inside certain human cells, it is less accessible than other pathogens to many components of the human immune system.[12] In the example of HIV-1, it appears that the virus is antigenically unstable and may mutate more rapidly than, for example, the influenza virus does. An additional challenge in designing a safe and effective anti-HIV-1 vaccine is that primary strains of HIV-1 are notoriously refractory to antibody neutralization.[20]

NEW VACCINES IN DEVELOPMENT

Many new vaccines are in preclinical development and clinical trials, including many vaccines with targets other than pathogens. For example, a vaccine currently in clinical trials for Alzheimer's disease involves administration of the amyloid ⊠-peptide to evoke a lasting immune response against it—there is increasing evidence that this peptide may be the culprit in the plaque formations seen in Alzheimer's patients.[51]

Other vaccines in clinical trials show promise in protecting against cocaine abuse/addiction and nicotine addiction. Succinylnorcocaine has been conjugated to the recombinant cholera toxin B and the conjugate administered to consenting former cocaine abusers in a human clinical trial. Highly specific IgG antibodies resulted, with an average binding affinity of 2.5×10^{-8} M to succinylnorcocaine, but these antibodies were no longer detected a year after vaccination.[28] It has been theorized that one or more booster injections are needed to generate and maintain a long-term immune response against cocaine.[28]

Human papilloma virus (e.g., HPV-16 and HPV-18) can cause several forms of cancer, including cervical and anogenital cancer. The HPV vaccines currently in human clinical trials are directed not only against HPV but also against cancer as a putative anticancer modality of treatment in patients already infected with HPV.[16] Limited early data from these trials indicate that most of the HPV vaccines being tested are "clinically effective."[7]

DNA vaccines have been and continue to be investigated. For example, DNA vaccines against malaria and HIV-1 have been researched, developed, and tested in human clinical trials. Although their safety has been acceptable, the potency has been "disappointing."[33] DNA vaccines have been administered as naked DNA by intravenous or intramuscular injection, via recombinant viral vectors and nonpathogenic bacterial hosts, and by ex vivo methods whereby some of a patient's cells are removed, the DNA of interest is transfected into these cells, and the treated cells are then returned to the patient.[1]

Merck (West Point, Pennsylvania) is conducting human clinical trials of an anti-HIV-1 vaccine utilizing replication-deficient (E1 and E3 deleted) adeno-virus-5 (Ad5) viral vectors containing a synthetic codon–optimized HIV-1 *gag* DNA sequence. Similar studies in rhesus monkeys with the same Ad5 viral vector containing synthetic codon–optimized SIV *gag* demonstrated a "reduced level of peak viremia" postinfection, minimal loss of CD4+ cells, and control of plasma levels of virus to nearly undetectable levels, i.e., "levels close to the assay detection limits (500 viral copies per mL)" of blood.[55] The results of this study suggest that such a vaccine may not prevent infection—indeed, low levels of virus were detected months later—but it may prevent illness. Virtually all the control monkeys that received either no vaccine construct or empty Ad5 vector got so sick (e.g., low CD4+ counts, high levels of viremia, opportunistic infections, weight loss, and chronic diarrhea) that they were euthanized.[55]

A cautionary note with regard to what may become one of the first DNA vaccines with demonstrable efficacy is that another group of researchers, also conducting immunodeficiency virus vaccination studies with rhesus monkeys, found that a single mutation in *gag* can lead to "eventual vaccine failure by viral escape in cytotoxic T cells." One of six rhesus monkeys vaccinated with a plasmid containing *env* and *gag* along with interleukin-2 had a viral mutagenic event in which a single nucleotide change in the *gag* sequence led to deteriorating health and death: "lower CD4+ cell counts, higher levels of viremia, clinical disease progression, and death from AIDS-related complications."[3]

Effective DNA vaccines are a new realm in the world of vaccines. A safe and effective vaccine against the HIV-1 virus—even if successful in only half of people vaccinated—would save millions of lives every decade. A proposal for the design and development of better DNA anti-HIV-1 vaccines is that multiple *gag* sequences (as opposed to a single sequence) be used in the vaccine construct to generate immunity against multiple known instances of the *gag* gene sequence.[3] There is typically a new influenza virus vaccine each year because the influenza coat mutates rapidly—certain mutations prevent last year's vaccine from being effective against this year's predominant strain. It may someday be discovered that the same is true of HIV-1 and other antigenically unstable pathogens, and that vaccines will therefore need to be updated as these pathogens mutate. We need to stay at least one step ahead.

CONCLUSION: PAST, PRESENT, AND FUTURE

The smallpox vaccine—the first vaccine to be discovered—is perhaps the best vaccine success story to date. Polio, although still not totally eradicated, has been sharply curtailed.[54] Vaccines continue to save millions of lives each decade, and research continues in pursuit of safe and effective vaccines for pathogens (e.g., malaria, HIV-1, and tuberculosis) that still kill so many people each year. Research efforts are also underway to discover vaccines to protect against newly emerging pathogens and bioterror threats. Severe acute respiratory syndrome (SARS)—another candidate for vaccine development—is an example of a newly discovered virus that has a high ratio of kill to infection. A coronavirus has been discovered and associated with SARS.[43]

The top 30 killers include diseases caused by communicable agents such as lower-respiratory-tract infections (e.g., influenza and pneumonia), diseases that cause diarrhea (e.g., cholera, giardiasis, and rotavirus), tuberculosis, measles, malaria, tetanus, pertussis, and HIV-1.[39] Several vaccines in current use could be better and more widely utilized. If distribution and administration of polio vaccines were more efficient, polio might be eradicated. It took nearly 200 years after Jenner's vaccination investigations for smallpox to be eradicated; it should not take 200 years for polio to be eradicated.

Rotavirus infection is a cause of diarrheal disease in infants and children. It often results in premature death, typically caused by the dehydration and/or malnutrition that can result from excessive and prolonged diarrhea. A vaccine exists for rotavirus, but it appears to have an "incidence of intussusception, estimated to be approximately 1 case per 10,000 immunized children."[41]

There is hope for commercially available vaccines to protect against non-pathogenic causes of disease and death, including many types of cancer. With regard to optimal biodefenses, it is challenging to speculate what bioterroristic agents might be used, and when. The best biodefenses may be to discover, develop, and test—at least for safety and for presence of antibodies after vaccination—as many vaccines as possible, against as many pathogens as possible, as quickly as possible.

REFERENCES

1. Alpar, H. O. and V. W. Bramwell. 2002. Current status of DNA vaccines and their route of administration. *Crit Rev Ther Drug Carrier Syst* 19: 307–383.
2. Arya, S. C. 2000. Efficient vaccination strategy against typhoid fever. *Vaccine* 18: 2321–2322.
3. Barouch, D. H., J. Kunstman, M. J. Kuroda, J. E. Schmitz, S. Santra, F. W. Peyerl, G. R. Krivulka, K. Beaudry, M. A. Lifton, D. A. Gorgone, D. C. Montefiori, M. G. Lewis, S. M. Wolinsky, and N. L. Letvin. 2002. Eventual AIDS vaccine fail-

ure in a rhesus monkey by viral escape from cytotoxic T lymphocytes. *Nature* 415: 335–339.

4. Barquet, N. and P. Domingo. 1997. Smallpox: The triumph over the most terrible of the ministers of death. *Ann Intern Med* 127: 635–642.

5. Baxby, D. 1999. Edward Jenner's inquiry: A bicentenary analysis. *Vaccine* 17: 301–307.

6. Bell, B. P. 2002. Hepatitis A vaccine. *Sem Pediatr Infect Dis* 13: 165–173

7. Berry, J. M. and J. M. Palefsky. 2003. A review of human papillomavirus vaccines: From basic science to clinical trials. *Front Biosci* 8: S333–345.

8. Bornside, G. H. 1981. Jaime Ferran and preventative inoculation against cholera. *Bull Hist Med* 55: 516–532.

9. Brandt, L., J. Feino-Cunha, A. Weinreich-Olsen, B. Chilima, P. Hirsch, R. Appelberg, and P. Andersen. 2002. Failure of the *Mycobacterium bovis* BCG vaccine: Some species of environmental mycobacteria block multiplication of BCG and induction of protective immunity to tuberculosis. *Infect Immun* 70: 672–678.

10. Brennan, M. J. 2000. Moving new vaccines for tuberculosis through the regulatory process. *Clin Infect Dis* 30: S247–249.

11. Carlsson, R. M., B. A. Claesson, E. Fagerlund, N. Knutsson, and C. Lundin. 2002. Antibody persistence in five-year-old children who received a pentavalent combination vaccine in infancy. *Pediatr Infect Dis J* 6:535–541.

12. Carvalho, L. J., C. T. Daniel-Ribeiro, and H. Goto. 2002. Malaria vaccine: Candidate antigens, mechanisms, constraints and prospects. *Scand J Immunol* 56: 327–343.

13. Cassidy, W. M. 2001. Adolescent hepatitis B vaccination: A review. *Minerva Pediatr* 53: 559–566.

14. Cummings, M. L. 2002. Informed consent and investigational new drug abuses in the U.S. military. *Account Res* 9: 93–103.

15. Fine, P. E. and J. A. Clarkson. 1987. Reflections on the efficacy of pertussis vaccines. *Rev Infect Dis* 9: 866–883.

16. Frazer, I. 2002. Vaccines for papilloma virus infection. *Virus Res* 89: 271–274.

17. Galazka, A. and F. Gasse. 1995. The present status of tetanus and tetanus vaccination. *Curr Top Microbiol Immunol* 195: 31–53.

18. Greenstreet, R. 1983. Adjustment of rates of Guillain-Barrè syndrome among recipients of swine flu vaccine, 1976–1977. *J Roy Soc Med* 76: 620–621.

19. Groschel, D. H. and R. B. Hornick. 1981. Who introduced typhoid vaccination: Almroth Write or Richard Pfeiffer? *Rev Infect Dis* 3: 1251–1254.

20. Ho, D. D. and Y. Huang. 2002. The HIV-1 vaccine race. *Cell* 110: 135–138.

21. Huebner, R. C., M. A. Gray, J. Holt, X. D. Wu, and J. P. Mays. 2000. Characterization of a recombinant outer surface protein A (OspA) vaccine against Lyme disease. *Dev Biol* (Basel) 103: 163–173.

22. Jenner, E. 1798. *An Inquiry into the Causes and Effects of the Variolae Vaccinae*. London: Low.

23. Jilg, W., B. Lorbeer, M. Schmidt, B. Wilske, G. Zoulek, and F. Deinhardt. 1984. Clinical evaluation of a recombinant hepatitis B vaccine. *Lancet* ii:1174–1175.

24. Jones, T. 2002. Varivax (Merck and Co). *Curr Opin Investig Drugs* 3: 54–57.

25. Kaper, J. B., J. G. Morris Jr., and M. M. Levine. 1995. Cholera. *Clin Microbiol Rev* 8: 48–86.

26. Kemper, A. R., M. M. Davis, and G. L. Freed. 2002. Expected adverse events in a mass smallpox vaccination campaign. *Eff Clin Pract* 5: 84–90.

27. Korger, G., U. Quast, and G. Dechert. 1986. Tetanus vaccination—tolerance and prevention of side effects. *Klinische Wochenschrift* 64: 767–775.

28. Kosten, T. R., M. Rosen, J. Bond, M. Settles, J. S. Roberts, J. Shields, L. Jack, and B. Fox. 2002. Human therapeutic cocaine vaccine: Safety and immunogenicity. *Vaccine* 20: 1196–1204.

29. Lang, J., S. Gravenstein, D. Briggs, B. Miller, J. Froeschle, C. Dukes, V. Le Mener, and C. Lutsch. 1998. Evaluation of the safety and immunogenicity of a new, heat-treated human rabies immune globulin using a sham, post-exposure prophylaxis of rabies. *Biologicals* 26: 7–15.

30. Lathrop, S. L., R. Ball, P. Haber, G. T. Mootrey, M. M. Braun, S. V. Shadomy, S. S. Ellenberg, R. T. Chen, and E. B. Hayes. 2002. Adverse event reports following vaccination for Lyme disease: December 1998–July 2000. *Vaccine* 20: 1603–1608.

31. Lee, C. Y., R. B. Tang, F. Y. Huang, H. Tang, L. M. Huang, and H. L. Bock. 2002. A new measles mumps rubella (MMR) vaccine: A randomized comparative trial for assessing the reactogenicity and immunogenicity of three consecutive production lots and comparison with a widely used MMR vaccine in measles primed children. *Int J Infect Dis* 6: 202.

32. Leroux-Roels, G., T. Cao, A. De Knibber, P. Meuleman, A. Roobrouck, A. Farhoudi, P. Vanlandschoot, and I. Desombere. 2001. Prevention of hepatitis B infections: Vaccination and its limitations. *Acta Clin Belg* 56: 209–219.

33. Liu, M. A. 2003. DNA vaccines: A review. *J Intern Med* 253:402–410.

34. MacLennan, I. C., C. Garcia de Vinuesa, and M. Casamayor-Palleja. 2000. B-cell memory and the persistence of antibody responses. *Philos Trans R Soc Lond B Biol Sci* 355: 345–350.

35. Marianneau, P., M. Georges-Courbot, and V. Deubel. 2001. Rarity of adverse effects after 17D yellow-fever vaccination. *Lancet* 358: 84–85.

36. Meyer, K. F., G. Smith, L. E. Foster, J. D. Marshall Jr., and D. C. Cavanaugh. 1974. Plague immunization. IV. Clinical reactions and serologic response to inoculations of Haffkine and freeze-dried plague vaccine. *J Infect Dis* 129: S30-S36.

37. Moore, S. A., E. G. Surgey, and A. M. Cadwgan. 2002. Malaria vaccines: Where are we and where are we going? *Lancet Infect Dis* 2: 737–743.

38. Moylett, E. H. and I. C. Hanson. 2003. Immunization. *J Allergy Clin Immunol* 111: S754–765.

39. Murray, C. J. and A. D. Lopez. 1997. Global mortality, disability, and the contribution of risk factors: Global Burden of Disease Study. *Lancet* 349: 1436–1442.

40. Nolan, T., M. S. Lee, J. M. Cordova, I. Cho, R. E. Walker, M. J. August, S. Larson, K. L. Coelingh, and P. M. Mendelman. 2003. Safety and immunogenicity of a live-attenuated influenza vaccine blended and filled at two manufacturing facilities. *Vaccine* 21:1224–1231.

41. Offit, P. A. 2002. The future of rotavirus vaccines. *Sem Pediatr Infect Dis* 13: 190–195.

42. Pearce, J. M. 2002. Louis Pasteur and rabies: A brief note. *J Neurol Neurosurg Psychiatry* 73: 82.

43. Peiris, J., S. Lai, L. Poon, Y. Guan, L. Yam, W. Lim, J. Nicholls, W. Yee, W. Yan, M. Cheung, V. Cheng, K. Chan, D. Tsang, R. Yung, T. Ng, and K. Yuen. 2003.

Coronavirus as a possible cause of severe acute respiratory syndrome. *Lancet* 361: 1319–1325.

44. Pellizzari, R., O. Rossetto, G. Schiavo, and C. Montecucco. 1999. Tetanus and botulinum neurotoxins: Mechanism of action and therapeutic uses. *Philos Trans R Soc Lond B Biol Sci* 354: 259–268.

45. Plotkin, S. A. and E. A. Mortimer, eds. 1994. *Vaccines*. Philadelphia: WB Saunders.

46. Poland, G. A. and R. M. Jacobson. 2001. The prevention of Lyme disease with vaccine. *Vaccine* 19: 2303–2308.

47. Pugachev, K. V., S. W. Ocran, F. Guirakhoo, D. Furby, and T. P. Monath. 2002. Heterogeneous nature of the genome of the ARILVAX yellow fever 17D vaccine revealed by consensus sequencing. *Vaccine* 20: 996–999.

48. Rajewsky, K. 1996. Clonal selection and learning in the antibody system. *Nature* 381: 751–758.

49. Rivera, A., J. C. Orengo, A. L. Rivera, C. Rodriguez, E. Calderon, J. Rullan, H. Yusuf, and L. Rodewald. 2002. Impact of vaccine shortage on diphtheria and tetanus toxoids and acellular pertussis vaccine coverage rates among children aged 24 months—Puerto Rico, 2002. *MMWR Morb Mortal Wkly Rep* 51: 667–668.

50. Robb-Nicholson, C. 2002. By the way, doctor. I'm 50 years old, and like everybody else my age, I was vaccinated against smallpox before first grade. Do I still have some immunity? *Harv Womens Health Watch* 10: 8.

51. Robinson, S. R., G. M. Bishop, and G. Munch. 2003. Alzheimer vaccine: Amyloid-beta on trial. *Bioessays* 25: 283–288.

52. Ropp, S. L., Q. Jin, J. C. Knight, R. F. Massung, and J. J. Esposito. 1995. PCR strategy for identification and differentiation of small pox and other orthopoxviruses. *J Clin Microbiol* 33: 2069–2076.

53. Rosenstein, N. E., M. Fischer, and J. W. Tappero. 2001. Meningococcal vaccines. *Infect Dis Clin North Am* 15: 155–169.

54. Senior, K. 2002. Polio eradication on track despite problems. *Lancet Infect Dis* 2: 321.

55. Shiver, J. W., T. M. Fu, L. Chen, D. R. Casimiro, M. E. Davies, R. K. Evans, Z. Q. Zhang, A. J. Simon, W. L. Trigona, S. A. Dubey, L. Huang, V. A. Harris, R. S. Long, X. Liang, L. Handt, W. A. Schleif, L. Zhu, D. C. Freed, N. V. Persaud, L. Guan, K. S. Punt, A. Tang, M. Chen, K. A. Wilson, K. B. Collins, G. J. Heidecker, V. R. Fernandez, H. C. Perry, J. G. Joyce, K. M. Grimm, J. C. Cook, P. M. Keller, D. S. Kresock, H. Mach, R. D. Troutman, L. A. Isopi, D. M. Williams, Z. Xu, K. E. Bohannon, D. B. Volkin, D. C. Montefiori, A. Miura, G. R. Krivulka, M. A. Lifton, M. J. Kuroda, J. E. Schmitz, N. L. Letvin, M. J. Caulfield, A. J. Bett, R. Youil, D. C. Kaslow, and E. A. Emini. 2002. Replication-incompetent adenoviral vaccine vector elicits effective anti-immunodeficiency-virus immunity. *Nature* 415: 331–335.

56. Smith, G. L. and G. McFadden. 2002. Smallpox: Anything to declare? *Nature Rev Immunol* 2: 521–527.

57. Smith, H. A. 2000. Regulation and review of DNA vaccine products. *Dev Biol* (Basel) 104: 57–62.

58. Steinhoff, M. and D. Goldblatt. 2003. Conjugate Hib vaccines. *Lancet* 361: 360–361.

59. Sweeney, J. A., Sumner, J. S. and J. P. Hennessey Jr. 2000. Simultaneous evaluation of molecular size and antigenic stability of PNEUMOVAX 23, a multivalent pneumococcal polysaccharide vaccine. *Dev Biol* (Basel) 103: 11–26.

60. Takahashi, H., V. Pool, T. F. Tsail, and R. T. Chen. 2000. Adverse events after Japanese encephalitis vaccination: review of post-marketing surveillance data from Japan and the United States. The VAERS Working Group. *Vaccine* 18: 2963–2969.
61. Von Seefried, A., J. H. Chun, J. A. Grant, L. Letvenuk, and E. W. Pearson. 1984. Inactivated poliovirus vaccine and test development at Connaught Laboratories Ltd. *Rev Infect Dis* 6: S345–349.
62. Willerford, D. M., W. Swat, and F. W. Alt. 1996. Developmental regulation of V(D)J recombination and lymphocyte differentiation. *Curr Opin Genet Dev* 6: 603–609.
63. Winau, F. and R. Winau. 2002. Emil von Behring and serum therapy. *Microbes Infect* 4: 185–188.

APPENDIX 3

■ ■

PERSONAL BIODEFENSES

Geoffrey Zubay

Although the emphasis in this book is on defenses that should be undertaken by the government, current indications are that the government is more likely to react than to initiate defensive measures. For example, it seems unlikely that respirators will be made available to the general public until after we have suffered a major attack using gas or an aerosolized pathogen. It would therefore be wise for individuals to take precautionary measures and to advise others to do the same. The following is a list of measures to be taken and key items that individuals might procure to protect themselves and family members.

1. Because most serious pestilences can be aerosolized, we must have protection against unfiltered air during a bioterrorist attack. Direct contact with outside air should be avoided in the first 1–2 days after an attack. Second, and of the utmost importance, at least one respirator should always be on hand. This means either carrying it around if you have only one or keeping one at home and the other at your workplace. A half mask covers the nose and mouth; a full mask protects the eyes as well. The mask should have replaceable filters that do not permit the passage of bacteria such as anthrax. Several filter replacements should be on hand for each mask. Consult the National Institute for Occupational Safety and Health (NIOSH) at 800-356-4674 or http://www.cdc.gov/niosh for information on proper masks and filters.

2. A portable radio will help you stay informed. Maintain a supply of extra batteries.

3. A flashlight is important, in case of a power outage. Again, keep extra batteries on hand.

4. In an emergency, elevators or other forms of transportation that rely on electricity should be not be used.

5 For nourishment, water is the most important commodity. Enough should be kept on hand to satisfy your needs for at least 2 weeks, as well as food stores, especially canned food that can be eaten without cooking.

6. You should have a good idea where your family members are at all times. Also, have a plan in the event of a situation in which you cannot communicate with one another.

7. In an attack, roads will probably be jammed, so don't attempt to go anywhere by car, especially if you live in a city. Since it may be impractical to stay where you are, have a plan for getting to a safe location promptly and efficiently.

8. A special whole-body suit would be helpful to protect against serious contact poisons such as nerve gas, but they are expensive. Watch for the availability of approved suits in the near future.

9. It has been highly publicized that plastic sheeting and duct tape could be used to seal off a room against the outside air. Keep in mind, however, that you need fresh air to breathe properly. Another complication is that many buildings have circulating air that is replenished by outside air. It would be advantageous to know the location of switches that control the circulating air so that they could be turned off during an emergency.

10. If you take special medicines, try to maintain a 2-month supply to protect against short supplies.

11. Always wash your hands with soap before eating. Soap will kill pathogens that are membrane-enclosed.

APPENDIX 4

■ ■

INFORMATION RESOURCES ON BIOTERRORISM

Kathleen Kehoe

The word *terrorism* was first used in 1795 to describe the reign of terror that gripped Paris when the Jacobins came into power.[24] This revolutionary faction executed scores of citizens in an effort to consolidate its control over French society. We now use the term to denote attacks launched by extremist groups to achieve political and social change. The magnitude of fear and panic inflicted on civilian populations is attributable to the unpredictable, irrational, and arbitrary character of the attacks.

Bioterrorism has yet to appear in many English-language dictionaries. A search of the Lexis-Nexis database revealed that the term first appeared in American newspapers in 1997. It refers to the use of not only disease organisms as weapons but also chemical weapons such as nerve gases and mustard gas. There is a long history of military use of biological and chemical weapons. The sixth-century (B.C.E.) Assyrians are believed to have poisoned water sources of their enemies with a toxic fungus.[26] In the fourteenth century, Tartar armies are said to have thrown the corpses of plague victims over their enemies' walls in an effort to start epidemics. An example of chemical warfare is the use of mustard gas by the German army on Allied troops in World War I. The 1925 Geneva Protocol outlawed chemical weapons in warfare. In 1972, the production, stockpiling, and use of biological weapons were outlawed by an international convention.[30] Military application of these weapons may be constrained, but terrorist groups have adopted their use because they provide an inexpensive alternative to costly and difficult-to-obtain conventional weapons.

In the past 20 years, numerous atrocities against civilians have been committed by terrorists in different parts of the world. In 1984, a group of religious cult members released the nerve gas sarin in the Tokyo subway, killing and injuring 750 people.[25] Another incident was the mailing of anthrax-containing letters in several regions of the United States in 2001. Twenty-two people were infected and five died as a result of contact with spores in the envelopes.[29] In view of these and similar occurrences, there is increasing interest in the mechanisms and methods of terrorists. Such information is useful in preparing for bioterrorist emergencies. This appendix reviews reliable information sources on biological weapons. Ancillary sources of information on chemical and radioactive weaponry are also provided.

An extensive body of information on bioterrorism and biological weapons is available to the public. It can be found in government reports, books, newspapers, magazines, and scientific journals, as well as on the Internet. Unfortunately, because there is also incorrect, unverified, and misleading information—both in print and online—care must be taken to select accurate and unbiased sources. The most reliable of these are, not surprisingly, government agencies, international health organizations, university research centers, and scientific societies. These resources are discussed below, along with commercially published reference books that warrant consideration. Website addresses are listed at the end of the text immediately before the references.

U.S. GOVERNMENT AGENCIES

CENTERS FOR DISEASE CONTROL AND PREVENTION

The Centers for Disease Control and Prevention (CDC) are responsible for national oversight of public health, especially in the area of infectious diseases. The CDC website[5] is the best first stop for background information on bioterrorism and biological weapons. The content is available in both English and Spanish and is frequently updated (see figure A4.1). It provides information and links to additional resources on pathogens and infectious diseases. Access to CDC materials on bioterrorism is initiated by clicking on "Emergency Preparedness and Response" and then selecting the subsection "Agents, Diseases and Threats," where the bacteria, viruses, and toxic substances most likely to be used by terrorists are listed (see figure A4.2).

The CDC classifies pathogens as Category A, B, or C, depending on the degree of virulence and contagiousness. Organisms on the Category A list, which have the highest likelihood of being chosen by terrorists, include *Bacillus anthracis* (anthrax), *Clostridia botulinum* (botulism), *Yersinia pestis* (plague), *Variola major* (smallpox), *Francisella tularensis* (tularemia), and viral hemorrhagic fever agents. The most notorious pathogens in this last group are the Marburg and Ebola viruses. The Category B organisms are also dangerous but more difficult to use as weapons; e.g., they do not spread easily among populations. Ex-

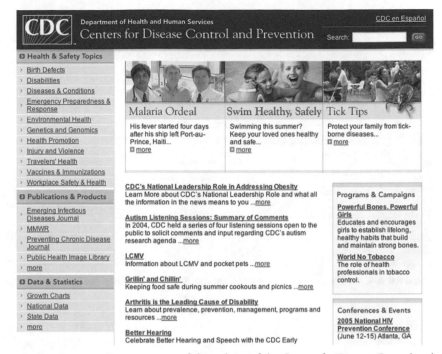

FIGURE A4.1. Opening screen of the website of the Centers for Disease Control and Prevention.

amples are the species of the genus *Brucella* (brucellosis), *Vibrio cholerae* (cholera), *Coxiella burnetti* (Q fever), *Escherichia coli* 0157:H7 (a particularly toxic member of the *E. coli* group), *Salmonella typhi* (typhoid fever), and *Shigella dysenteriae* (shigellosis).

The Category C list contains newly identified pathogens such as those that cause severe acute respiratory syndrome (SARS), hantavirus, and West Nile virus. Selecting an organism's name from the list provides summaries of the agent's life cycle, modes of infection and transmission, the diseases it causes, and methods of treatment and prevention, along with images of the pathogens and/or visible symptoms. References to key articles are provided for each organism. There are also links to the CDC journals *Emerging Infectious Diseases* and the *Morbidity and Mortality Weekly Report*. The user may examine tables of contents or search journal issues by keyword or author. The CDC recently added entries for the chemical weapons ricin, sarin, VX, and sulfur mustard.

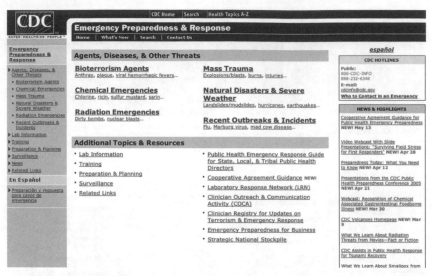

FIGURE A4.2. The Emergency Preparedness and Response section of the CDC website.

THE NATIONAL LIBRARY OF MEDICINE

The National Library of Medicine (NLM) is a division of the National Institutes of Health. The nation's repository for all biomedical publications, it maintains an electronic catalog called LocatorPlus.[19] Another, more focused NLM website, MedlinePlus,[13] is especially useful for health consumers. It provides a broad range of medical references and is available internationally. The MedlinePlus homepage (see figure A4.3) provides a starting point for access to the site's resources, which include medical databases, the *Adam Health Illustrated Encyclopedia*,[1] the *Merriam-Webster Medical Dictionary*,[16] and a directory of medical organizations. There is an excellent guide for judging the reliability of Internet references called *The MedlinePlus Guide to Healthy Web Surfing*.[14] People unfamiliar with biomedical literature should read the guide before using the Internet for health research.

Medline, originally called Index Medicus, is the most important database maintained by the NLM. It is available as a free resource under the name PubMed, and is the best source of articles on bioterrorism. It indexes 4600 journals from more than 70 countries and contains over 12 million citations and abstracts. PubMed is accessed through either the National Center for Bio-

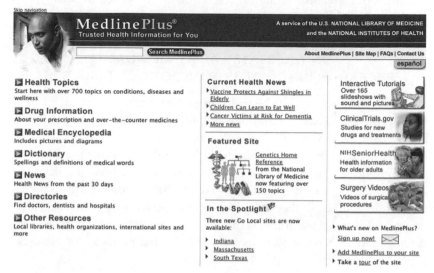

FIGURE A4.3. Opening screen of the MedlinePlus website.

medical Information[20] or MedlinePlus as discussed above. PubMed's coverage continually expands with the addition of records from previous years as well as new records. The database contains the past 50 years of biomedical literature—references from 1950 to 1966 were added in December 2004. It includes enhancements unavailable in other versions of Medline; for example, citations to articles having common indexing terms are linked to one another, and these can be retrieved to enlarge a set of search results. Also, citations are linked to entries in other databases, such as GenBank and Toxline.

Using PubMed is not simple. Search statements require appropriate terminology and syntax. Before beginning, it is recommended that novices use the tutorial and read the "help notes." This explains the syntax required for a search as well as how to use the thesaurus to select the best search terms. For example, a search on "smallpox and vaccine" retrieved 3102 articles, certainly many more references than most people would need. Much of the information will be overly technical, and many entries will be redundant. Narrowing this search by adding the term "bioterrorism" reduced the yield to 241 articles. The results can be further reduced by using the Limits feature. A search on "smallpox vaccine and bioterrorism" limited to reviews in English that deal only with humans resulted in only 28 articles. The citations retrieved have the following format: title, author(s), journal title (often abbreviated), date of publication, volume number, issue number, and page numbers (sample: "Wright ME, Fauci

AS. Smallpox immunization in the 21st century: the old and the new. JAMA 2003 June 25; 289(4): 3306–8").

UNITED STATES ARMY MEDICAL RESEARCH INSTITUTE FOR INFECTIOUS DISEASES

Selected military resources on bioterrorism are now available to the public. The U.S. Army Medical Research Institute for Infectious Diseases (USAMRIID), in Fort Detrick, Maryland, is the leading laboratory for the U.S. biological defense research program. Its website[21] features links to other sites that provide references on nuclear, biological, and chemical warfare sites. Journal articles and books written by USAMRIID staff are linked to the site. Among these is the latest electronic edition of the *Medical Management of Biological Casualties Handbook*,[31] which contains a history of biological warfare, criteria for distinguishing deliberate attacks from natural disease outbreaks, and protocols for managing battlefield casualties.

U.S. FOOD AND DRUG ADMINISTRATION

Still other federal resources can be found through two U.S. Food and Drug Administration (FDA) websites. The Center for Food Safety and Nutrition[9] has compiled an online handbook of foodborne pathogenic microorganisms that cause epidemics of food poisoning, whether deliberate or accidental. The Center for Drug Evaluation Research[8] publishes information on vaccines and drugs for the treatment of victims of terrorist attacks and provides guidelines on preventive measures.

INTERNATIONAL HEALTH ORGANIZATIONS

The best source of international information on bioterrorism and infectious diseases is the Communicable Disease Surveillance and Response (CSR)[22] section of the World Health Organization (WHO) website. WHO monitors epidemics and disease outbreaks around the globe. Fourteen diseases are considered epidemic threats, and WHO has designated four of them potentially effective biological weapons: anthrax, plague, smallpox, and the Ebola virus. The CSR site contains a section, Preparedness for Deliberate Epidemics, that includes reports on public health responses to biological and chemical weapons. There are also United Nations resolutions and background papers on responses to bioterrorist attacks on civilians. The site links to specialized WHO publications, including the *Bulletin of the World Health Organization*, *Disease Outbreak News*, and electronic issues of the *Weekly Epidemiological Review*. The *Review* can be explored, by keyword or author, back to 1996. Earlier issues (to 1926) are available in pdf format. The CSR website also includes specialized monographs, such as *Smallpox and Its Eradication*.[27] The entire WHO library, starting in 1948, can be searched online in the WHO Library Catalog.[23]

UNIVERSITY RESEARCH CENTERS

Several universities have centers for research on bioterrorism and biosecurity. The Center for Civilian Biodefense Strategies at The Johns Hopkins School of Public Health was the first. It has recently been divided into two entities: 1) the original center, as part of a new Institute for Global Health and Security (IGHS) at Johns Hopkins, and 2) a new Center for Biosecurity (CBS), which has relocated to the University of Pittsburgh Medical Center.

The IGHB website is not yet available, but the CBS has set up an excellent new site.[4] It contains links to major organizations doing work on bioterrorism, fact sheets on pathogenic bacteria and viruses, bibliographies for full-text articles, news bulletins, book reviews, and a calendar of meetings on bioterrorism and infectious diseases. Selected transcripts of relevant congressional testimony and guidelines for public responses to terrorism threats are posted on the site. It also contains the first peer-reviewed journal on the topic, the *Biosecurity Bulletin* (formerly *The Biodefense Quarterly*)

The Saint Louis School of Public Health supports two bioterrorism research centers: the Center for the Study of Bioterrorism and the Center for Emerging Infections. Together, they maintain a website[7] containing a section on chemical warfare agents and radiological weapons. Chemical weapons are organized by the body system they affect, for example, blood agents (arsinides), nerve agents (sarin and the organophosphates), vesicants (blistering agents such as mustard gas), and pulmonary agents (chlorine and phosgene). "Radiological weapons" refers to the deliberate radiation poisoning of a population by "dirty bombs," nuclear blasts caused by reactor meltdowns, and actual bomb attacks in war.

The University of Minnesota maintains the Center for Infectious Disease Research and Policy. Its website[6] contains monthly records of bioterrorism events since September 11, 2001. These records are compiled from news sources such as *The New York Times*, Associated Press, and Reuters. The center focuses on public health preparedness, with particular attention to the safety of the nation's food supply. The website contains references on food biosecurity measures aimed at preventing the deliberate introduction of biological or chemical agents into commercially prepared foods. Other resources on the site concentrate on agricultural biosecurity, which refers to the infection of livestock, poultry, fish, or crops with disease agents and other poisonous substances.

SCIENTIFIC SOCIETIES

Three websites offer resources of scientific organizations and societies. The National Academies of Science (NAS) is composed of four divisions: the National Academy of Science, the Institute of Medicine, the National Research Council, and the National Academy of Engineering. The website of the NAS contains an electronic library[18] that is a gateway to more than 3000 books and tens of thousands of Academy reports and documents. It can be explored through a powerful search mechanism not available on many of the websites discussed earlier. For example, it produces an ordering of documents relevant to an entered keyword, e.g., "anthrax." The article, book, or report having the most citations to the keyword is displayed, followed by the document with the next most citations, and so on, until there are no more documents containing the keyword. The documents themselves can be downloaded individually.

The Federation of American Scientists is a nonprofit organization founded by a group of scientists interested in influencing public policy. Principal concerns of the organization are nuclear disarmament, limitation of nonconventional weapons, and prevention of terrorist attacks. The Federation website[10] provides a wealth of information on disarmament and arms treaties, weapons of mass destruction, and biological, chemical, and nuclear terrorism.

The website of the American Society for Microbiology[3] hosts a Biological Weapons Resource Center. A pdf version of the book *Biosafety in Microbiological and Biomedical Laboratories*[33] is available here. There are links to legislation on bioterrorism enacted since 2001. The site contains regulations and guidelines for handling and shipping biological agents, and other technical information for laboratory specialists.

REFERENCE BOOKS

Heightened public interest in bioterrorism has resulted in the publication of a growing number of reference books and bibliographies. These can be located by consulting library catalogs and *Books in Print* as well as electronic catalogs of the Library of Congress[12] and the NLM (LocatorPlus)[19] mentioned earlier. These two libraries have the nation's most extensive collection of books relevant to bioterrorism. *Books in Print* is a comprehensive listing of available books published in the United States. It can be searched by title, author, or subject.

Encyclopedia articles are a good starting point for gathering background information on bioterrorism and related topics. The *Encyclopedia of Microbiology*[32] and the *Encyclopedia of Virology*[28] are valuable resources. A number of electronic reference works on microbiology and virology are available to the public. The McGraw-Hill Companies have created a website—Harrison's Online[11]—that includes nine chapters from a standard internal medicine textbook. The chapters cover pathogens, the diseases they cause, patient treatment, vaccines, and preventive measures. The pharmaceutical firm Merck and Co. has deposited the bioterrorism portion of *The Merck Manual of Diagnosis and Therapy* on its website.[2] One of the Internet's best sites is All the Virology on the World Wide Web.[22] This is a gateway to dictionaries, course notes, information summaries, and full-text articles. Another online compilation is the Medical Library Association's bioterrorism bibliography and resources.[17] This website provides references selected for their excellence by medical librarians.

CONCLUSION

The advantages to the public of these resources cannot be gainsaid. However, there is increasing concern in some government circles about the sensitive nature of information pertaining to bioterrorism. Indeed, there is a trend toward withdrawing some of these materials from public access, including information on government websites.[34] Even scientific organizations such as the American Society for Microbiology are advocating that researchers withhold information from their publications in the interest of national security. It is not clear to what extent future legislation might limit what investigators may publish.

WEBSITES

1. *Adam Health Illustrated Encyclopedia.* http://www.nlm.nih.gov/medlineplus/encyclopedia.html. Accessed 12/28/04.
2. *All the Virology on the WWW.* http://www.tulane.edu/˜dmsander/garryfavweb.html. Accessed 12/ 28/04.
3. American Society for Microbiology. http://www.asm.org. Accessed 12/28/04.
4. Center for Biosecurity UPMC. http://www.upmc-biosecurity.org. Accessed 12/28/04.
5. Centers for Disease Control and Prevention. http://www.cdc.gov. Accessed 1/08/04.
6. CIDRAP. http://www.cidrap.umn.edu/cidrap Accessed 12/28/04.
7. CSB and EI. http://www.bioterrorism.slu.edu. Accessed 12/28/04.

8. FDA CDER. http://www.fda.gov/cder/drugprepare. Accessed 12/28/04.
9. FDA CFSAN. http://www.cfsan.fda.gov. Accessed 12/28/04.
10. Federation of American Scientists. http://www.fas.org. Accessed 12/28/04.
11. *Harrison's Online*. http://www.accessmedicine.com. Accessed 12/28/04.
12. Library of Congress Catalog. http://www.loc.gov. Accessed 12/28/04.
13. MedlinePlus. http://www.nlm.nih.gov/medlineplus. Accessed 12/28/04.
14. MedlinePlus *Guide to Healthy Web Surfing*. http://www.nlm.nih.gov/medlineplus/healthywebsurfing.html. Accessed 12/28/04.
15. *Merck Manual of Diagnosis and Therapy*, 17th ed. 1999. http://www.merck.com/pubs/mmanual/section13/chapter157/157c.htm. Accessed 12/28/04.
16. *Merriam-Webster Medical Dictionary*. http://www.nlm.nih.gov/medlineplus/mplus-dictionary.html. Accessed 12/28/04.
17. MLA. http://mlanet.org/resources/caring/resources.html. Accessed 12/28/04.
18. NAS *Searchable Webshelf Collection*. http://www.nap.edu/firstresponders. Accessed 12/28/04.
19. NLM LocatorPlus. http://www.locatorplus.gov. Accessed 12/28/04.
20. PubMed. http://ncbi.nlm.nih.gov/entrez. Accessed 12/28/04.
21. USAMRIID. http://www.usamriid.army.mil. Accessed 12/28/04.
22. WHO (CSR). http://www.who.int/csr. Accessed 12/28/04.
23. WHO Library. http://www.who.int/library/database. Accessed 12/28/04.

REFERENCES

24. Britannica Student Encyclopedia. *Chemical and Biological Terrorism*. Online: http://search.eb.com/ebi/article?eu=295426. Accessed Dec. 26, 2003.
25. Broad, W. J. 1998. Sowing death: A special report; how Japan germ terror alerted world. *New York Times*, May 26.
26. Christopher, G. W., T. J. Cieslak, J. A. Pavlin, and E. M. Eitzer Jr. 1997. Biological warfare: A historical perspective. *JAMA* 278: 412–417.
27. Fenner, F. and D. M. Henderson. 1988. *Smallpox and Its Eradication*. Geneva: World Health Organization.
28. Granoff, A. and R. G. Webster. 1999. *Encyclopedia of Virology*, 2nd ed. San Diego: Academic Press.
29. Gursky, E., T. V. Inglesby, and T. O'Toole. 2003. Anthrax 2001: Observations on the medical and public health response. *Biosecur Biodef* 1: 97–108.
30. Kaplan, M. M. 1999. The efforts of WHO and Pugwash to eliminate biological and chemical weapons—a memoir. *Bull World Health Organ* 77: 97–108.
31. Kortepeter, M. and G. W. Christopher, eds. 2003. *Medical Management of Biological Casualties Handbook*, 5th ed. Frederick, Maryland: USAMRIID.
32. Lederberg, J., ed. 2000. *Encyclopedia of Microbiology*, 2nd ed. San Diego: Academic Press.
33. Richmond, J. Y. and R. W. McKinn. 1999. *Biosafety in Biological and Biomedical Laboratories*. Atlanta: Centers for Disease Control and Prevention.
34. Rindskopf-Parker, E. and L. Gielow-Jacobs. 2003. Government controls of information and scientific inquiry. *Biosecur Bioterror* 1: 83–87.

Acetylcholine (Ach) Neurotransmitter that functions at vertebrate neuromuscular junctions and at many neuron–neuron synapses in the brain and peripheral nervous system.

Acetyl CoA An acetyl group linked to coenzyme A; formed during oxidation of pyruvate, fatty acids, and amino acids.

Acidosis A metabolic condition resulting from the accumulation of acid or the depletion of bicarbonate in the blood and tissues.

Acquired immunity The ability of an individual to produce specific antibodies in response to antigens to which the body has previously been exposed based on the development of a memory response.

Active transport The transport of solute molecules across a membrane against a concentration and/or electrical gradient; it requires a carrier protein and the input of energy.

Acyclovir A synthetic purine nucleoside derivative with antiviral activity against herpes simplex virus.

Adenylate cyclase The enzyme that catalyzes the formation of cyclic-3′,5′-adenosine monophosphate (cAMP) from ATP.

Adhesions Thin protein structures that mediate the adhesion of *E. coli* to epithelial surfaces and contribute to the virulence of the bacteria.

Adjuvant A drug ingredient that aids or modifies the action of the main ingredient.

Aerobic bacterium A bacterium that requires oxygen for growth and survival.

Agglutination reaction The formation of an insoluble immune complex by the cross-linking of cells or particles.

Albuminuria Abnormally high level of albumin, a plasma protein, in the urine.

Allele One of two or more alternative forms of a gene located at the corresponding site (locus) on homologous chromosomes.

α-helix A helical folding pattern found in protein backbones. α-helices are characterized by having 3.6 amino acid residues per turn.

Anabolism The synthesis of complex molecules from simpler molecules with the input of energy.

Anaerobic bacterium A bacterium that does not require oxygen for maintenance or growth.

Anaphylaxis An immediate hypersensitivity reaction following exposure of a sensitized individual to the appropriate antigen. Mediated by reagin antibodies, chiefly IgE.

Anoxic Without oxygen; anaerobic.

Antibody Large Y-shaped proteins produced by higher organisms (e.g., humans, monkeys, rodents) in response to an antigen. They can be of a variety of immunoglobulins and bind with high specificity to their target (e.g., antigens, toxins) to neutralize it.

Antigen Any substance that stimulates the immune system; a substance (e.g. foreign protein, toxin) that provokes an immune response; bacteria and viruses are common sources of antigens.

Antigenic determinant (epitope) The part of an antigen molecule that binds to the antigen-binding region of an antibody.

Antigenic drift Major antigenic change, typically seen in influenza viruses and other pathogens, that occurs because of accumulated genetic mutations and recombinations; process of antigenic change that can cause gene reassortment between an animal and a human strain; a gradual and cumulative process of antigenic changes that become apparent only with time.

Antigenic shift Genetic mutation caused by the addition of new genes that produces new strains of influenza viruses.

Antisense RNA A single-stranded RNA with a base sequence complementary to a segment of another RNA molecule that can specifically bind to the target RNA and inhibit its activity.

Apical membrane The inner or upper surface of transporting epithelial cells that in vertebrates is separated from the basolateral membrane by a ring of tight junctions that prevents mixing of proteins and substances between the two membranes.

Apoptosis (programmed cell death) A genetically encoded sequence of actions, normal in many cells, that leads to the cells' death.

Arthralgia Sharp, severe joint pain; arthritis.

ATP (adenosine triphosphate) The main source of readily available biochemical energy.

Auxotroph A cell or microorganism that grows only when the medium contains a specific nutrient or metabolite that is not required by the wild type; commonly, a microbial mutant strain that cannot synthesize an essential metabolite as judged by its inability to grow in a minimal medium.

Bacillus The genus of the organism that causes anthrax infection. It also refers to the rod-shaped structure of vegetative bacteria in the *Bacillus* genus. All members of the genus, in addition to being rod-shaped bacteria, form endospores when environmental conditions deteriorate.

Bacteriophage A virus whose host is a bacterium; a virus that replicates within bacterial cells.

Bacteriostatic Inhibiting the growth and reproduction of some types of bacteria without necessarily killing them.

Basolateral surfaces The outer or bottom surface of transporting epithelial cells that in vertebrates is separated from the apical membrane by a ring of tight junctions that prevents mixing of proteins and substances between the two membranes.

B cells Bone marrow–derived immune cells that synthesize and secrete antibodies. B cells are an integral part of the humoral immune system.

Bioinformatics A field of study that extracts biological information from large data sets such as sequences, protein interactions, and microarrays; also includes the area of data visualization.

Biopreparat The program/organization sponsored by the government of the former Soviet Union with the goal of developing biological weapons as well as treatments and countermeasures for such weapons.

Biovars Subdivisions of a serovar.

Broad-spectrum drugs Chemotherapeutic agents that are effective against many kinds of pathogens.

C3/C5 convertases Components of the complement system, a major factor in the body's defense mechanisms against foreign pathogens.

Calmodulin Ubiquitous Ca^{2+}-binding protein whose binding to other proteins is governed by changes in intracellular Ca^{2+} concentration. Its binding modifies the activity of many target enzymes and membrane transport proteins.

cAMP-dependent protein kinase (cAPK) Cytosolic enzyme that is activated by cAMP and functions to regulate the activity of numerous cellular proteins; also called protein kinase.

Capsule A micropolysaccharide outer shell that envelops some bacteria.

Carbohydrate General term for certain polyhydroxyaldehydes, polyhydroxyketones, or compounds derived from these, usually having the formula $(CH2O)n$.

Carcinogen A chemical or physical agent that can cause cancer when cells or organisms are exposed to it.

Catabolism The breakdown, or degradation, of large energy-rich molecules within cells.

Cell adhesion molecules (CAMs) Proteins that protrude from the surface of the plasma membrane and form loops or other appendages that the cells use to grip one another and the surrounding connective-tissue fibers.

Chaperone Proteins that prevent misfolding of a target protein (molecular chaperones) or actively facilitate its proper folding (chaperonins).

Chelator A chemical compound that is capable of removing heavy metals, such as lead or mercury, from the bloodstream. An example is EDTA.

Chemokine family receptor A receptor capable of binding chemokines, thereby activating white blood cells.

Chemotaxis The pattern of microbial behavior in which the microorganism moves toward chemical attractants and away from repellents.

Cholesterol A lipid containing the four-ring steroid structure with a hydroxyl group on one ring; a major component of many eukaryotic membranes and precursor of steroid hormones.

Cholinergic synapses Synapses that are activated or released by acetylcholine.

Chromatin Complex of DNA, histones, and nonhistone proteins from which eukaryotic chromosomes are formed.

Cilia Threadlike appendages extending from the surface of some protozoa that beat rhythmically to propel them; cilia are membrane-bound cylinders with a complex internal array of microtubules, usually in a 9 + 2 pattern.

Class I MHC molecule MHC molecule involved with antigen presentation and recognition; found on the surfaces of nearly all nucleated cells. Class I MHC molecules with their bound foreign antigens are recognized exclusively by T lymphocytes carrying the CD8 marker.

Class II-associated antigen Antigen derived from extracellularly synthesized proteins such as bacterial proteins and soluble exotoxins that react with class II MHC molecules.

Class II MHC molecule MHC molecule involved with antigen presentation and recognition; expressed only on B lymphocytes, macrophages, dendritic cells, and endothelial cells; composed of two glycoprotein chains, alpha and beta, that are noncovalently associated. Class II MHC molecules with their bound antigen are recognized exclusively by $CD4^+$ cells (T-helper cells).

Clathrin Protein that assembles into a polyhedral cage on the cytoplasmic side of a membrane so as to form a clathrin-coated pit, which buds off to form a clathrin-coated vesicle.

Clone A group of genetically identical cells or organisms derived by asexual reproduction from a single parent.

Cloning vector A DNA molecule that can replicate and is used to transport a piece of inserted foreign DNA, such as a gene, into a recipient cell. It may be a plasmid, package, or cosmid.

Coccobacillus A bacillus that is short and round.

Coenzyme A loosely bound cofactor that often dissociates from the enzyme active site after product has been formed.

Collagen A protein that forms fibrils of great tensile strength; a major component of the extracellular matrix and connective tissues.

Commensal strains (apathogenic commensal strains) Relating to *E. coli* strains that either symbiotically exist in humans as intestinal flora or are incapable of causing disease because they are unable to colonize within the intestines and/or do not possess key virulence factors, such as laboratory *E. coli* strains.

Complement Group of proteins normally present in plasma and tissue fluids that participate in antigen–antibody reactions, allowing reactions such as cell lysis.

Complement fixation The binding of complement to an antigen–antibody complex so that the complement is unavailable for subsequent reactions.

Conformation The precise shape of a protein or other macromolecule in three dimensions resulting from the spatial location of the atoms in the molecule. The conformation of proteins is most commonly determined by x-ray crystallography. Small changes in the conformations of some proteins greatly affect their activity.

Conjugation The temporary union of two bacterial cells during which one cell transfers part or all of its genome to the other.

Conjunctival Relating to the clear membrane that coats the outer surface of the eye.

Coryza Excess mucus production.

Cyanosis A bluish coloring of the skin and/or mucous membranes caused by inadequate oxygenation of the blood.

Cytokine Any of numerous secretions, small proteins (e.g., interferons, interleukins) that bind to cell-surface receptors on certain cells to trigger their differentiation or proliferation. Some cytokines, also called lymphokines, function to regulate the intensity and duration of the immune response.

Cytopathic effect Deterioration or degeneration in cells caused by the multiplication of a virus.

Cytotoxic Toxic to cells.

Cytotoxic T cells (CTLs) Class of lymphocytes that have the phenotype CD4⁺ or CD8⁺ and kill cells.

Denaturation Drastic alteration in the conformation of a protein or nucleic acid due to disruption of various noncovalent bonds caused by heating or exposure to certain chemicals.

Diauxic growth A biphasic growth pattern or response in which a microorganism exposed to two nutrients initially uses one of them for growth and then alters its metabolism to make use of the second.

Diuresis Increased urine excretion.

DNA library Collection of cloned DNA molecules consisting of fragments of the entire genome (genomic library) or of DNA copies of all the mRMAs produced by a cell type inserted into a cloning vector.

Dysphagia Difficulty in swallowing.

Dyspnea Difficulty in breathing associated with lung or heart problems and causing shortness of breath; "air hunger."

Ecchymosis Hemorrhaging from ruptured blood vessels into the subcutaneous tissue, causing a purple coloring of the skin.

Edema An excessive accumulation of serous fluid in tissue spaces or a body cavity.

Electrolytes Substances that dissociate into ions, which are thus capable of conducting electric current.

ELISA (enzyme-linked immunosorbent assay) Assay in which the antibody against the protein of interest is immobilized on an inert solid such as polystyrene. The solution being assayed is applied to the antibody-coated surface and the resulting protein antibody is further reacted with a second protein-specific antibody to which an easily assayed enzyme has been linked.

Endemic Describes a disease that is constantly present in a given area; for example, malaria has been constantly present (to varying degrees) for a prolonged period of time in sub-Saharan Africa.

Endocarditis Infection of the endocardium or heart valves caused by bacteria or, in the case of intravenous drug abusers, fungi.

Endocytosis Uptake of extracellular materials by invagination of the plasma membrane to form a small membrane-bounded vesicle.

Endopeptidase (protease) An enzyme that catalyzes the hydrolysis of internal peptide linkages, resulting in the degradation of a protein.

Endoplasmic reticulum The extensive array of internal membranes in a eukaryotic cell involved in coordinating protein synthesis.

Endosome A membranous vesicle formed by endocytosis.

Endosperm cell Cell making up the nutrient-rich tissue within the seed of a flowering plant surrounding the embryo.

Endotoxin A molecule (mostly lipopolysaccharides) associated with the cell wall of gram-negative bacteria. The toxicity of most endotoxins is associated with the lipid component of the molecule. Compared with exotoxins, endotoxins have a relatively low potency and low specificity because they lack the specificity of proteins.

Enteropathogen An organism that causes disease in the intestinal tract.

Enterotoxin A toxin specifically affecting the cells of the intestinal mucosa, causing vomiting and diarrhea.

Enthalpy (H) Heat; in a chemical reaction, the enthalpy of the reactants or products is approximately equal to their total bond energies.

Entropy (S) A measure of the degree of disorder or randomness in a system; the higher the entropy, the greater the disorder.

Enzootic Describes a disease prevalent among certain animals in a certain area.

Eosinophils White blood cells that are important in allergic response and in combating internal parasite infestations..

Epinephrine The primary hormone secreted by the adrenal medulla; important in preparing the body for "fight or flight" responses and in regulation of arterial blood pressure; adrenaline.

Episome A genetic particle, such as a plasmid, usually found in bacterial cells, that can exist either autonomously in the cytoplasm or as part of a chromosome.

Epistaxis Bleeding from the nose.

Epithelial tissue A functional grouping of cells specializing in the exchange of materials between the cell and its environment; it lines and covers various body surfaces and cavities and forms secretory glands.

Epitope A chemical structure capable of eliciting an immune cell response and of being specifically recognized by molecules of immune recognition (e.g., antibodies, T-cell receptors); usually refers to a portion (i.e., the exact region) of the surface of a foreign protein to which components of the immune system (e.g., antibodies, T-cell receptors) bind with specificity.

Epizootic Describes a disease temporarily prevalent among certain animals in a certain area (as opposed to enzootic).

Erythema Redness due to dilation and congestion of capillaries, a sign of inflammation or infection.

Eubacteria Class of prokaryotes that constitutes one of the three distinct evolutionary lineages of modern-day organisms; phylogenetically distinct from archaea and eukaryotes.

Eukaryote An organism whose cells(s) contain a distinct, membrane-bound nucleus.

Excitatory postsynaptic potential (EPSP) A small depolarization of the postsynaptic membrane in response to neurotransmitter binding, thereby bringing the membrane closer to threshold.

Executioner caspase A type of protein that coordinates the execution phase of apoptosis by cleaving multiple structural and repair proteins; the caspases are a family of cysteine proteases that together cause apoptosis.

Exocytosis Release of intracellular molecules contained within a membrane-bounded vesicle by fusion of the vesicle with the plasma membrane of a cell. This is the process whereby most molecules are secreted from eukaryotic cells.

Exon Segments of a eukaryotic gene that reaches the cytoplasm as part of a mature mRNA, rRNA, or tRNA molecule.

Exotoxins Toxins that are protein in nature and, rather than being associated with the cell wall as endotoxins are, are diffusible and in most cases secreted by the cell. In some cases, however, exotoxins are released only when a bacterial cell is lysed.

Facultative anaerobe An organism, such as a bacterium, that lives in the absence of atmospheric oxygen but can also function in the presence of oxygen.

FAD (Flavin adenine dinucleotide) A coenzyme that participates in oxidation reactions by accepting two electrons from a donor molecule and two H^+ from the solution.

Fatty acid A hydrocarbon chain that has a carboxyl group at one end; a major source of energy during metabolism and precursors for synthesis of phospholipids.

FDA (U.S. Food and Drug Administration) A federal agency established in the 1930s with primary responsibility to permit or not permit the sale and marketing of individual drugs on a case-by-case basis in the United States of America. The European Union's equivalent is the EMEA, the European Medicines Agency (http://www.emea.eu.int).

F factor The fertility factor, a plasmid that carries the genes for bacterial conjugation and makes its *E. coli* host cell the gene donor during conjugation.

Fibronectin An extracellular multiadhesive protein that binds to other matrix components, fibrin, and cell-surface receptors of the integrin family. It functions to attach cells to the extracellular matrix and is important in wound healing.

Fimbriae Hair-like particles extending from the exterior membrane of the bacterium that enable it to be mobile and confer adhesion particles.

Flagellum A thin, threadlike appendage on many prokaryotic and eukaryotic cells that is responsible for their motility.

F′ plasmid An F plasmid that carries bacterial genes and transmits them to recipient cells where the F′ cell carries out conjugation; the transfer of bacterial genes in this way is often called sexduction.

Frameshift mutation A type of mutation that causes a change in the three base sequences read as codons, i.e., a change in the phase of transcription arising from the addition or deletion of nucleotides in numbers other than three or multiples of three.

Ganglioside Any of a group of galactose-containing cerebrosides found in the surface membranes of neurons.

Gene Physical and functional unit of heredity that carries information from one generation to the next.

Gene array A small slide (usually glass or silicon) that has small pieces of DNA arrayed on its surface.

Glucagon A peptide hormone produced in the α-cells of the pancreas that triggers the conversion of glycogen to glucose by the liver.

Glucocorticoids The adrenocortical hormones that are important in intermediary metabolism and in helping the body resist stresses; primarily corticol.

Glucose Six-carbon monosaccharide that is the primary metabolic fuel in most cells.

Glycolysis The anaerobic conversion of glucose to lactic acid by use of the Embden-Meyerhof pathway.

Glycoprotein A protein–carbohydrate complex.

Golgi complex (Golgi apparatus) Stacks of membranous structures in eukaryotic cells that function in processing and sorting of proteins and lipids destined for other cellular compartments or for secretion.

Gram's stain procedure Differential staining procedure in which bacteria are classified as gram-negative or -positive, depending on whether they retain the primary stain when treated with a decolorizing agent; the staining procedure reflects the underlying structural differences in the cell walls of gram-negative and -positive bacteria.

GTP (guanosine 5′-triphosphate) A nucleotide that is a precursor in RNA synthesis and plays a special role in protein synthesis, signal-transduction pathways, and microtubule assembly.

GTPase superfamily Group of guanine nucleotide–binding proteins that cycle between an inactive state with bound GDP and an active state with bound GTP. These proteins—including G proteins, Ras proteins, and certain polypeptide-elongation factors—function as intracellular switch proteins.

Haptophore The group on a toxin responsible for target cell binding; this group exercises its activity immediately after entry into the organism.

Helix loop–helix motif A structural motif found in many proteins that consists of a polypeptide region containing two segments of ⊠-helix connected by a loop region of varying length and contour.

Hemagglutination The agglutination of red blood cells by antibodies.

Hemagglutinin The antibody responsible for a hemagglutination reaction.

Hematogenous Having to do with blood.

Hemolysis The destruction or dissolution of red blood cells.

Hfr strain A bacterial strain that donates its genes with high frequency to a recipient cell during conjugation because the F factor is integrated into the bacterial chromosome.

High-throughput Describes methods that produce large volumes of data and can process many samples quickly. Robots and computerized data collection are common in high-throughput processes.

Horizontal gene transfer The process of passing genes from one species to another. The mechanism for this is largely unknown, although pathogens and transposons are suspected causes.

Hormone A long-distance chemical mediator that is secreted by an endocrine gland into the blood, which transports it to its target cells.

Humoral immunity A form of immunity whereby B lymphocytes and plasma cells produce antibodies to foreign agents, such as antigens, and stimulate T lymphocytes to attack them (cellular immunity).

Humoral response An immune response that is mediated by antibodies.

Hydrophobic Not interacting effectively with water, as oil. Literally, "water-averse."

Hyperglycemia Elevated glucose concentration in the blood.

Hyponatremia A condition characterized by excessive sodium loss due to dehydration.

Hypoperfusion Describes decreased blood flow through an organ.

Hypotension Low blood pressure.

Hypovolemia Decreased blood volume.

IgG (immunoglobulin G) The largest fraction of the body's immunoglobulins and a major antibody that circulates through the body; an immunoglobulin that has a major role in protecting the body against systemic microbial infections.

Immunoassays An investigative procedure that measures and identifies an antigen via reaction with antibody.

Immunohistochemical staining Staining technique utilizing labeled antibodies to localize and identify immunoreactive substances.

Induration The hardening of a normally soft tissue or organ, especially the skin.

Inoculation A deliberate exposure to an antigen or pathogen (or modified antigen or pathogen) that stimulates the development of immunity. A typical route of inoculation is injection via syringe and needle. Also applies to administration orally (e.g., a polio vaccine and a new vaccine for cholera) and via a nasal spray (e.g., FluMist for influenza vaccination).

Insertion sequence A small piece of DNA that moves mobile genetic elements.

Insulin A protein hormone produced in the β-cells of the pancreas that stimulates uptake of glucose into muscle and fat cells and with glucagon helps to regulate blood glucose.

Interferon A chemical released from virus-invaded cells that provides nonspecific resistance to viral infections by transiently interfering with replication of the same or unrelated viruses in other host cells.

Interleukin-1 (IL-1) A soluble cytokine secreted by monocytes, macrophages, and accessory cells that is involved in the activation of T and B lymphocytes. Its biological effects include replacing the macrophage requirement for T-cell activation as well as having many effects on other cell types. It is often involved in inflammatory and immune-system responses.

Interleukin-2 (IL-2) A cytokine formed by T lymphocytes that stimulates the growth of T lymphocytes and cytokine production by T cells.

Intestinal flora The bacteria and other microorganisms that normally inhabit the intestines.

Intron A noncoding intervening sequence in a split or interrupted gene that codes for RNA missing from the final RNA product.

In vitro A reaction or process taking place in an isolated cell-free extract; sometimes used to distinguish cells growing in culture from those in an organism. Literally,. "in glass."

In vivo In an intact cell or organism. Literally, "in life."

Ion channel A transmembrane protein complex that forms a water-filled channel across the phospholipid bilayer, allowing selective ion transport down its electrochemical gradient.

Isoelectric point (pI) The pH of a solution at which a dissolved protein or other potentially charged molecule has a net charge of zero.

Kaposi's sarcoma A skin cancer commonly associated with AIDS victims.

Kinase An enzyme that catalyzes the transfer of a group, usually a phosphate group, from a high-energy phosphate molecule (e.g., ATP) to a substrate.

Kligler iron agar (KIA) test A type of biochemical screening test that can be bought as a kit and used to rule out pseudomonas and certain enterobacteriaceae when making confirmatory diagnosis of *V. cholerae.*

Km A parameter that describes the affinity of an enzyme for its substrate and equals the substrate concentration that yields the half-maximal reaction rate.

Lateral gene transfer The movement of genetic material between bacteria other than by inheritance from a parent species (through cell division).

Lectin A term used to describe plant proteins that recognize and bind to specific carbohydrate groups on proteins or cell membranes.

Lentivirus Any of a group of animal viruses that cause diseases having an unusually long incubation period. HIV is a lentivirus.

Leukocytosis Condition of abnormally elevated white-blood-cell count.

Leukopenia Condition of abnormally low white-blood-cell count.

Ligand Any molecule other than an enzyme substrate that binds tightly to another molecule or metallic cation.

Lumen Cavity enclosed by an epithelial sheet (in a tissue) or membrane (in a cell).

Lymphadenopathy A chronic, abnormal enlargement of the lymph nodes.

Lymphatic system The interconnected system of spaces and vessels between body tissues and organs by which lymph circulates throughout the body and mediates immune responses.

Lymphokine A biologically active protein secreted by activated lymphocytes, especially sensitized T cells. It acts as an intercellular mediator of the immune response and transmits growth, differentiation, and behavioral signals.

Lysine iron agar (LIA) test A biochemical test that is useful for screening out aeromonas and non-cholera-causing vibrio species which, unlike *V. cholerae*, do not decarboxylate lysine. *V. cholerae* produce a purple slant and butt in the LIA test, whereas other species induce a yellow butt and/or a red slant.

Lysosome Small membrane-bounded organelle having an internal pH of 4–5 and containing hydrolytic enzymes, which aid in the digestion of material ingested by phagocytosis and endocytosis.

Macrophages Mononuclear phagocytes; large, actively phagocytic cells found in the spleen, liver, lymph nodes, and blood; important factors in nonspecific immunity.

Major histocompatibility complex (MHC) Proteins, found on almost all cells in the body, that are responsible for presenting processed foreign protein antigens to T helper cells or cytotoxic T cells; they were first identified as the main determinants of tissue or graft rejection in transplantation from one individual to another. There are two classes of MHC molecules: I and II.

Malaise A general feeling of illness, accompanied by restlessness and discomfort.

MAP kinase Protein kinase activated in response to cell stimulation by many growth factors; mediates cellular responses by phosphorylating specific target proteins.

Mass spectrometer (MS, also known as Mass Spec) A machine that can determine with exquisite accuracy an exact weight for a compound, molecule, peptide, or protein by deflecting ions into a small opening and measuring ion current with an electrometer. MS can often use the weight to identify a chemical, toxin, peptide, or protein. MS/MS, also known as tandem mass spec, provides amino acid sequence information on a peptide or protein, facilitating its identification.

Medium The food source for growing cells (e.g., bacterial and eukaryotic cells) in vitro.

Membrane potential A separation of charges across the membrane; a slight excess of negative charges lined up along the inside of the plasma membrane and separated from a slight excess of positive charges on the outside.

Meningismus Pain due to irritation of the layers (meninges) surrounding the brain and spinal cord.

Metalloprotease An enzyme dependent on the presence of a certain metal element that cleaves protein bonds.

Miasma A poisonous atmosphere, once thought to rise from swamps and putrid matter and cause disease.

Microhematuria Microscopic amounts of blood in the urine.

Mitogen-activated protein kinase kinase (MAPKK) family Family of proteins that, when activated by mitogen signals, phosphorylates members of the MAPK family of proteins. It is thought to be part of a protein cascade involved in pushing the cell into the cell cycle.

Monocistronic A term used to define mRNAs that encode only one polypeptide chain.

Monoclonal antibody An antibody produced from a clone of hybridoma cells that makes only antibody molecules with a singular specificity.

Monocytes Large, circulating, phagocytic white blood cells that constitute 3 to 8% of the white cells in humans.

mRNA (messenger RNA) The template for protein synthesis.

Mutagen A chemical or physical agent that induces mutations.

Mutation A permanent heritable change in the nucleotide sequence of a chromosome.

Myalgia Diffuse muscle pain.

Myocardium The muscular tissue of the heart.

NAD$^+$ (nicotinic adenine dinucleotide) A coenzyme that participates in oxidation reactions by accepting two electrons from a donor molecule and one H$^+$ from the solution.

Na$^+$-K$^+$ pump A carrier that actively transports Na$^+$ out of the cell and K$^+$ into the cell.

Neutrophil A granular, amoeboid leukocyte with a heavily lobed nucleus that phagocytoses and destroys pathogens. The most abundant leukocyte in the blood, it is also involved in the inflammatory response. Often called a polymorphonuclear neutrophilic leukocyte.

NfκB A mammalian transcription factor that is translocated to the nucleus of immune-system cells in response to extracellular signals.

Odynophagia Difficulty or pain with swallowing.

Oliguria A urine output insufficient to excrete the daily produced load; associated with acute renal failure.

Oncogene A gene whose product is involved in either transforming cells in culture or inducing cancer in animals. Most oncogenes are mutant forms of normal genes (protooncogenes) involved in the control of cell growth or division.

Open reading frame A sequence of nucleotides that contains no termination codons.

Operon A group of contiguous genes that are coordinately regulated by two *cis*-acting elements, a promoter and an operator.

Organelle Any membrane-limited structure found in the cytoplasm of eukaryotic cells.

Oxidation Loss of electrons from an atom or molecule, as occurs when hydrogen is removed from a molecule or oxygen is added; the opposite of reduction.

Oxidation potential The voltage change when an atom or molecule loses an electron.

Oxidative phosphorylation The synthesis of ATP from ADP using energy made available during electron transport.

Oxygen debt The oxygen, in excess of the oxygen that is immediately available, that an organism needs to consume after a period of metabolic exertion.

Pandemic A widespread outbreak of disease, usually involving more people than an epidemic.

Papule A small, solid, elevated, and usually inflamed area of the skin that is not filled with pus; in smallpox, develops from a macule.

Passive immunity Short-term immunity brought about by the transfer of preformed antibody from an immune subject to a nonimmune subject.

Pathogen A microorganism (e.g., a virus, bacterium, fungus, parasite, or protozoon) capable of producing disease.

Pathogenicity island A location on the bacterial genome where genes for virulence factors tend to congregate.

Peplomers Protruding spikes of a virus that affect pathogenicity of a particular viral strain.

Pericardium The membranous, fluid-filled sac that encircles the heart and the roots of the major blood vessels exiting from the heart.

Periplasmic space The space between the plasma membrane and the outer membrane in gram-negative bacteria, and between the plasma membrane and the cell wall in gram-positive bacteria.

Peritoneal dialysis A procedure in which bodily wastes are removed via a tube filled with solution and inserted into the peritoneum (tissue lining the abdominal cavity).

Petechiae Small, reddish-purple spots on the skin due to hemorrhages in or under the skin, often indicating a low platelet count.

pH The logarithm to the base 10 of the reciprocal of the hydrogen ion concentration; $pH = \log 1/H^+$ or $pH = -\log H^+$.

Phagocytosis The process by which one cell (often a macrophage) engulfs extracellular material such as an invading bacterium or virus.

Phenotype The observable characteristics of a cell or organism as distinct from its genotype.

Phospholipids The major class of lipids in biomembranes, usually composed of two fatty-acid chains esterified to two of the carbons of glycerol phosphate, with the phosphate esterified to one of various polar groups.

Phylogenetic Relating to the evolutionary history of a species or a specific taxonomic group.

pKa A unit of measure of the relative ease with which an amino acid releases its dissociable protons.

Plasmid A circular DNA duplex that replicates autonomously in bacteria. Plasmids differ from viruses in that they never form infectious nucleoprotein particles.

Platelet (thrombocytes) Fragments of megakaryocytes only 2 to 4 μm in diameter. Their main functions are to plug leaks in small blood vessels and to release substances essential for the initiation of blood clotting.

Polyadenylation A posttranscriptional modification that characterizes a mature mRNA molecule ready for translation by the human cell translation machinery; the addition of 100–200 adenosine residues to the 3′ end of an mRNA transcript. The poly(A) tail aids in the export of mature mRNA from the nucleus, stabilizes the transcript by protecting the coding region from degradation, and serves as a recognition signal for the ribosome.

Polymer adjuvant A synthetic chemical polymer used in formulations with an immunogenic peptide or peptides comprising a vaccine in order to boost or contribute to the vaccine's ability to stimulate the immune response.

Polymerase chain reaction (PCR) A laboratory technique used to amplify specific DNA sequences. It can be used as a diagnostic tool to confirm the identify of an organism by attempting to amplify DNA known to be in a certain species.

Polyprotein A protein that is cleaved after synthesis to produce several functionally distinct proteins.

Prodrome Early symptom, or class of symptoms, that indicates the onset of a disease.

Promoter The region of the gene that signals RNA polymerase binding and the beginning of transcription.

Protease An enzyme (e.g., endopeptidase or exopeptidase) that catalyzes the hydrolytic breakdown of proteins into peptides or amino acids by proteolysis.

Proteinuria Abnormally high levels of protein in the urine.

Proteome The sum total of an organism's proteins.

Pseudogene A segment of DNA that resembles a gene but cannot be transcribed.

Pustule Small, inflamed elevation of the skin filled with pus; in smallpox, develops from papule.

Quarternary structure In a protein, the way in which the different folded subunits interact to form the multisubunit protein.

Ras protein A GTP-binding protein that functions in intracellular signaling pathways and is activated by ligand binding to receptor tyrosine kinases and other cell-surface receptors.

Recombination The animal formation of new genetic combinations, by crossing over (or independent assortment).

Redox potential (E) The relative tendency of a pair of molecules to release or accept an electron. The standard redox potential is the redox potential of a solution containing the oxidant and reductant of the couple at standard concentrations.

Reductive evolution Process undergone by small, asexual populations, in which deleterious substitutions and deletions accumulate over time in an irreversible manner.

Regulator A region of DNA (considered to be either part of a gene or a separate coding region) that controls the expression of a gene through direct or indirect processes.

Regulon A situation in which two or more spatially separated genes are regulated by a common molecule.

Restriction enzyme A protein that recognizes specific DNA sequences and cleaves DNA at those recognized sequences.

Restriction site String of nucleotides recognized by a restriction enzyme; restriction enzyme recognition site.

Reticuloendothelial system A part of the body's natural defense system consisting of reticuloendothelial cells spread throughout the body. These cells filter out and destroy bacteria, viruses, and foreign substances and destroy worn-out or abnormal cells and tissues.

Retrovirus A virus that contains two single-strand linear RNA molecules per virion and reverse transcriptase.

Reverse transcription Mechanism for RNA synthesis in which the RNA viruses use their RNA genome as a template for an RNA-directed DNA polymerase, the reverse of normal information flow within a cell.

R factors Plasmids bearing one or more drug-resistant genes.

Ribosome Small cellular particles made up of ribosomal RNA and protein. Together with mRNA, they are the site of protein synthesis.

Ribozyme An RNA molecule with catalytic activity.

RNA interference (RNAi), also known as siRNA (short inhibitory RNA) Short dsRNA capable of inactivating genes by blocking the production of the encoded proteins. First discovered in the worm *C. elegans*, RNAi works in a wide range of species.

Scissile bond Bond capable of being cleaved easily.

Sebaceous glands Glands in the dermis of the skin that open into a hair follicle and produce and secrete sebum, concentrated most heavily in the face.

Second messenger An intracellular chemical that is activated by binding of an extracellular first messenger to a surface receptor site and triggers a preprogrammed series of biochemical events that result in altered activity of intracellular proteins to control a particular cellular activity.

Septicemia Condition in which an infectious agent is distributed throughout the body via the bloodstream; blood poisoning, the condition attended by severe symptoms in which the blood contains many bacteria.

Serotype A group of closely related microorganisms distinguished by a specific set of antigens.

Serovars Different forms or species of a bacterium.

Sialylation The addition of sialic acid side chains to a molecule, such as an oligosaccharide.

Signal peptides A short amino acid sequence that is translated as part of certain proteins that dictate its cellular localization. The signal peptide can either remain part of the protein or be cleaved during posttranslational modification.

Signal-transduction pathway A series of coupled intracellular events, triggered by binding of a signaling molecule to a receptor, that occur in a sequential fashion to convert an extracellular signal to a cellular response.

Soft palate The part of the roof of the mouth that is muscular and lacks bone.

Southern blotting Technique for detecting specific DNA sequences.

Spore An asexual reproductive or resting body that is resistant to unfavorable environmental conditions, capable of generating viable vegetative cells when conditions are favorable; resistant and/or disseminative forms produced asexually by certain types of bacteria by a process that involves differentiation of vegetative cells or structures, characteristically formed in response to adverse environmental conditions.

Strip immunoblot assay An assay that tests for the presence of antigen by reaction with antibodies; the reaction proceeds by physical "blotting" of antibody and antigen samples.

Systemic infection Describes an infection that spreads from a local site throughout the body via the bloodstream. Common symptoms include fever, chills, general malaise, and joint pain. Systemic infections can result in major tissue damage and serious infection and other complications in major organs. They can be life-threatening.

2D gel Two-dimensional gel electrophoresis used to separate macromolecules (e.g., proteins, peptides) based on isoelectric point (pI) in the first dimension and molecular weight in the second dimension. A protein has a unique molecular weight and isoelectric point.

Tachycardia Unusually high heart rate.

Tachypnea Rapid breathing.

Tautomer Compound related to another molecule by the chemical process of tautomerization.

T cells (lymphocytes) Immune cells derived from the thymus; thymus-derived lymphocytes; the major component of cell-mediated immunity. There are several types: cytotoxic T cells (killer T cells), helper T cells, and suppressor T cells. Lymphocytes make up 20–35% of white blood cells and use antibodies to identify and attack invading antigens and pathogens.

T delayed-type hypersensitivity cells (Tdth) A class of T effector cells that function in delayed-type hypersensitivity reactions.

Testosterone The male sex hormone, secreted by the Leydig's cells of the testes.

T helper (TH) cells A class of T cells with CD4 markers that enhance the activities of B cells in antibody-mediated immunity; T lymphocytes that act as effector cells and interact with other T cells, B cells, and macrophages to activate the immune response; generally $CD3^+CD4^+CD8^-$ cells.

Thrombocytopenia A blood disease characterized by an abnormally low level of platelets (platelet deficiency), resulting in the potential for increased bleeding and clotting difficulty.

Thyroxine The most abundant hormone secreted by the thyroid gland, important in the regulation of overall metabolic rate.

Toxicology Study of drug candidates (e.g., small molecules, chemicals) to determine whether they are poisonous to mammals. A drug candidate found to be free of toxicity would be a good candidate for efficacy trials.

Toxin A poisonous substance.

Transcription A process whereby one strand of a DNA molecule is used as a template for synthesis of a complementary RNA by RNA polymerase.

Transduction The transfer of bacterial genetic material from one bacterium to another by a bacteriophage.

Transformation The process of introducing a DNA oligomer (typically a recombinant plasmid) into a prokaryotic cell line (e.g., *E. coli* cells), quite typically for the purpose of expressing a protein of interest in large quantities before purifying that same protein. A widely used methodology to transform bacterial cells is to first treat them with calcium chloride, add the plasmid of interest, and heat-shock for a moment at 42°C.

Transposon A DNA segment that carries the genes required for transposition and moves about the chromosome; if it contains genes other than those required for transposition it may be called a composite transposon.

T-suppressor cells (Ts) A class of T cells that produce cytokines that depress the activities of B cells in antibody-mediated immunity and other T cells and macrophages in cell-mediated immunity; generally CD3$^+$CD4$^-$CD8$^+$ cells.

Tumor necrosis factor–α (TNF-α) Molecules, often described as cytokines, that normally have the ability to preferentially kill tumor cells. TNF-α has a wide range of proinflammatory responses that include causing localized edema.

Ubiquitin A small, highly conserved protein that becomes covalently linked to lysine residues in other intracellular proteins. Proteins to which a chain of ubiquitin molecules is added are usually degraded in a proteasome.

USAMRIID (U.S. Army Medical Research Institute of Infectious Diseases) Division of the U.S. military established in 1953 with the purpose of researching infectious diseases, especially those that could be used in a biological warfare or bioterrorism setting, in order to develop vaccines and other countermeasures.

Vaccine Protection of susceptible individuals from disease by the administration of a living modified agent, a suspension of killed organisms, or an inactivated toxin.

van der Waals interaction A weak, noncovalent attraction due to small, transient asymmetric electron distributions around atoms.

Viremia The presence of viruses or viral particles in the bloodstream.

Vasopressor An agent that induces contraction of the muscular vascular tissue, thereby increasing blood flow rate and pressure.

Vertical transmission The passage of a pathogen from mother to fetus during pregnancy or to baby during birth.

Virion A virus particle, typically characterized by nucleic acid enclosed in a protein shell or capsid, possibly further enclosed in a lipid envelope; the functional unit of a virus.

Western blot Proteins that have been separated by size using SDS-PAGE , transferred to a special type of paper (called a membrane), and probed with an antibody. Western blots are used to determine molecular weight, tissue distribution, and relative amount of a protein of interest.

X-ray crystallography A process that analyzes the x-ray diffraction patterns of crystals of a molecule in question (e.g., crystals of proteins) to determine its three-dimensional structure.

X-ray diffraction An investigative tool that utilizes x-rays to determine the three-dimensional structure of complex molecules such as proteins.

Zinc finger Several types of conserved DNA-binding motifs composed of protein domains folded around a zinc ion; present in several types of eukaryotic transcription factors.

Zoonosis A disease of animals that can be transmitted to humans.

INDEX

acetylcholine (ACh), 32, 33, 35
Acinetobacter, 220
acyclic nucleoside analog, 20, 21–22, 25
Acyclovir, 5, 21–22
addiction, 317
adenovirus, 310
aerosolization, 10; of anthrax, 130, 134–35, 149, 157, 158, 163–64; of antibiotics, 56; of botulism, 28, 31, 35, 36, 39, 40–41, 134–35; of Ebola, 59, 60, 70, 73; of flaviviruses, 23, 24, 25; of hantaviruses, 112, 123; natural, 112, 123, 124, 130; and personal defense, 325; of plague, 203, 205, 223, 224, 225–26; of SARS, 193, 194; secondary, 158, 164, 165; of smallpox, 240, 247; of tularemia, 43, 48, 49; of vaccines, 127–28
Afghanistan, 255
aflatoxin, 134
Africa, 6, 81; cholera in, 252, 255; Ebola in, 59–62, 61, 72; plague in, 199, 201; smallpox in, 230, 233, 249
Albert, Prince (England), 277
Alexander the Great, 277
Alibek, Kenneth, 55, 73, 132, 160, 248
al-Qaeda, 161
Alzheimer's disease, 317
amantadine, 98, 99
American Media, Inc. (AMI), 135
American Society for Microbiology, 334, 335
Amherst, Jeffrey, 232
Andrewes, Christopher, 82

animals. *See* birds; mammals; rodents
anthrax (*Bacillus anthracis*), 3, 39, 129–72, 296, 311, 327–28, 332; acquisition of, 161–62; availability of, 157–58, 162; CDC category of, 166, 328; cutaneous, 129, 133, 146–47, 151, 152; and decontamination, 131–32, 156–57; description of, 136–37; diagnosis of, 146–50, 151, 153; dormant spores of, 129, 131, 136, 137, 152, 154–57, 163; forms of, 129, 146; gastrointestinal (GI), 146, 147; gene transcription in, 138–39; history of, 129–36, 151; inhalational, 129, 131, 135, 146–49, 152, 164; mechanisms of, 140, 147; molecular biology of, 136–46; and other pathogens, 50, 52, 53, 55, 176, 290; processing of, 162–64; resistant strains of, 160–61; strains of, 135, 136, 154, 155; symptoms of, 146–49; threat status of, 158–61; treatment of, 132, 150–54, 160, 166, 167–68; virulence factors in, 129, 137; weaponization of, 129, 131–36, 156–66
Anthrax Prevention Act (England; 1919), 130
antibiotic resistance: of anthrax, 151–52, 160–61; of cholera, 268, 270; of plague, 222–23, 226–27; of salmonella, 288, 290; of tularemia, 50, 52, 54, 55, 56
antibiotics: for anthrax, 132, 147, 149–55, 157, 160, 166–67; for cholera, 254, 268, 270; for Ebola, 72; for flaviviruses, 20;